THE MAP AND THE MANUSCRIPT

THE MAP AND THE MANUSCRIPT

Journeys in the Mysteries of the Two Rennes

Simon M. Miles

IGNOTUM PRESS

First published in Great Britain
in 2022 by Ignotum Press
W2/211 Woodend Mill 2
Manchester Road
Mossley OL5 9RR
www.ignotumpress.co.uk

Copyright ©2022 by Simon M. Miles

10 9 8 7 6 5 4 3 2 1

The right of Simon M. Miles to be identified as the author of this work has been asserted by him in accordance with the Copyrights Act.

ISBN 978-1-7397347-0-1

All rights reserved. No part of this book may be reprinted or reproduced or used in any form or by any means without permission of the publishers.

Acknowledgements: Every effort has been made to obtain the necessary permissions with respect to copyright material. Full details of images and acknowledgements may be found in the List of Figures in Appendix II. For any omissions, please contact the publisher who will be pleased to make appropriate arrangements in any future editions.

Design, layout and cover by Ignotum Press.

"Le Rêve est une seconde vie."

Gérard de Nerval, *Aurélia*, Paris, 1855.

Table of Contents

Prologue: A Dream in Athens 1
Introduction 5

PART ONE
Identification

Chapter One: On the Path 15
Chapter Two: *Le Serpent Rouge* 41
Chapter Three: First Inklings 61
Chapter Four: Sightlines 67
Chapter Five: Sacred Geography 95
Chapter Six: Converging Circles 111
Chapter Seven: The Number of the Famous Seal 119

PART TWO
Orientation

Chapter Eight: The Zodiac of Rennes-les-Bains 131
Chapter Nine: The Cromlech of Rennes-les-Bains 143
Chapter Ten: Delphi, Apollo and the Python 167
Chapter Eleven: Confirmation from the Team 187

PART THREE
Solution

Chapter Twelve: Geographic Cryptography 209
Chapter Thirteen: The Riddle of the Parchments 241
Chapter Fourteen: The Arques Square 287

PART FOUR
Transmission

Chapter Fifteen: Grand Voyager of the Unknown 339
Chapter Sixteen: Dreams, Alchemy and the Omphalos 357
Chapter Seventeen: Imprint of a Seal 383
Chapter Eighteen: A Walk in the Woods 401
Chapter Nineteen: Reassembling the Scattered Stones 427
Epilogue: Coda to a Dream 437

APPENDICES

Appendix I: The Parchment Text Decipherments..........................445
Appendix II: List of Figures 465
Appendix III: The Text of *Le Serpent Rouge*..............................474
Appendix IV: Chronology of Key Texts......................................480
Appendix V: Bibliography..482
Expanded Table of Contents 489

"And, very circumspectly, he replied: 'In the manuscript, of course, there was also the map, or, rather, a precise description of the map, of the original.'"

Umberto Eco, *Foucault's Pendulum*, (London, Picador, 1988), p.552.

Prologue
A DREAM IN ATHENS

On a spring morning in 1958 in Athens, Professor Jean Richer awoke with a start from a vivid dream:

> "A statue of Apollo, a *kouros* (simultaneously evoking two statues preserved in the National Museum of Athens), appeared to me from behind, then, slowly, it turned clockwise one hundred and eighty degrees, to face me."¹

Somehow, he knew intuitively that the vision of Apollo as a *kouros* statue was related to a problem he had been turning over in his mind. It was a *"songe divinatoire"* or prophetic dream, presented to him, as he was to describe later, "not as a warning, but as a response to a clearly stated question."

He also knew exactly what he had to do next:

> "All that was needed was a map of Greece, a ruler and a compass to interpret this dream."

Richer was a French academic specialising in symbolist poetry and an authority on the nineteenth century author Gérard de Nerval. He was staying with his wife in a house on the slopes of Mt Lycabettus, a prominent hill which is one of the highest locations in the Greek capital. They had been spending time in Greece in recent years, teaching and exploring its history, culture and antiquities.

At some of the locations they had visited certain enigmatic details began to attract his attention. For example, at the entrance to the temple precinct of Delphi, the spiritual centre of ancient Greece, dedicated to Apollo, the first element of the complex which greets the

---
1   Jean Richer, *Delphes, Délos et Cumes : Les Grecs et le Zodiaque*, (Paris, Julliard, 1970), pp. 14, 15. (Passages translated from the original French.)

visitor is the Sanctuary of Athena Pronaia. Located on a narrow strip of land on the southern slope of Mount Parnassus, an Archaic Temple of Athena, goddess of wisdom and inspiration, was erected here in the seventh century BC.

It seemed curious to Richer that the pilgrim arriving at the site of the ancient oracle should find a sanctuary dedicated to Athena Pronaia, rather than to Apollo, the sun god himself. What was the reason for this, he wondered?

There was no obvious indication at Delphi itself or in any ancient texts that he could find to suggest any relevant connection which might provide a clue. He had become intrigued by this minor puzzle and resolved to find the answer:

> "I had posed a precise question, which was formulated as follows: Why does the traveller, arriving from Athens to Delphi, find at the entrance of the sacred site a sanctuary of Athena Pronaia? The answer came in the dream of a spring morning."[2]

In Richer's account of that night, he describes fumbling for the light-switch as he looked around for the required materials, and then:

> "Still half asleep, I took the first map of Greece that came to hand—which was the little map at the end of the *Guide Bleu*[3]. I drew the line Delphi – Athens. Oh, the surprise: when extended, it passed over Delos and Camiros, on the island of Rhodes, where is found the most ancient sanctuary of Apollo of that island."[4]

The statue of Apollo in the dream had inspired, perhaps almost compelled, Richer to get out of his bed and find a map of Greece, a ruler and a pencil. With the ruler touching Delphi and Athens, he drew the line on the map across the Peloponnese and the Greek Islands, and marvelled to see that it passed over Delos, of all places, the tiny island on which Apollo was said to have been born. It then continued on to another temple of Apollo on the island of Rhodes.

Richer had discovered, or rediscovered, an accurate alignment stretching across Greece, measuring over 500 km from end to end. It passed through four ancient sacred sites, each of which had associations with Apollo. The dream had brought a piece of hidden information

2 Ibid p. 14
3 A popular French tourist guide to Greece published by Hachette.
4 Jean Richer, *Delphes, Délos et Cumes*, (Paris, Julliard, 1970), p. 15.

Figure 1: Richer's dream.

from the past to light. Richer must have been elated by this discovery, as much by the wonder of the line itself as the manner in which the revelation had come to him.

Yet, while he knew somehow that the dream related in some way to his question, and that the insight of the Delphi-Athens-Delos-Camiros alignment must be relevant, it still did not, on its own, provide any clear answer as to why the statue of Athena Pronaia stood at the gates of Delphi. For that, he would have to wait a little longer.

The experience had nevertheless stirred his imagination. On waking from the dream that morning, he had also awakened to something else. It had been his first fleeting glimpse of a fragment of a forgotten knowledge, a remnant of a vast subtle science of the ancient world which had long vanished from view.

Without quite realising yet, he had taken the first step on a path to the rediscovery of what he would later call the "sacred geography" of Greece. In the years to come, it would inspire him to create and pursue entirely new fields of research.

Over time, Jean Richer was able to reassemble many of the elements of this lost antique system and through the discovery of a particular master key, to gain an extraordinary perspective and insight into the mindset of the early Greeks. He laid out his discoveries in a series of richly detailed volumes which offer a radical re-appraisal of the relationship between culture, myth and landscape in the ancient world.

It is true that, for a variety of reasons, these ideas have not gained much acceptance by other scholars, or even the general public, but they were to have a profound impact on Richer's thinking, his career and his life.

And eventually, the key that he discovered would also lead him to the answer to his original riddle.

INTRODUCTION

The Landscape in the Languedoc

THERE ARE two villages bearing the name of Rennes in the foothills of the northern slopes of the Pyrenees, in the Languedoc, southern France. Though they are twinned by name, the pair could not be more different in aspect.

Rennes-le-Château ("Rennes-the-Castle") perches high on a conical hilltop, with breathtaking panoramic views over a charming bucolic landscape of craggy, wooded mountains, green valleys with red soil, and spectacular limestone outcrops. It is visible from miles around, and its elevated position provides a superb vantage point from which to survey the surrounding countryside and the narrow winding roads which have carried traffic through the valleys for millennia.

A few kilometres eastward, Rennes-les-Bains ("Rennes-the-Baths") nestles deep in the cleft of a river valley, surrounded by dense oak forests which blanket a surrounding ring of imposing peaks. It is hidden away from the gaze of the passing world, accessible only by a single narrow winding road running north-south which hugs the west bank of the River Sals. The traveller approaching from the north passes between the dramatic heights of the impressive mountain Pech Cardou to the east and the ruins of the Templar Château Blanchefort on a towering rocky outcrop to the west, twin sentinels or guardians of the land that seem to watch silently over all who pass between them.

The villages are located on a plateau known as the Haute Vallée de l'Aude or High Valley of the River Aude. This region is dotted with settlements, connected by narrow winding roads, trails and footpaths. These mountains and valleys have been home to humans since earliest times who left their marks in many places which remain to this day. In the town square of Rennes-le-Château for example, the visitor can inspect a giant, monolithic block of limestone covered in ancient engravings, including crosses and shallow bowl-like indentations.

The area, now within the *département* of the Aude, in the Occitanie region, is situated at a major crossroads of trading and travel routes in this south-west corner of Europe. To the south lie the mountains of the Pyrenees, and Spain; to the east, the city of Perpignan and access to the Mediterranean; to the north, Carcassonne with its spectacular medieval walls and turrets; and to the north-west, the ancient city of Toulouse.

There is another valuable advantage offered by the Haute Vallée de l'Aude in this remote part of France: the surrounding mountain ranges form a barrier that creates a natural sanctuary tucked away in its own world. As a result, the Haute Vallée, as it is known, is a protected enclave that chance and nature have created to form an inviting and secure location in which to live, close to trade-routes, the plains and the ocean, and yet separated from them by natural barriers with limited, easily defended access routes. In each direction, access is only possible by a small number of roads snaking through the gorges carved through the ranges by ancient rivers. It is an ideal place to keep things safe, or to be busy with projects that might benefit from being away from the direct gaze of a curious world.

The landscape is truly spectacular, consisting of rugged peaks and ridges, interspersed with valleys, rivers and dramatic gorges. Geologically, it is classified as karst, a topography characterised by limestone or other soluble rocks which readily form caverns and drainage systems to carry water deep into the earth. These subterranean chambers slowly fill up over thousands of years of rains falling on the mountains. The massive underground lakes that result are then slowly heated by deep thermal sources and the water is eventually forced back to the surface under pressure, creating hot springs on the surface. People have been coming here for centuries, even millennia, to take the thermal baths in and around the village fed by the mineral rich streams from below.

The original inhabitants of the Pyrenees (or the Mountains of Fire) and the Iberian Peninsula came from Asia Minor, as did, later, the Cathar doctrines that were to take root and flourish here. In the first decades of the Christian era, the Greek geographer Strabo wrote that the area was inhabited by a Celtic tribe known as the Tectosages. He relates that they had originally emigrated from lands beyond the Alps and had been part of the loose confederation of Gauls who had invaded Delphi in 279BC. This was a defining moment in ancient history. At the

Battle of Thermopylae, under the leadership of their General Brennus, the Gallic tribes had achieved the unthinkable, and broken through into the inner sanctum of the Temple of Delphi itself. According to some reports, they sacked the treasury and carried off the great hoard of gold and treasure stored there, accumulated over centuries by the priesthood.[5]

Some of that treasure made its way back to the lands around what is now the city of Toulouse. The legend of the cursed gold of Toulouse grew up around this captured booty, bringing ill-luck to those who tried to obtain it for themselves. Here indeed is the seed from which, much later in history, the same legends would attach themselves to tales of buried riches around Rennes-le-Château.

It is easy enough to see what attracted the Celts to the area. There are abundant natural resources, good soil and many excellent locations for establishing settlements. The weather is delightful in summer and mild in winter compared to northern climes. There is wild boar for food, thermal springs for health and relaxation, and a sense of theatre in the landscape that is just as dramatic and appealing today as it undoubtedly was in earlier times.

When the Romans arrived, in the first century BC or earlier, they built elaborate bathhouses at Rennes-les-Bains to take advantage of the natural thermal springs, long reputed to have healing properties, which flow into the River Sals. Ruins of these remain to this day.

Waves of migration continued after the Romans were gone, including the Visigoths who invaded in the fifth century AD. In the ninth to twelfth centuries, the region became a haven for the Cathars, denounced as heretics by the Church. It was also the era of the Troubadours and the Templars, who left their marks in culture and landscape in so many indelible ways.

To this day, many of the châteaux built by the Templars and others remain, in various stages of ruin, throughout the region. Some are well known, like Montségur, where the Cathars were besieged in 1244, and many hundreds were burned at the stake as martyrs. It occupies a commanding position, atop a prominent "pog", on the northern slopes of the mountain range crowned by the Pic de Saint-Barthélemy some twenty-five miles to the west of the Haute Vallée.

Others are less well known, but equally awe-inspiring. Puislaurens Château is found a few miles due south of the Haute Vallée, a spectacular medieval defensive fort occupying a high peak at the head of a steep

---

5   Strabo, *Geography* Book IV, Chapter I.

river valley. There are many others, with names that are etched into local mythology: the Quéribus Château, where the Cathars had their final stand in 1255, Peyrepertuse Château, Puivert Château, famous for its Troubadours, to name but a few. The layers of history and culture are palpable, and while the modern world has inevitably intruded, it still does not completely displace or quench the sense of otherworldliness and mystery which effortlessly pervades the entire region.

And there is a mystery here.

The Affair of Rennes

Rennes-le-Château has become the epicentre of a global phenomenon, the site of a curious mélange of interlocking enigmas which have spawned more books than there are residents in the village. Rennes-les-Bains, too, has its mysteries, but it has held its secrets rather closer to its bosom, away from the spotlight.

In the late nineteenth century, each of these two villages had a priest whose unusual activities would give rise to much conjecture and even controversy. Like the villages themselves, the two priests have had quite different profiles in terms of their visibility in the years since.

Abbé Bérenger Saunière, appointed to the parish of Rennes-le-Château in the mid-1880s, has become the face of the so-called Affair of Rennes. When he arrived in the village, its ninth-century church, dedicated to Mary Magdalene, was in urgent need of repairs. As there were only limited funds available from the church or elsewhere for such purposes, he began with some small minor renovations. What happened next is subject to conjecture and set in train a sequence of enigmatic events. According to the testimony of eyewitnesses, Saunière discovered some objects in a hidden niche, under the altar in the church. They were said to include certain ancient parchments containing a coded message.

Whatever he found, and whatever the truth of the story, what is certain is that his fortunes changed, almost overnight, and in a short space of time he became very wealthy. He went from being a pauper priest with no funds available for basic repairs to having what appeared to be almost unlimited sums at his disposal. He proceeded to undertake major renovation and building work in the church and surrounding buildings. He entertained a stream of visitors, dined on the finest food and wine, and travelled to destinations around Europe.

From this minor local mystery, the main branch of the narrative of the Affair expands and branches out, or splinters if you prefer, until it becomes entangled in a dizzying range of interconnected topics. The story of Saunière's acquired riches is well documented, if controversial; indeed, one can still visit Rennes-le-Château today and see the results of his building programme. Not everything, though, is quite as it seems. Certain elements of the story have been created out of thin air and inserted into the narrative to create mystification. It is exceedingly difficult to separate fact from fiction in coming to grips with this mystery.

Abbé Henri Boudet, the *curé* of Rennes-les-Bains, has remained the lesser-known of the pair. His career was not as outwardly spectacular as that of his fellow priest, but he too was a custodian of secrets. He spent considerable time wandering the local countryside, collecting relics of the Celts who had inhabited the land, and exploring the geology and geography. He wrote a very strange book entitled *La Vraie Langue Celtique et le Cromleck de Rennes-les-Bains*, published according to the title page in 1886, which has baffled generations of readers and led to considerable speculation.

The entire history of the Affair has been thoroughly scrutinised by a small army of writers and researchers for decades. A veritable cottage industry has sprung up of books, magazines, websites, online discussion forums, conferences and movies devoted to exploring the tangled web of mysteries, real and fictional, centred on Rennes-le-Château and Rennes-les-Bains and its surrounding landscape rich in history, culture and myth. It provided the inspiration for one of the biggest selling books of all time, Dan Brown's 2003 mystery thriller *The Da Vinci Code,* and the Hollywood movie franchise which followed. From humble beginnings, this sliver of local history from an obscure village in southern France, the Affair of Rennes, has gone viral, and is known around the world.

It has to be acknowledged that the topic has also attracted a slightly controversial reputation over the years as a magnet for unreliable history, dubious theology and questionable theories. Certainly, there is no escaping the fact that there have been many false, misleading and outright fictional stories projected onto the core narratives of the Affair of Rennes.

But is that all there is to it? Or is there something here more sophisticated than just a simple hoax or deception? Is it possible that beneath layers of obfuscation and deflection something true and genuine lies concealed? Have some of the false elements woven into the story been introduced as a subterfuge to keep an authentic secret safe?

The twentieth century American science fiction writer Philip K. Dick is said to have coined the idea of the "fake fake", an authentic object, work, or situation that has been made to appear as if it is a worthless forgery. This intriguing notion might be a useful concept to keep in mind when navigating through the layers making up the Affair of Rennes. Is it a "fake" or a "fake fake", and how could we tell the difference? These are questions to be explored in the pages that follow.

It must be stressed at the outset, however, that this book is unlike anything previously published on the Affair of Rennes, in several respects. To begin, many of the usual themes which appear in such works are almost entirely missing from this one. This is definitely not a treasure hunt. It has nothing to do with gold or riches, or any physical object of monetary or even historic value. I have never had any interest in how Saunière acquired his apparent wealth and will offer no insights into this question.

On the other hand, this book contains things found nowhere else: true solutions to some of the core mysteries at the heart of the Affair. If that should seem an unlikely, even preposterous, claim to make, I can only agree. The only justification I can offer is that the evidence is here, in your hands, for you to be able to evaluate and decide for yourself.

And there's something else. I have found something. It has left only faint traces but these are enough. It can be measured, observed, photographed and explored. There is an abundance of credible witnesses who testify to its existence. It has been described in a range of sources and texts but, curiously, the specific details have very often been veiled.

It can be expressed in a single, simple image, and yet it has taken a book-length treatment to begin to document. It is a kind of time capsule, a relic of our ancient past which has been carefully hidden for ages, patiently waiting for the time when it will re-emerge into view. Its description has been deposited with great care in certain places and in various formats to preserve its memory. An intricate game has been played across decades, even centuries, to preserve the knowledge and simultaneously conceal it. By such means, it has propagated across time, and has survived into our own era.

But how could I describe such a thing, and how would I share it? The only way, I decided, was to start at the beginning. Here it is then. This is the story of my adventure in the mysteries of the Affair of Rennes: how it began, where it has taken me, and what I found along the way. I would like to invite you to accompany me as I retrace the route I have taken into this tangled wood, along paths where few feet have previously passed, all the way to a surprising, and hopefully satisfying, final destination.

No prior knowledge of the topic will be assumed or needed in order to undertake this journey. All of the necessary background context, maps, compasses and tools required for navigation will be provided. Admittedly, I must caution the reader: the path is quite lengthy, and somewhat convoluted. There will be some difficult stretches, perhaps, but there will also be plenty of fine views along the way to keep us going. In the end, when we arrive at our goal, I hope that it will have been more than worth the effort required to reach it.

PART ONE

Identification

Chapter One
ON THE PATH

MY FIRST encounter with the Affair of Rennes is etched in my memory. It was around 1977, and I was a teenager, growing up in Adelaide, South Australia. Turning on the television one late Sunday afternoon, I stumbled upon a documentary. It recounted a very strange tale about a priest in a poor remote village in the south of France in the late nineteenth century, who had apparently come into huge wealth after the discovery of some ancient, coded manuscripts in his church during renovations. I was drawn in and became fascinated, though it was difficult to follow, partly because I had missed the beginning. When it was over, I did not know quite what to make of it. I was not even sure whether it was fiction or non-fiction. Despite my confusion, or perhaps because of it, this obscure tale took root in my imagination and I never forgot it.

Years later, browsing in a bookshop in Melbourne, I came across a slim volume entitled *The Holy Place: The Mystery of Rennes-le-Château – Discovering the Eighth Wonder of the Ancient World*.[6] Instantly, I recognised the cover image from the documentary, a painting by Poussin of shepherds gathered around a tomb. A light went on! This was the curious story I remembered watching on television that afternoon. I purchased the book, took it home and read it with great interest.

I learned that the programme I had seen all those years earlier was a BBC production, from their *Chronicle* series, called "The Priest, the Painter and the Devil".[7] It had been made in 1974 by the author of the book I had found, Henry Lincoln, who had gone on to co-author the

---

6   Henry Lincoln, *The Holy Place: The Mystery of Rennes-le-Château – Discovering the Eighth Wonder of the Ancient World*, (London, Jonathan Cape, 1991).
7   BBC Chronicle documentary, "The Priest, the Painter and the Devil", (1974).

international best-seller *The Holy Blood and the Holy Grail*[8] (1982) and *The Messianic Legacy* (1986), with Richard Leigh and Michael Baigent.[9] This new work of Lincoln's, published in 1991, was quite different to these previous volumes, however.

The Holy Place was a deceptively compact book, illustrated with many maps, charts, photographs and other images. It touched on the topics I recalled from the documentary but most of the focus was on an entirely new aspect to the story.

Lincoln claimed to have made a remarkable discovery on the 1:25,000 scale map of the landscape around Rennes-le-Château. He had found a complex network of alignments between various mountain peaks, churches, châteaux, and other ancient sites, consisting of perfectly straight lines passing through three or more of such locations over distances of several miles. He had also tried to discern some traces of order amongst the chaotic profusion of lines. The complex, he claimed, had been laid out using English measures, specifically poles and miles, and many of these alignments were arranged into geometric forms, including grids, circles, and even a regular pentagon of mountain peaks.

Though I had read many books on topics in ancient history and earth mysteries, this work was quite unlike anything I had encountered before, and I found Lincoln's account of his discovery curiously engaging. More than anything, it was the sheer number of alignments he claimed were present that grabbed my attention. Even if only a small fraction of the claims turned out to be true, it seemed to me quite a remarkable situation. Unfortunately, however, it was not possible to properly verify Lincoln's results using the maps included as the reproductions were simply too small.

By the time I had finished the book, my curiosity was burning, and I had many unanswered questions. Were the alignments genuine? How had they been created, by whom and when? Lincoln himself admitted that he was struggling to understand what he had found:

> "As I approach the end of my survey, I am conscious that we are rather at the beginning of a long road towards an understanding of what has come to light. Already I am struggling with the huge number of questions which the discovery has raised."[10]

---

8   Henry Lincoln, Michael Baigent, Richard Leigh, *The Holy Blood and the Holy Grail*, (London, Jonathan Cape, 1982).
9   Henry Lincoln, Michael Baigent, Richard Leigh, *The Messianic Legacy*, (London, Jonathan Cape, 1986).
10 Henry Lincoln, *The Holy Place*, op. cit. p. 154.

I was impressed by this admission that the mystery remained unsolved. Lincoln also appealed for assistance:

> "There is a need for a far greater expertise to be brought to bear than I possess in order to reveal the Holy Place in all its majesty."[11]

I agreed that his discovery deserved further study and attention, but it all depended of course on whether the claims of alignments and geometry as laid out in the book were valid and accurate. Someone would have to go to the trouble of checking. I kept thinking about this book and wondering if anyone would take up the challenge. One day, I decided I would.

It seemed like a worthwhile exercise. If the claims in the book held up to scrutiny, it might open the material for further investigation, and if they did not, well at least I would know. Either way, it would hopefully make for an interesting side project. And so, my adventure began.

The Map

My first step was obtaining a copy of the 1:25,000 scale map of the area.[12] In those pre-internet days of the early 1990s and living in Australia, this meant posting an order to the Institut Géographique National (IGN) in Paris and waiting patiently for the package to be delivered.

When it arrived, I had the map mounted on thick board and laminated, just as Lincoln had suggested. I obtained a good long ruler and some felt-tip pens and was ready to start my new project. Little did I realise how far it was going to take me.

The map offered a fascinating window into an exotic, far-off landscape on the other side of the world. It depicted the isolated, secluded, protected region known as the Haute Vallée de l'Aude, or High Valley of the River Aude. There was something mesmerising about the strange place names and the sense of faded history that it effortlessly exuded. I slowly began to familiarise myself with its contents and the local geography.

The Aude is one of seven major rivers which descend from the Pyrenees into France. It flows first northwards across the area covered by the map, then turns east to empty into the Mediterranean at Narbonne, whilst the other six rivers flow to the west into the Atlantic. As it passes

---

11   Henry Lincoln, *The Holy Place*, op. cit. p. 153.
12   IGN map number 2347OT Quillan. All IGN maps including this one are now freely available online at https://www.geoportail.gouv.fr/carte.

through the Haute Vallée, carving out valleys and creating spectacular vertical gorges, it is joined by other smaller rivers fed by underground springs including the Sals, the Blanque and the Rialsesse.

The landscape is home to many villages, some very small, connected by an intricate network of roads, tracks and mountain passes. The map shows more than a hundred of these, each with its own church, marked with a distinctive cross sign. In addition, there are at least fifty châteaux in the surrounding region, many dating from the Templar and Cathar era between the ninth and twelfth centuries. There are also many peaks, lookouts, trails, cemeteries, ancient stones, caves and underground passages depicted. But did it also contain hidden geometry, concealed in the distribution of the churches and châteaux and mountain peaks? This was the question, and now that I had the tools, I could begin to tackle it.

The initial challenge was simply to check if the alignments Lincoln had cited were indeed present. I wanted to see with my own eyes if these lines did in fact pass through the sites he had listed, and if so, what degree of accuracy was involved, and whether the distances between points were as regular as the book claimed.

I decided to proceed methodically on my "audit" of Lincoln's book. I would work slowly through *The Holy Place*, page by page, from the beginning and verify for myself every claim, every line, every measure, every angle. There was a lot of material to check, but slowly a picture began to emerge.

To begin with the question of alignments, there was no doubt that most of the lines that Lincoln described held up to scrutiny. There were admittedly a few minor exceptions, some stray loose points and one or two typographical errors, but even eliminating all of these, an impressive body of accurate alignments remained.

Many of these were quite remarkable. There were numerous examples of four or more sites very precisely aligned over relatively short distances. These incorporated both natural landscape features and châteaux, churches and other significant locations. I soon became convinced that Lincoln was on to something. Even if I discarded every questionable alignment there remained an impressive array of solid, tight examples. Inevitably, these positive results set me thinking about the whole question of intentionality. Were these observable geometrical relationships deliberately and consciously planned by ancient engineers? Or were they just the result of random distribution?

One approach to resolving such issues might be to calculate how many alignments could be expected by chance for a given area with a certain number of sites, and then compare this figure to what has been found. It seemed to me, however, that such an approach is only valuable up to a point. All it can do is offer a probability of an expectation. It cannot settle the question for any given alignment. I wanted to find a better approach to evaluating whether individual alignments were the result of conscious intervention in landscape.

It occurred to me that it might be more useful to consider a wider range of factors in addition to the mere observation that four points, for example, fell on a straight line. For example: does the alignment point in a significant direction, perhaps along a major compass angle, or towards a significant astronomical direction such as the midwinter sunrise or the rising or setting of a particular star? Does the alignment interact with other landscape features through which it passes? What cultural or historical traces might it have it left? Does it interact with other lines? Are there any parallel alignments nearby? Do multiple alignments pass through the same location, marking a "node"?

These seemed like valuable considerations. If a line satisfied more than one of these other conditions, then the possibility that it had arisen through deliberate intent would surely be strengthened.

So, I began to think about the question of intention from the perspective of these additional qualities. I started looking not just for straight lines connecting points, but for evidence of other attributes that might help to eliminate the possibility of chance as an explanation for their origin.

And, indeed, as I slowly progressed, I was able to find satisfying examples which met multiple criteria of order. There were many interactions to be observed between the lines and the landscape. Some alignments coincided closely with the paths of rivers. Others passed through natural lookouts or *points de vue* as they are known in France. There were lines which passed through multiple churches or châteaux and terminated on significant peaks at either end, which could have offered vantage points for convenient sighting. There were sites at which two, three or more alignments converged, forming node points that were much less likely to arise by chance. There were significant angles between lines. There was much to sift through and think about.

Lincoln had also noticed several sets of alignments which ran parallel to each other, or at right angles, suggesting the possibility of a common purpose governing the direction of these lines. Some of these parallel clusters also appeared to interact with each other, forming grid-like patterns. In some instances, they shared sites in common. I found the idea of these grids strangely fascinating, and paid close attention to these claims.

Not all the material in *The Holy Place* survived close scrutiny though, by any means. One of the major claims was the presence of a regular pentagram, described by Lincoln as "exact", marked by five mountains. However, on close inspection, the peaks did not align precisely with a theoretically perfect geometrical pentagram but deviated by errors of up to 3–4 millimetres on the 1:25,000 map, or 75 to 100 metres in the landscape. For me, this was simply too large to be acceptably described as "exact" as Lincoln had claimed. Furthermore, the peaks were by no means the most prominent on the map, with the fifth being particularly insignificant. The proposed pentagram of five mountains was neither very exact nor particularly remarkable.

Unfortunately, it must be said that this same issue affected other geometric forms which he had proposed. The "Esperaza hexagram", for example, looks impressive on the page. On close inspection, several of the alignments are inaccurate, and some are completely incorrect. With these removed, the "hexagram" dissolves. I began to sense that in his eagerness to bolster his thesis, Lincoln was willing to tolerate some "fudging".

But did this matter? Even if I put to one side all the claims in the book which were not very precise, or invalid, there remained an impressive list of observations Lincoln had made which *were* accurate. It was clear to me that his problem was being somewhat overwhelmed with material, rather than straining to make a case. I was left with the sense that there was some genuine underlying phenomenon here, but that Lincoln was somewhat out of his depth when it came to the details. In fact, he himself said as much:

> "I confess that the geometry is now becoming far too sophisticated for my amateur's grasp. (...) Perhaps the simplicity I am seeking lies elsewhere in the recesses of the Temple. (...) For me however, the essential task remains the hunt for the elementary first steps of the construction, the basis of the design."[13]

Overall, despite my reservations on much of his geometry, I concluded that many of the alignments which Lincoln had found were valid and they often involved multiple factors which at least suggested

---

13   Henry Lincoln, *The Holy Place*, op. cit. p. 140.

the possibility of intentionality. I was intrigued and slowly became hooked on my new project. Soon, exploring the map of this area on the other side of the world to where I lived became a slight obsession.

The 45° alignment

Against this background, there was one alignment which made a particularly strong impression on me. It was approximately seven miles in length and passed through a total of five significant ancient sites—including the church of Rennes-les-Bains itself – all of which were dated to the thirteenth century or earlier.

This line is found in Lincoln's book, but it had first been described as part of a larger cluster of alignments by another researcher and author, David Wood, in his book *Genisis*[14]. However, I noticed an aspect that neither of them seemed to have remarked upon: the alignment runs due north-east at a bearing of very close to 45°.

This observation struck me as highly significant, coming in addition to the high quality of the alignments between the sites. It suggested the possibility that the original architects and builders of the structures intentionally arranged the spatial relationships between the locations.

Here is the list of the five sites, in order from south-west to north-east:
- St-Just-et-le-Bézu Church: dating from the ninth century AD, this small church lies at the centre of the village, which is itself surrounded by dramatic ridges and valleys.
- Lavaldieu: Site of a Templar Commanderie from at least eleventh century AD. Now remains in ruins.
- Rennes-les-Bains Church: at centre of village, dating from at least eleventh century AD.
- Montferrand-le-Château: Now lying in ruins, the château is located high on the flank of Pech Cardou in a tiny village.
- Château d'Arques: An impressive tall, narrow square tower within a walled rectangular compound. Dating from at least eleventh century AD and possibly much earlier.

Château d'Arques

The Château d'Arques is a unique and remarkable structure. Its physical appearance is quite unlike any of the other châteaux which may be found in the area. For a start, it does not occupy a high point, or strategic lookout, but sits in a large expanse of flat open ground nestled between surrounding peaks. Architecturally, it is very curious: a tall,

---
14   David Wood, *Genisis*, (Tunbridge Wells, The Baton Press, 1985).

narrow tower with a square cross-section and impressive turrets. As its four sides are oriented to the cardinal directions, the 45° alignment neatly traverses the tower from its south-west to north-east corners. The rooflines mark these diagonals in the design of the Château itself.

The Château sits within a large, rectangular compound formed by high stone walls. In the south-west corner is another building of square cross-section, which also features diagonal axes marked in the roof lines. Very remarkably, the 45° alignment through the Château d'Arques passes perfectly through this building and the south-west corner of the compound, as shown in Figure 4. Notice that the Château itself does not fall on the diagonal through any of the other corners of the compound, thus strongly suggesting the alignment at 45° through the south-west corner, the guardhouse, and the tower is intentional.[15]

Windows in the turrets located on the corners of the Château d'Arques allow the observer to direct their gaze naturally along the 45° alignment, to the south-west, over the corner building and the corner of the compound. The eye is then drawn beyond to look directly at a large prominent rocky outcrop on the flank of Pech Cardou, which happens to mark the path of the alignment as it crosses the ridge.

It was noteworthy to find five structures, from the same historical era, falling on an exact alignment, over a short distance, but to also be oriented at a bearing of 45° and marked in this manner by the Château d'Arques complex, seemed to me a strong indicator of conscious, co-ordinated intervention by the builders of these structures to achieve the result. But why? What purpose would it serve? What would be the point? Whatever it was, it must have been worth it, as even just to achieve this one single alignment would have been an impressive feat. This was not flat terrain. This was rugged landscape, of wooded peaks and river valleys, and a challenging environment in which to produce such high quality of surveying.

This 45° alignment easily met my criteria for conscious intervention. There were multiple overlapping consistent signals which pointed to a deliberate origin for this blatantly geometrical form.

Other books had also been written on this strange subject of landscape geometry around Rennes-le-Château.[16] Without going into detail on these, as it would take the story too far off track, they each provided

---

15   Lucien Bayrou, *Le Château d'Arques*, (Carcassonne, Centre d'Archéologie Médiévale du Languedoc, 1988).
16   For example, David Wood, *Genisis*; David Wood and Ian Campbell, *Geneset*, (Tunbridge Wells, The Baton Press, 1994); Paul Schellenberger and Richard Andrews *The Tomb of God*, (London, Little, Brown and Co, 1996).

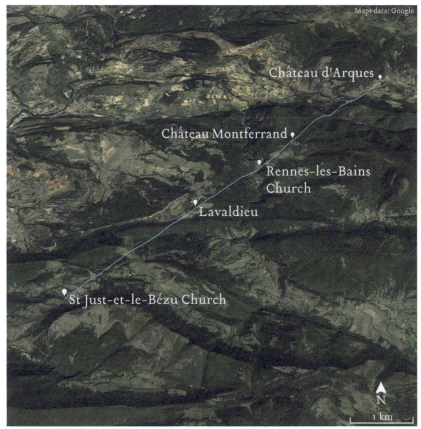

Figure 2: The 45° alignment.

a mix of intriguing material and additional frustration. Some described new alignments which I found were accurate and impressive, and added to my catalogue of material built up from Lincoln's book and my own work. Yet, at the same time, each author laboured hard to present theories that failed to convince me.

All their issues, in fairness, could be traced to the underlying difficulty which challenges everyone who engages with this material: what is it, exactly, that we are looking for? What *is* the mystery? The question I slowly began to formulate in my mind was this: what was the original, underlying reason why this place became the site of these speculations in the first place? Somehow, I wanted to find a way to strip away all the false narratives, distortions and distractions that have been projected over the events. I wanted to get to the foundation layer of bedrock certainty if it existed.

Figure 3: The imposing tower of the Château d'Arques, looking towards the west, with Pech Cardou sloping up to the left in the background.

Spurred on by my dissatisfaction with these other accounts, I pushed on slowly with my own approach, exploring the map, the alignments and their relationships. There was something here which I could not quite grasp and yet could not quite let go. The map became covered in a forest of alignments. What did it all mean? I was becoming lost in the woods. Eventually, I knew I had to start thinking about the problem in a different way.

Meridians in the Mountains

An idea occurred to me. Suppose, I thought, that some at least of these alignments are the result of genuine historical interventions in landscape, even if it was only a small number. In that case, one of these alignments must, by definition, have been the first, or earliest one marked out. Now, if there was such an initial alignment, I began to wonder, what kind of alignment might we expect it to be? What factors or qualities might characterise the very first alignment?

Thinking about it in these terms, a sensible answer readily suggested itself: a good potential candidate for such an original, primary line might be a meridian line, or due north-south line, as this is and always has been the prime direction which must first be established to fix

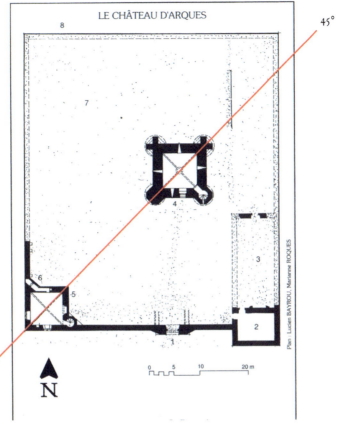

Figure 4: Ground plan of the Château d'Arques compound. The 45° alignment co-incides with the diagonals of both the square "guardhouse" in the south-west corner of the compound and the château itself. (Image: Lucien Bayrou and Marienne Roques, Le Château d'Arques, (Carcassonne, Centre d'Archéologie Médiévale du Languedoc, 1988) p.18)

location. It was, if nothing more, a suggestion worth consideration: if any alignments existed, then there must have been an initial alignment, and this might have been a meridian.

The act of marking a meridian is a fundamental – perhaps *the* fundamental – human intervention in the world. It is the primary expression of our relationship to physical space. To define a meridian requires an understanding of the relationship between sun and earth and observer. It is the first act in establishing one's location and is the precondition for all competent navigation. It lies at the root of our conception of time, as well as space. In this sense, it can be considered as the primal expression of human consciousness. Nature does not mark meridians; if we should find one, it can only have arisen through conscious intent.

So, I set out to find any examples of meridians on the map. I had in mind looking for north-south alignments passing through the peaks of mountains or ridges, and through churches or châteaux. I had no great expectations of success, but it seemed like an easy idea to check and a worthwhile exercise.

I oriented my long transparent ruler with the left side of the map so that it was aligned north south. Then I began to move it slowly and carefully across the surface, keeping it parallel, while searching for any instances where it might pass over more than one peak or church or other noteworthy location falling on the same meridian.

I had made my way painstakingly to about half-way across the map when I found what I was looking for: two peaks falling on the same north-south line. The first was Le Sarrat Rouge, (the Red Hill), the local highest point of the plateau on the west bank of the River Sals.

The second was a very distinctive local highpoint and popular lookout called La Pique, which culminated in an outcrop of jutting rock at the highest point of a large triangular ridge. My ruler showed a perfect meridian between these two marked positions, Le Sarrat Rouge and La Pique.

As I traced the line further to the south, I noticed that it passed over two parallel ridges or crests which ran approximately east-west across the landscape. It took me a few moments to realise that in both cases, the meridian crossed on the highest marked spot height of each crest. There were four local high points in close proximity falling with unerring precision on the same meridian, namely $2°17'50''$E.

It was far more impressive than I had dared to imagine, consisting of four distinct peaks on an impeccable, razor-sharp meridian alignment. Even more remarkable to me was the fact that the peaks ascended in height, from north to south. This implied that if one were to view the scene looking southward from a convenient vantage point further to the north, on the same meridian, one might expect to see the four peaks "stacked", one above the other. When I checked, I found that there was indeed such a suitable vantage point to the north, a long high ridge which ran east-west across the landscape and acted as a natural northern border of the plateau. It offered a perfect view of the peaks and valleys to the immediate south. On the map, I saw that the closest place name to the point where the meridian crossed the sighting ridge bore the uncanny name: *L'Homme Mort*. The Dead Man.

La Pique Meridian

I decided to name the meridian after the remarkable landmark that is La Pique. As an aside, the French phrase *la pique* does not mean "peak" as one might expect, but rather denotes a thin rod or lance, corresponding to the term "pick" in English, as in a "tooth-pick". Here are the positions on the La Pique meridian, from north to south, on longitude 2°17′50″E.

- L'Homme Mort (706 m): Vantage point on long ridge running east-west to the north of the Rennes plateau.
- Le Sarrat Rouge (569 m): ("The Red Hill"), a small, symmetrical, well-defined peak and local high point. Meridian passes exactly though marked high point.
- La Pique (582 m): Prominent lookout highpoint at culmination of a distinctive triangular limestone plateau. Meridian passes through marked high point.
- La Serre Calmette (843 m): Ridge which crosses landscape west-to-east. Meridian crosses on marked highest point on crest.
- Les Crêtes d'al Pouil (1,037 m): Parallel ridge which also crosses landscape west-to-east. Meridian again crosses on marked highest point on crest.

I was utterly amazed by what I had stumbled upon. I kept on checking it, over and over, almost in disbelief, but there was no mistake. I had found a meridian marked in the landscape.

This was a major moment. Alignments are one thing. Yet, no matter how impressive any given example might appear, it is difficult to exclude completely the possibility that it may be purely a result of random forces. But an alignment comprising four or more positions which also runs precisely north-south is another matter entirely. In such a case, we have independent, defined conditions of order both being satisfied simultaneously, all but eliminating chance as an explanation.

I had only searched half of the map, so I still had further work to do. I resumed moving the ruler across the map again, continuing to the east of the La Pique meridian. It only took a few minutes before I stumbled upon a second example. And this one, if anything, was even better than the first.

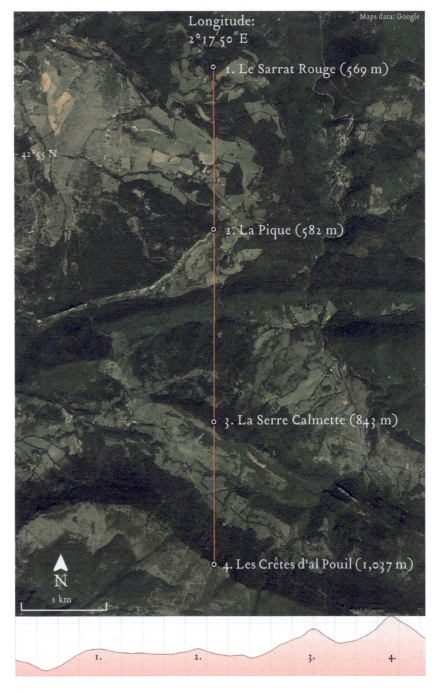

Figure 5: The La Pique Meridian. Shown from above and in elevation profile.

The Pech Cardou Meridian

The highest peak on the plateau immediately surrounding the two Rennes is Pech Cardou. At 795 metres tall, it is dwarfed by the higher mountains of the Pyrenees to the south, but it is nevertheless an imposing sight visible from miles around. It is found to the east of the River Sals, a short distance north of Rennes-les-Bains.

When I reached the summit of Pech Cardou on the map, I leaned in to inspect the path traced by the edge of the ruler. To my utter astonishment, it passed exactly through a total of eight peaks to the south. These nine positions all fell on a precise bearing of 180°.

Incredibly, there was also a tenth peak on the same meridian, to the north of Pech Cardou. Again, this offered a convenient high vantage point from which one might, at least in principle, view the entire meridian. In this case, the viewpoint was a spot height marked 802 m, named Planditou on the map, the highest point of the same long ridge on which was located L'Homme Mort, the vantage point for the La Pique meridian, further to the west. Ten "peaks", (which included several highest points of ridges, or crests) fell without deviation on the precise 180° line ruled through the summit marker of Pech Cardou on the map.

Furthermore, proceeding south from Pech Cardou, the eight peaks also rose in order of ascending height, from north to south. In theory, according to the map, this implied that to an observer on the summit of Cardou the peaks would appear, again, as if they were stacked one above the other, stretching away to the south and culminating in the towering peak of Pech dels Escarabatets at 1,190 m. The meridian falls on longitude 2°19'40"E. Here is the list of the ten locations.

- Planditou (806m): Highest point on long ridge running east-west, with commanding views over the Rennes area to the south, and Pyrenees beyond.
- Pech Cardou (795 m): Highest mountain in the local area.
- Col Doux (450 m): Small peak south of Cardou.
- Soula de la Carbonnière (515 m): Highest marked point on ridge crest running east to west across landscape.
- Notch on La Garosse (590 m): Marked spot height on ridge.
- Unnamed summit (802 m): Marked spot height.
- Col du Vent (825 m): This is a V-shaped "groove" in the high, steep, ridge which runs east-west across the landscape.

- Serrat dels Avets (858 m): Marked spot height.
- Pech de la Quière (1,029 m): Highest point on ridge, with sharp break in the crest at that point, as if the western side of point "demolished".
- Pech dels Escarabatets (1,190 m): Peak of major mountain.

Notice that one of the "peaks" (Col du Vent) is in fact a low point on a ridge rather than a high point, but far from breaking the pattern, it is the exception that proves the rule.

Col du Vent is the site of a V-shaped "groove" in the crest, where a narrow track crosses over the ridge permitting vehicles and walkers to pass from one side to the other. The only alternative is a long drive around the ridge in one direction or the other. The road has clearly existed on this same path for a very long time. The track, the crest and the meridian all coincide in physical space, a remarkable conjunction which cannot have arisen by chance.

After I had caught my breath at the wonder of the Pech Cardou meridian, I took stock of what had just happened. I had found two meridians, one of five peaks and the other of ten, within a short distance of each other, both perfect and exquisite.

Any lingering possibility that these might have arisen naturally by the play of random geological forces, or simply by chance, completely evaporated. There could be no doubt about it. If one meridian on its own was powerful evidence for intentional intervention in landscape by humans, then two of them, so close together and so accurate, must surely be considered as close to definitive proof.

The game was afoot.

The observation that the bearings of at least three alignments – the two meridians and the 45° alignment – were set to significant compass angles, was a decisive breakthrough and took the argument for intentionality to the next level. It was one thing to find four churches in a straight line, but it was always possible that such arrangements might occur by chance, given a large enough sample. If such alignments were also oriented to regular compass angles however, then the factor of chance as an explanation was much more difficult to argue. I realised that this was an excellent test to apply to these observations. Perhaps there were also further similar examples to be found. This gave me all the encouragement I needed to persist, and I redoubled my efforts.

So many questions occurred to me. I now had little doubt that, in some sense, human activity had been involved in creating these lines, but what could that mean? Can a man move a mountain? Can he nudge

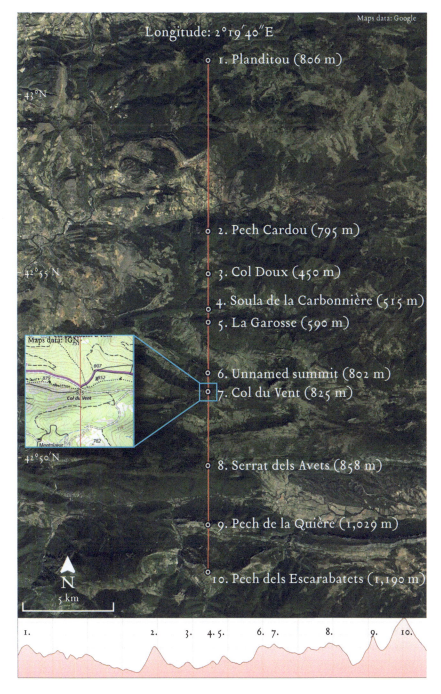

Figure 6: *The Pech Cardou meridian. Shown from above and in elevation profile.*

a peak, a little to the east, or a little to the west? In any case, I soon realised that the physical act of carving or otherwise altering the position of the landscape was not even the most challenging aspect of such an operation.

The real difficulty would have been in calibration. How could ancient man have reliably determined these precise compass bearings? How does one lay out an exact north-south line? How does one mark a 45° bearing over several miles of mountainous terrain? I began to consider these questions as I continued to pore over the map.

Units of Measure

Lincoln also grappled with the question of the units of measure that might have been used to lay out the geometry. He suggested that significant sites in the landscape were separated by distances of whole numbers of miles and gave a long list of examples. I went through and methodically checked each length by carefully measuring the distances on the map. Many were inexact. Lincoln's case that the geometry was laid out in miles in the landscape, or in poles as he also suggests, left me unconvinced.

Nevertheless, I thought these questions about measure were useful to ask. If these alignments were indeed genuine interventions in the landscape by some "intelligent agents", then there must have been some measurement system employed. What was it? The question of the units of length that might have been used continued to occupy my thoughts.

It also made me start to think about a related and much simpler question: what did they use for a map? Any building work requires a plan. There must have been some equivalent to our modern map, though obviously not something laid out on paper, far less on a computer screen. So, what did they use instead? How could they have designed the geometry in the landscape, executed it or communicated about it without some shared resource equivalent to modern cartography?

I was interested in the possibility that significant distances in the landscape geometry, if they were intentionally laid out, might resolve to whole round numbers when expressed in their original measures. After all, this is why, even today, particular units are chosen and used: for convenient handling.

To test for this, I was in the habit of routinely converting any of Lincoln's distances that he claimed to be significant into various ancient measures to see what might turn up. One day something did. It was to have a profound impact on my understanding of the nature of the

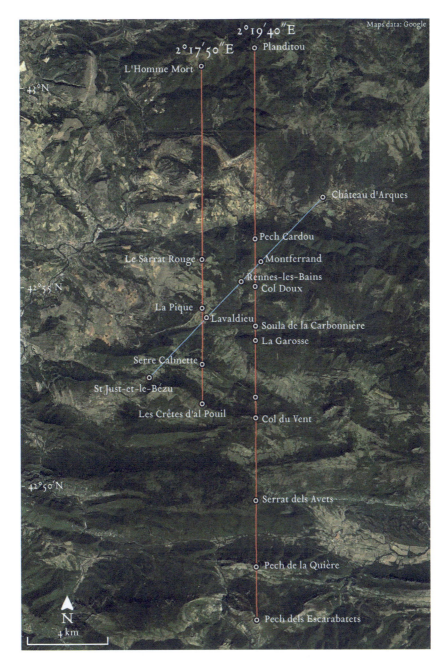

Figure 7: The two meridians of La Pique (on the left) and Pech Cardou (on the right), with the 45° alignment, viewed from above. The "sighting ridge" appears at the top, with the locations of sighting positions L'Homme Mort and Planditou.

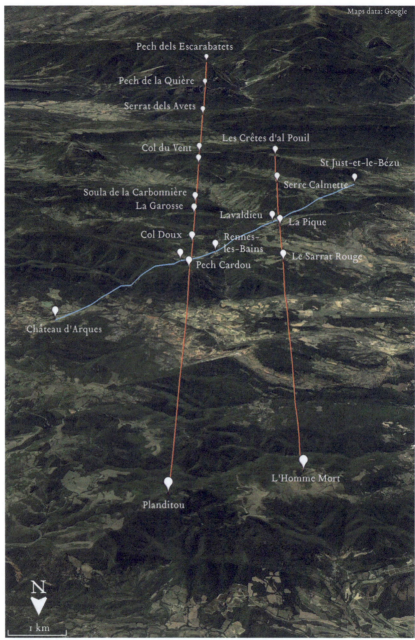

Figure 8: The two meridians of Pech Cardou (on the left, or east) and La Pique, (on the right, or west), in perspective view, looking towards the south. The 45° alignment is also shown. The sighting ridge is in the foreground, showing the vantage points of Planditou and L'Homme Mort.

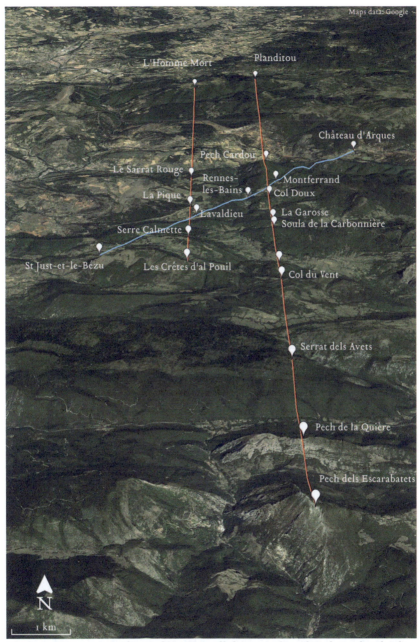

Figure 9: The two meridians of La Pique (left, or west) and Pech Cardou (right, or east) viewed in perspective looking north. The peak of Pech dels Escarabatets can be seen at lower right, the final summit on the Pech Cardou meridian.

relationship between the map, the landscape and the mind of the landscape engineer. It was to prove even more uncanny as time moved on.

In *The Holy Place*, Lincoln discussed an alignment which he called the "Bugarach Baseline", previously identified by David Wood in his book *Genisis*. It begins at Bugarach Church. Running slightly west of north, it passes over the highest marked point on a ridge called La Soulane, then through Montferrand Château, Pech Cardou summit, the Château Serres and a location called Combe Loubière, where three trackways meet. This alignment easily met my basic criteria for consideration as potentially intentional: it included six valid ancient sites or peaks, in good alignment[17] over a relatively short distance, and shared a site in common with the 45° alignment, Montferrand Château.

It also lay at a right angle to another important alignment in Lincoln's scheme called the "Sunrise Line". This ran from the Campagne-sur-Aude Church, through Rennes-le-Château Church, Château Blanchefort and Arques Church. Lincoln made an observation about the length of the Bugarach Baseline which caught my attention. He pointed out that the distance between Bugarach Church and Combe Loubière was divided precisely into three equal segments by La Soulane and by the point where it crosses the Sunrise Line.

I checked and the claim was accurate enough. The length on the map between Bugarach Church and Combe Loubière was 457 mm. It was divided into three lengths of 152 mm (to the nearest mm) by La Soulane and the Sunrise Line. This seemed like a promising observation. Three equal lengths suggested the possibility of a repeating measure. Was this length significant?

Multiplying 457 mm by the scale factor of 25,000 gave a distance of 11,425 metres in the landscape. I proceeded to convert this figure into other units of measure, ancient and modern, as I had done many times. Ideally, I was looking for some suitable unit of measure which expressed the length in a whole round number and was also readily divided into three. I soon found one that met these requirements neatly: the total overall length was very close to 450,000 inches[18], which divided cleanly into three sections of 150,000 inches each. This was a satisfying result.

Now I needed to check these new values on the map to make sure that I had performed the calculations correctly. In order to do

---

17   Château Serres is slightly to one side of the line, but it does pass through the grounds. The remaining five sites are accurately aligned.
18   457 mm x 25,000 = 11,425 metres = 449,803 inches. This differs from 450,000 inches by less than 200 inches in the landscape.

so, I needed to convert this distance in the landscape back into a length on the map, again using the 1:25,000 scale.

This time the calculator was not necessary: 150,000 divided by 25,000 is simply 6. A distance of 150,000 inches in landscape corresponds to 6 inches on a 1:25,000 scale map. All this time I had been working exclusively with a ruler marked in millimetres. Now I scrambled to find one marked in inches, which was no easy feat in a country that had converted to the metric system decades earlier. Eventually I found one in the back of a drawer and was able to confirm the measure of the line segments on the map. They were indeed each six inches, for a total length of the Bugarach Baseline of eighteen inches.

Suddenly, the result hit me like a thunderclap! All this time I realised I had only been considering the idea of distances in whole round numbers in suitable units *in the landscape*. It had never occurred to me to think about whole round measures *on the map*.

After all, why would it? Any measures on the map will be, of course, dependent on the scale of the map, which in this case has been dictated by the choice of the cartographers at the IGN. But what if, by some astounding coincidence, the ancient landscape engineers themselves had also worked with a map of scale 1:25,000?

What if they had employed a combination of this map scale and the inch measure to allow them to convert readily between distances in the landscape and such a map? Of course, this could not have been a paper map, far less a digital one, but as a concept this seemed to me to be an immensely powerful idea. In that moment of insight, from which I think I have never really retreated, or recovered, I saw the possibility for an interconnected system of representation of landscape which does not require a physical map.

After all, consider the simple implication of all that we are contemplating here. If indeed there was ancient intervention in landscape, a conscious programme undertaken by sentient agents, presumably humans, then there surely must have been a map. How could any design be inscribed at large in landscape without one? What form then could such a map have taken?

The essential difficulty that I had always run up against in considering how all of this could have been achieved was the challenge of planning, executing, discussing and managing such landscape work without the convenience of the paper map. I reasoned that there must have been some kind of functional equivalent. Now I could begin to see for the first time what such a map might be made of: geometry.

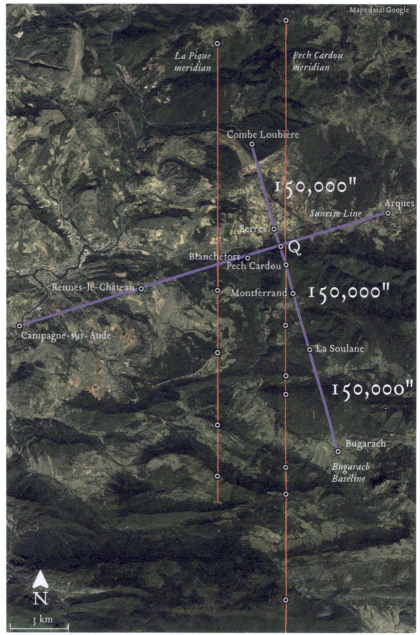

Figure 10: The Sunrise Line and the Bugarach Baseline (purple) with the La Pique and Pech Cardou meridians. The Bugarach Baseline is 450,000 inches in length from Combe Loubière to Bugarach Church, divided into three equal segments of 150,000 inches by point Q and La Soulane.

All of this came to me in a giddy flash as the inch ruler lined up with the lengths of the line segments. This was not the same as saying I had found the details of such a system. In fact, I definitely had not. But deep in my intuition, something had stirred. Once it had occurred to me, I could not easily put aside this notion of a mental "tool kit" employed by the builders comprising geometry, scale and measure by which to conceive and describe and communicate the geometry embedded in the landscape.

Consider a map comprised of a design which is held in the mind's eye, the inner imagination, the picture forming capability inherent in human consciousness. Suppose that it consists of a geometrical figure, which can be visualised rigorously, line by line, within the mind, according to some well-defined set of directions. Now permit it to be associated with a measure and scale. The entire mental construct is formed and held in the inner vision of the geometer.

The set of instructions can then be verbally described and transmitted to other geometers. It can also be recreated in detail in an accurate format in the outer world if suitable materials are available. In this manner, an equivalent to our modern concept of the map might have been created.

The more I thought about it, the more I became intrigued by the realisation that whatever form of map or geometry the ancients used, and there surely must have been some, it must have had a scale. Could that scale have been 1:25,000, identical to the one we use today? Was the inch the unit of measure employed to lay out the geometry, so that one inch on the map represented 1:25,000 inches in the landscape? Was this the basis of a map that could be carried around internally, in memory, without the need for it to be printed on paper, but permitting enough accuracy and record of form to allow even quite detailed calculations? This was a question I began to think about. There was no turning back.

I was starting to feel that I was making modest progress. I had discovered the meridians and marvelled at the properties of the 45° alignment. There were some first tentative ideas about measure, and even scale. I had concluded that there was indeed something real, something intentional and something very ancient embedded into this landscape.

It was woven into the positions of the mountain peaks themselves, and yet it was also assimilated to the placement of structures dating to the Middle Ages and even earlier.

By now, it was the second half of the 1990s and the era of the rise of the internet. I discovered online a thriving world of research, websites, documents and forums dedicated to the Affair of Rennes and even launched my own first tentative site on my work in 1998.

Through these, inevitably, I had slowly begun to familiarise myself a little more with other aspects of the complicated sprawling mass of topics which made up the mystery. It was not with any particular expectation that they might shed light on the landscape geometry, but just out of curiosity and for background and context.

Frankly, I did not intend to engage too deeply with this other material. My strategy was to focus on the geometry. After all, I thought confidently, what could the tale of Saunière and his riches, or the bizarre decorations he installed in this church, or the coded parchments he supposedly found, or any of the other myriad facets to the story have to do with the entirely separate question of landscape geometry?

It was during this period that I first came across references to a poem called *Le Serpent Rouge* connected to the Affair. I must have read it somewhere, but I cannot say that this strange obscure work made a particularly strong impression on me initially. I did not for a moment imagine that it would play any part in what I was looking for. As it happened, I could not have been more wrong.

Chapter Two
LE SERPENT ROUGE

The Pamphlet

IN EARLY 1967, a small, crudely printed pamphlet with the full title *Le Serpent Rouge: Notes sur Saint Germain des Prés et Saint Sulpice de Paris* was deposited in the Bibliothèque nationale in Paris, as is legally required of every document published in France. It contained an enigmatic, three-page prose-poem, also with the same name, which has resisted interpretation for more than half a century.

What is it about? Who wrote it? Why was it published? This intriguing poem has been subject to considerable discussion, analysis and speculation, but these fundamental questions have never been satisfactorily answered.

Le Serpent Rouge was one of a stream of occasional publications that have come to be known as the *Dossiers Secrets*[19], produced over the course of several years during the mid to late 1960s in France by a small group of serious esoteric pranksters who called themselves the Priory of Sion. The exact purpose, membership and even existence of this group has been the subject of much speculation. Nevertheless, we can be reasonably certain that its activities were directed by the redoubtable Pierre Plantard, an elusive and mysterious character about whom much has been written.

Within the *Dossiers Secrets*, this series of strange, privately published writings, the Priory of Sion wove a complex multi-layered narrative, straddling the boundary line between fiction and history, which took as its starting point an obscure item of local French lore, the Affair of Rennes-le-Château, and used it as a platform on which to build a nest

---

[19] For a list of these texts, including titles and dates of publication, please see the end of the Bibliography in Appendix V.

of mystifications. In this elaborate game, which was in many ways typical of a certain playful strand of French literary production in the twentieth century, Plantard was assisted to some degree by at least two other colleagues, namely Gérard de Sède and Philippe de Chérisey. Both were accomplished writers who were active in various Parisian literary and esoteric coteries.

The Poem

Several articles on the mystery of Saunière and his unexplained wealth had appeared in local press in the south of France in the late 1950s, but it was not until the publication in 1967 of Gérard de Sède's *L'Or de Rennes*[20] that the affair was brought to the notice of a wider public. Based on source material from the *Dossiers Secrets*, this book offered a detailed narrative of how certain mysterious parchments were found by Saunière during renovations of the church, including reproductions of the presumed originals. It tells of his travels to Paris to seek help in deciphering them and his apparent discovery of treasure that followed his return from the capital. It describes his impressive building programme, his travels and friendships with famous names of the era, and his subsequent troubles with the Catholic Church when he failed to answer demands from the local bishop to account for the source of his wealth.

While *Le Serpent Rouge* certainly appeared within the context of the Affair of Rennes-le-Château as part of the Priory of Sion's *Dossiers Secrets*, the poem itself gives the distinct impression that it stands alone, somewhat separate and apart from the other material, as if it came from a different source. *Le Serpent Rouge* sets its own textual agenda and establishes a tone quite distinct from the rest of the Priory output. It contains clear references to the wider Rennes narrative, but it is quite unlike anything else which came from the pen of Plantard or his usual stable of writers and partners. *Le Serpent Rouge* does not pursue the same questions as *L'Or de Rennes*. It has nothing to say for example about Saunière or hidden treasure of any kind.

It is curious to note that *Le Serpent Rouge* is cited in the bibliography of *L'Or de Rennes* and yet is not actually mentioned anywhere in the text! For some reason, de Sède must have wanted to establish a connection between the poem and his book on Rennes-le-Château,

---

20 Gérard de Sède, *L'Or de Rennes ou La Vie Insolite de Bérenger Saunière*, (Paris, Julliard, 1967), (republished the following year as *Le Trésor Maudit de Rennes-le-Château* and translated by Bill Kersey in 2001 as *The Accursed Treasure of Rennes-le-Château*, Surrey, DEK Publishing.

both appearing in that crucial year of 1967 for our story. This reticence to discuss the poem openly, which we will encounter again, hints at the strange relationship between *Le Serpent Rouge* and the rest of the mystery.

Le Serpent Rouge is a short work, but it contains a wealth of learning and knowledge. It also offers a passageway into something else, something vast, like an antechamber which leads on to larger rooms beyond. It is a series of riddles, a guidebook, a map and an esoteric encyclopaedia. This story about an obscure French poem of the 1960s has much wider implications and ramifications.

The pamphlet in which the poem appeared was itself a very unusual document of just fifteen pages. The title page featured a fictitious coat of arms depicting a crudely drawn horse's head, a vase and a strange Rosicrucian seal. The names of three authors were cited on the cover, complete with their home addresses but, as we will see, these men almost certainly were not involved in any way whatsoever with the creation of the work.

On the second page was printed a short note to the reader including a quotation from a nineteenth-century French astronomer Abbé Moreaux.

> "Before reading the lines that follow, the reader is asked to remember that: '...after a long sleep, the same hypotheses are revived, no doubt returning with new and richer clothes, but the foundation remains the same and the new mask in which they are arrayed could not deceive the man of science...'"[21]

Below the quotation appeared a line drawing by the French esoteric writer Oswald Wirth, depicting a seated figure contemplating some blocks at his feet, to which a variation on a quotation from the poem has been added: *"découvrir une à une les soixante-quatre pierres..."*, or "discover one by one the sixty-four stones...". These two introductory pages were followed by the poem of *Le Serpent Rouge,* which occupied a further three pages. The full text of the poem both in the original French and in English translation may be found in Appendix III at the end of this book.

The remaining ten pages of the pamphlet were padded out with a haphazard collection, almost like a scrapbook, of various items: a few clippings of photocopied articles on historical aspects of Paris, some

---

21   *Le Serpent Rouge*. p. 2. translated.

(falsified) genealogies of the Merovingian kings, a selection of historical maps of France, a plan of the interior of the church of St Sulpice and various other images and random excerpts.

Written in a dreamlike, stream-of-consciousness style, *Le Serpent Rouge* is comprised of thirteen stanzas of dense, cryptic prose. The poem unfolds a narrative of a journey undertaken by an unnamed hero, around an unknown landscape, through various trials, which culminates in an encounter with an enormous red serpent, the *serpent rouge*.

This broad outline is easy enough to trace, but beyond this, the details are shrouded in ambiguity. Things are never quite made plain or fully resolved. Without knowing who is undertaking the journey, or where it takes place, or what the nature of the testing might be, or the identity of the *serpent rouge*, the poem has remained opaque to understanding.

Its language draws on an eclectic range of sources that encompass elements of French history, literature, astrology, art, alchemy, ecclesiastical architecture and fairy tales. It is replete with references to a host of highly obscure items which would tax the knowledge of the most dedicated esotericist.

And yet, despite the near-impenetrable tangle of words and the temptation to suspect it might all just be a jumble of meaningless nonsense, *Le Serpent Rouge* conveys a strong impression of coherence, in the same way, for example, that a surrealist painting can create an impeccable internal logic. Somehow, it manages to make deep sense on its own terms and conjures up a complete (if somewhat baffling) miniature world in its thirteen brief stanzas.

For all its relative obscurity, and the difficulties it presents in understanding, there can be little doubt that it is a serious work from an accomplished writer. But who was it, if it did not come from the pen of Plantard, or any of the other members of his Team? It has attracted high praise and significant interest over the years and inspired many attempts to understand and explain it, but to the best of my knowledge, no one has ever put forward a plausible suggestion as to the identity of the author.

Le Serpent Rouge remains a riddle, an unsolved esoteric puzzle whose origin and meaning have continued to elude the critics, the researchers and the conspiracy buffs for more than fifty years. It seems to be saying something important, but the nature of the message hovers tantalisingly out of focus, on the periphery of clear vision. The basic questions remain unanswered, and it has remained resolutely locked to understanding. If there are keys to be found that would enable the decoding of the poem, they have apparently not yet been discovered.

+ LE SERPENT ROUGE +

NOTES SUR SAINT GERMAIN DES PRÉS ET SAINT SULPICE DE PARIS
===

par

PIERRE FEUGÈRE
    LOUIS SAINT-MAXENT
        GASTON DE KOKER

****

Les exemplaires de cet ouvrage sont en vente à :

PONTOISE,	ARGENTEUIL,	ERMONT,
chez PIERRE FEUGÈRE	chez L. SAINT-MAXENT	chez G. DE KOKER
9, Rue des Cordelliers	63, Bd. Jean Allemanne	118, R. de Sannois

----

PONTOISE - 17 JANVIER 1967 - 3 Frs

Figure 11: Front cover of Le Serpent Rouge.

- 2 -

LE SERPENT ROUGE

* *

NOTES SUR SAINT GERMAIN DES PRES ET SAINT SULPICE DE PARIS

par PIERRE FEUGERE, LOUIS SAINT-MAXENT & GASTON DE KOKER

Avant de lire les lignes qui suivent,
Au lecteur de daigner se souvenir qu'

" ...après un long sommeil, les mêmes hypothèses
ressuscitent, sans doute nous reviennent-elles
avec des vêtements neufs et plus riches, mais
le fond reste le même et le masque nouveau
dont elles s'affublent ne saurait tromper
l'homme de science..."

Abbé Th. MOREUX
Directeur de l'Observatoire
de Bourges, page 10, du livre L'ALCHIMIE MODERNE.

...DÉCOUVRIR UNE AUNE
LES SOIXANTE QUATRE PIERRES...

PONTOISE - 17 JANVIER 1967 - 3 FRS

Figure 12: Le Serpent Rouge, page 2.

LE SERPENT ROUGE
NOTES SUR SAINT GERMAIN ET SAINT SULPICE DE PARIS

Avant-Propos

Comme ils sont étranges les manuscrits de cet Ami, grand voyageur de l'inconnu, ils me sont parvenus séparément, pourtant ils forment un tout pour celui qui sait que les couleurs de l'arc-en-ciel donnent l'unité blanche, ou pour l'Artiste qui sous son pinceau, fait des six teintes de sa palette magique, jaillir le noir.

Cet Ami, comment vous le présenter ? Son nom demeura un mystère, mais son nombre est celui d'un sceau célèbre. Comment vous le décrire ? Peut-être comme le nautonier de l'arche impérissable, impassible comme une colonne sur son roc blanc, scrutant vers le midi, au-delà du roc noir.

Dans mon pèlerinage éprouvant, je tentais de me frayer à l'épée une voie à travers la végétation inextricable des bois, je voulais parvenir à la demeure de la BELLE endormie en qui certains poètes voient la REINE d'un royaume disparu. Au désespoir de retrouver le chemin, les parchemins de cet Ami furent pour moi le fil d'Ariane.

Grâce à lui, désormais à pas mesurés et d'un oeil sûr, je puis découvrir les soixante-quatre pierres dispersées du cube parfait, que les Frères de la BELLE du bois noir échappant à la poursuite des usurpateurs, avaient semées en route quand ils s'enfuirent du Fort blanc.

Rassembler les pierres éparses, oeuvrer de l'équerre et du compas pour les remettre en ordre régulier, chercher la ligne du méridien en allant de l'Orient à l'Occident, puis regardant du Midi au

Figure 13: Le Serpent Rouge, page 3.

Nord, enfin en tous sens pour obtenir la solution cherchée, faisant station devant les quatorze pierres marquées d'une croix. Le cercle étant l'anneau et couronne, et lui le diadème de cette REINE du Castel

Les dalles du pavé mosaïque du lieu sacré pouvaient-être alternativement blanchies ou noires, et JESUS, comme ASMODEE observer leurs alignements, ma vue semblait incapable de voir le sommet où demeurait cachée la merveilleuse endormie. N'étant pas HERCULE à la puissance magique, comment déchiffrer les mystérieux symboles gravés par les observateurs du passé. Dans le sanctuaire pourtant le bénitier, fontaine d'amour des croyants redonne mémoire de ces mots : PAR CE SIGNE TU le VAINCRAS.

De celle que je désirais libérer, montaient vers moi les effluves du parfum qui imprégnèrent le sépulcre. Jadis les uns l'avaient nommée : ISIS, reine des sources bienfaisantes, VENEZ A MOI VOUS TOUS QUI SOUFFREZ ET QUI ETES ACCABLES ET JE VOUS SOULAGERAI, d'autres : MADELEINE, au célèbre vase plein d'un baume guérisseur. Les initiés savent son nom véritable : NOTRE DAME DES CROSS.

J'étais comme les bergers du célèbre peintre POUSSIN, perplexe devant l'énigme : "ET IN ARCADIA EGO..." ! La voix du sang allait-elle me rendre l'image d'un passé ancestral. Oui, l'éclair du génie traversa ma pensée. Je revoyais, je comprenais ! Je savais maintenant ce secret fabuleux. Et merveille, lors des sauts des quatre cavaliers, les sabots d'un cheval avaient laissé quatre empreintes sur la pierre, voilà le signe que DELACROIX avait donné dans l'un des trois tableaux de la chapelle des Anges. Voilà la septième sentence qu'une main avait tracée : RETIRE MOI DE LA BOUE, QUE JE N'Y RESTE PAS ENFONCE. Deux fois IS, embaumeuse et embaumée, vase miracle de l'éternelle Dame Blanche des Légendes.

Commencé dans les ténèbres, mon voyage ne pouvait s'achever qu'en Lumière. A la fenêtre de la maison ruinée, je contemplais à travers les arbres dépouillés par l'automne le sommet de la montagne. La croix de crête se détachait sous le soleil du midi, elle était la quatorzième et la plus grande de toutes avec ses 35 centimètres! Me voici donc à mon tour cavalier sur le coursier divin chevauchant l'abîme.

Figure 14: Le Serpent Rouge, page 4.

- 6 -

Vision céleste pour celui qui se souvient des quatre oeuvres de Em. SIGNOL autour de la ligne du Méridien, au choeur même du sanctuaire d'ou rayonne cette source d'amour des uns pour les autres, je pivote sur moi-même passant du regard la rose du P à celle de l'S, puis de l'S au P... et la spirale dans mon esprit devenant comme un poulpe monstrueux expulsant son encre, les ténèbres absorbent la lumière, j'ai le vertige et je porte ma main à ma bouche mordant instinctivement ma paume, peut-être comme OLIER dans son cerceuil. Malédiction, je comprends la vérité, IL EST PASSE, mais lui aussi en faisant LE BIEN, ainsi que CELUI de la tombe fleurie. Mais combien ont saccagé la MAISON, ne laissant que des cadavres embaumés et nombres de métaux qu'ils n'avaient pu emporter. Quel étrange mystère recèle le nouveau temple de SALOMON édifié par les enfants de Saint VINCENT.

Maudissant les profanateurs dans leurs cendres et ceux qui vivent sur leurs traces, sortant de l'abime où j'étais plongé en accomplissant le geste d'horreur : " Voici la preuve que du sceau de SALOMON je connais le secret, que REINE de cette REINE j'ai visité les demeures cachées. " A ceci, Ami Lecteur, garde toi d'ajouter ou de retrancher un iota ... Médite, Médite encore, le vil plomb de mon écrit peux contient peut-être l'or le plus pur.

Revenant alors à la blanche coline, le ciel ayant ouvert ses vannes, il me sembla près de moi sentir une présence, les pieds dans l'eau comme celui qui vient de recevoir la marque du baptême, me retournant vers l'est, face à moi je vis déroulant sans fin ses anneaux, l'énorme SERPENT ROUGE cité dans les parchemins, salée et amère, l'énorme bête déchainée devint au pied de ce mont blanc, rouge de colère.

Mon émotion fut grande, "RETIRE MOI DE LA BOUE" disais-je, et mon réveil fut immédiat. J'ai omis de vous dire en effet que c'était un songe que j'avais fait ce 17 JANVIER, fête de Saint SULPICE. Par la suite mon trouble persistant, j'ai voulu après réflexions d'usage vous le relater en conte de PERRAULT. Voici donc Ami Lecteur, dans les pages qui suivent le résultat d'un rêve m'ayant bercé dans le monde de l'étrange à l'inconnu. A celui qui PASSE de FAIRE LE BIEN !

Octobre 1966
l'Auteur,
LOUIS SAINT-MAXENT

Figure 15: Le Serpent Rouge, page 5.

In the final stanza, it is revealed that the action has all taken place within a dream. Ultimately, it is the logic of the dream-world that governs this strange universe and points the way to the path of understanding.

Where Does the Action of the Poem Take Place?

The geographical setting of *Le Serpent Rouge*, as befitting a dream, is fragmented, or layered, so that things seem to be taking place simultaneously in different spaces. One of these layers, at least, is clearly identified. It is the ancient Roman Catholic church of St Sulpice in Paris, the second largest in Paris, dating from the thirteenth century.

This is confirmed both by the full title of the pamphlet itself, as well as by the multiple unambiguous references in two stanzas – Virgo and Scorpio – to specific people, art works and architectural features connected to the church.

The other is an unnamed wooded landscape, where the walk itself takes place, which can nevertheless also be positively identified. It is undoubtedly set in the area around Rennes-les-Bains, as it contains several clearly recognisable references to local landmarks, a topic which will be explored further in Part Two. The reference to the meridian of St Sulpice in the Scorpio segment seems to hint at some connection between the two locations in which the action of the poem is set, both in the landscape and in the Paris church.

There is something else to notice about the spatial geography in which the poem takes place. The hero, and the journey, proceed around a zodiac, as indicated by the stanza headings. What is this doing here?

What is the Nature of the Zodiac in the Landscape?

As printed in the original pamphlet, each of the thirteen stanzas of *Le Serpent Rouge* is marked with a small glyph representing one of the signs of the zodiac. In this way, the poem is arranged in a zodiac format, with one paragraph for each of the signs. Hence, as the hero moves from one stanza to the next, he passes through the various zodiac signs. The zodiac is, of course, usually considered to be found amongst the stars of the night sky. How then are we to understand the poet's notion of a zodiac overlaid or embedded in a landscape in the south of France?

There are some unusual aspects to this zodiac, in addition to its being somehow on the land rather than in the sky. As there are thirteen stanzas, a further sign additional to the usual twelve is required to match the structure of the poem to the zodiac. This is fulfilled by the inclusion of the sign of Ophiucus between Scorpio and Sagittarius to form a thirteen-sign format.

One further curiosity is that the first stanza of the poem is allocated to Aquarius, rather than Aries, which is traditionally considered the first sign of the zodiac. After Aquarius, the allocation to the stanzas proceeds in the usual order: the second is matched with Pisces, the third to Aries, the fourth to Taurus, and so on.

The sequence continues around the poem and concludes with the tenth stanza paired with Scorpio, the eleventh with Ophiucus, the twelfth with Sagittarius, and the thirteenth and final stanza is linked with Capricorn.

The Riddles

In addition to these unresolved issues relating to its origin and location, there is a sequence of specific questions which are posed by *Le Serpent Rouge* itself. Some of these take the form of actual riddles explicitly inserted within the text, while other queries are rather suggested or implied. There are descriptions of people and places and things which are full of specific detail, yet curiously veiled at the same time. It becomes apparent that identities within the poem have been omitted or erased, and there is an implicit challenge to the reader to find them out.

The core outline of the narrative of the poem, and some of the key questions which it poses, can be summarised as follows.

The stanzas are numbered, and I have added their zodiac signs in French and English as indicated by small glyphs in the original text.

1. *Verseau* /Aquarius
   "How strange are the manuscripts of this friend, the grand voyager of the unknown."

The opening sentence of the first stanza and the poem presents an unnamed person, *"cet ami"* or "this friend", who is also the "grand voyager of the unknown". Who is he? Why is he referred to by these names? What are these "strange manuscripts" of his? There are also a number of references to colours. Why are these significant?

2. *Poissons* /Pisces
   "This friend, how would you know him? His name remains a mystery, but his number is that of a famous seal."

The unnamed figure from the first stanza appears again. Now the reader is presented with a riddle and challenged to work out the mystery of his name. His number, we are told, is that of a "famous seal". What does this mean? Here is the master question at the heart of the poem. The revelation of the secret identity of the "grand voyager of the unknown" will unlock the hidden layers of *Le Serpent Rouge*.

3. *Bélier* / Aries

> "In my initiatory journey, I tried to hack with my sword
> a way through the tangled vegetation of the woods. ...
> Desperate to find the way, the parchments of this Friend
> were for me, the thread of Ariadne."

In the third stanza, we meet the narrator, the central character of the poem. Again, he is not named, but simply refers to himself in the first person. He tells us that he is undertaking a *"pèlerinage éprouvant"*[22], or an "initiatory journey", through a wooded landscape, and that the "manuscripts" of the "friend" from the first two stanzas have been the *fil d'Ariane*, or thread of Ariadne, by which he has found his way in this labyrinth. He states the goal of his "pilgrimage", which is to reach the place of the "sleeping beauty", a mysterious queen of a lost kingdom.

4. *Taureau* / Taurus

> "with measured steps and a sure eye, I am able to discover
> the sixty-four scattered stones"

The fourth, fifth and sixth stanzas present a series of clues that seem to be cryptic instructions of a geometric nature, involving sixty-four "stones". Whatever these might be, some ordering amongst them has been lost, and the poem challenges the reader to find it and restore it.

5. *Gémeaux* /Gemini

> "Gathering together the scattered stones, working with
> the square and compass to put them back in order, looking
> for the line of the meridian going from East to West, then
> looking from the South to the North"

Restoring the order of the scattered stones apparently requires geometric tools, and involves a search for a meridian.

6. *Cancer* /Cancer

> "The stones of the mosaic pavement of the sacred place
> could be alternately white or black."

The description of these stones suggests a chessboard pattern of black and white squares.

---

22 The word is misspelled *péleŕinage* in the poem.

7. *Lion*/Leo
>"ISIS: queen of the benevolent springs."

The narrator arrives at a spring, the place of the queen, also known as Isis, or Madeleine.

8. *Vierge*/Virgo
>"when the four horsemen jumped, one horse had left four hoofprints on the rock."

An enigmatic image of hoofprints on a rock amidst references to Poussin, Delacroix and the church of St Sulpice in Paris.

9. *Balance*/Libra
>"The cross of the crest stood out under the midday sun, it was the fourteenth and the tallest of all at 35 centimetres!"

The journey around the landscape continues. What are these crosses on crests? Why are there fourteen of them and why the very specific dimension?

10. *Scorpion*/Scorpio
>"the Meridian line ... I turn around looking from the rose of the P to that of the S, then from the S to the P"

A reference to the famous meridian in St Sulpice, Paris, which is marked with brass inlaid letters, a P and an S.

11. *Ophiucus*/Ophiucus
>"Here is the proof that the seal of SALOMON I know the secret ... the base lead of my writing contains perhaps the purest gold."

This stanza contains certain curious remarks addressed directly to the reader. It speaks of a secret involving the Seal of Salomon, or Solomon. There are also several apparent references to alchemy.

12. *Sagittaire*/Sagittarius
>"Then returning to the white hill ... facing me I saw, unrolling endlessly his coils, the huge RED SERPENT quoted in the parchments."

The climax of the action of the poem takes place in his penultimate stanza. The narrator arrives back where he began, at the "white hill", where he has an encounter with an enormous red serpent, the *serpent rouge* of the poem's title. Who or what is this serpent? Where is this white hill? What does the encounter represent? Why is the beast described in these terms? The author is presenting the reader with an enigma. We cannot understand this poem if we do not know the identity and role of the *serpent rouge* itself.

13. *Capricorne* /Capricorn
> "I awoke immediately. I have omitted to tell you in fact
> that this was a dream that I'd had."

The final stanza brings the journey and poem to their conclusion. The narrator awakes; it was all a dream.

To summarise the story: the action of *Le Serpent Rouge* unfolds in the woods, hills and valleys of an unnamed landscape, with the progress calibrated by a 13-sign zodiac, beginning with Aquarius, including Ophiucus and ending with Capricorn.

Within this zodiac framework, the narrator, also unnamed, assisted by the manuscripts of his "friend", the "grand voyager of the unknown" undertakes a pilgrimage or initiatory journey. After various adventures, the traveller comes face to face with an enormous red serpent whose identity is equally shrouded in uncertainty.

There is a tension here between coherence and ambiguity. We can grasp the basic outline easily enough, but we understand nothing because all identities are blurred. This is without even considering the various cryptic riddles and obscure references sprinkled throughout the poem.

It is not just the contents of the poem which are shrouded in mystery. There are also some very puzzling aspects surrounding the circumstances of its publication and appearance. Perhaps the strangest of these concerns the real identities of the three men who were named on the cover, and whether these were indeed the genuine authors.

Who Wrote the poem?

The title page lists the authors as Pierre Feugère, Louis Saint-Maxent and Gaston de Koker. None of the three are mentioned anywhere else in the Priory of Sion materials, nor do they appear to have any other literary or esoteric footprint.

On the face of it, the case for their collective authorship seems straightforward enough. The original catalogue slip for the pamphlet at the Bibliothèque nationale shows that *Le Serpent Rouge* was submitted to the library for inclusion on 15 February 1967. It also lists the same three names as the authors and is signed by one of them, Pierre

Feugère. Six weeks later, on 20 March 1967, the slip was stamped to denote that the pamphlet had been officially accepted for deposition, and a catalogue number was allocated. So, the library catalogue document is consistent with the title page, and it even carries the signature of one of the men.

Yet there is a serious problem. In 1978, a French researcher, Franck Marie, investigating the identities and backgrounds of the three authors, made an unsettling discovery. He learned that three men with (almost) identical names had died in separate instances of apparent suicides by hanging, all within the Paris Metropolitan area, during the same 24-hour period over 6 and 7 March 1967.

If these three men who died were indeed the authors of *Le Serpent Rouge*, then their deaths took place during the six-week period after the pamphlet was submitted to the Bibliothèque nationale, and before it was accepted for deposit. These circumstances obviously raise disturbing questions.

Franck Marie interviewed the families of the men and was informed that none of them had any interest or involvement in literary, occult or esoteric activities whatsoever, nor did they seem to be known to each other.[23] Their untimely deaths in similar circumstances on the same weekend seemed to be nothing but a macabre coincidence, which had apparently been co-opted for some reason by the pamphlet publisher.

On further investigation, Marie was able to establish some details with a degree of confidence. The three men had definitely existed, and their deaths had indeed taken place on the weekend of March 6/7, according to the death certificates that Marie obtained. These confirmed the details of their names and addresses, with some minor but curious discrepancies.

If we assume that the death certificates as reproduced in Marie's report are accurate, then for each of the three men there is precisely one error in their full name and address as they are given on the cover of *Le Serpent Rouge*.

In addition, the discrepancy occurs in a different place in each case. For M. Feugère, his street name is cited incorrectly as rue des Cordelliers (rather than rue des Cordeliers). For M. Saint-Maxent, the house number is incorrect (53, rather than 33). Finally, for M. DeKoker, his first name is incorrectly given as Gaston (rather than Gustave). These carefully choreographed errors do not seem to suggest carelessness,

---

23   Franck Marie, *P.F... le Serpent Rouge preuves*, (Malakoff, Editions S.R.E.S. Vérités Anciennes, 1979).

but, on the contrary, a certain meticulous, or even obsessive, attention to detail.

Franck Marie's real breakthrough came when he was able to prove that the library catalogue slip had been forged and the details falsely backdated. He was able to identify the typewriter on which the deposit slip was typed as the same one used to type the poem itself. Further, it is the same typewriter used elsewhere to create the *Dossiers Secrets* and almost certainly belonged to Pierre Plantard. He may have typed out the poem, but this of course is not the same as having authored it. It does seem likely though that Plantard was involved in the mystification surrounding the named authors as he was responsible, if Franck Marie was correct, for creating the false deposit slip.

His conclusion was that the person or persons behind the production of the *Le Serpent Rouge* pamphlet learned of the three deaths by hanging, which had taken place around the time when the pamphlet was ready for deposition, in the second or third week of March 1967, and decided, for some unknown reason, to create the false impression that the men were the three co-authors. They accomplished this by obtaining a blank deposit slip from the Bibliothèque nationale and filling it in with the dead men's names as the authors of the pamphlet. They then backdated it, to February 15, and, somehow, managed to have it authorised with the correct official stamps, and entered into the records of the library, without the ruse being detected. On March 20, it was officially approved and became part of the collection, completing a very bizarre sequence of events.

We can be confident, therefore, thanks to the diligent research of Franck Marie, that the three named men had nothing whatsoever to do with the publication of *Le Serpent Rouge*. What could possibly have motivated the perpetrators to go to such lengths to create mystification?

Perhaps more importantly: if these three named men were not responsible for the poem, as it appears, then the identity of the true author remains unknown. Who was it? Why did the author choose to remain anonymous and even to have his identity concealed behind those of the three men who had been found dead?

Summary

With that, we conclude this overview of the main problems posed by the poem *Le Serpent Rouge*, intended as a brief sketch, or map, of where we are heading. These are the questions I will tackle in the pages that follow. It should be made clear that I will certainly not aim to unpack

```
                                    0/
         RÉGIE                                                    19
       DU DÉPÔT LÉGAL              N° d'enregistrement
                                  DL_20 3 1967 _ 04 _97

                                 Cadre réservé à l'Administration
```

Je soussigné : (PIERRE FEUGERE

Représentant légal

demeurant à (PONTOISE

rue (des Cordelliers n° 9

agissant en qualité d'Éditeur, déclare avoir adressé ce jour à la Régie du Dépôt Légal à la Bibliothèque Nationale à Paris, en
 1 exemplaires, l'ouvrage désigné ci-dessous, accompagné de la présente déclaration en triple exemplaire :

Auteurs PIERRE FEUGERE,
         LOUIS SAINT MAXENT,
         GASTON DE KOKER.

Titre : LE SERPENT ROUGE
       NOTES SUR ST. GERMAIN DES PRES ET
Texte imprimé par : ST. SULPICE DE PARIS
                    LES AUTEURS

Planches imprimées par :

Broché par : LES AUTEURS

Achevé d'imprimer le : 17 JANVIER 1967
                       à PONTOISE

Format : 21 x 27

Prix : 3 FRANCS

Date de mise en vente ou en distribution : JANVIER 1967

Chiffre déclaré du tirage : CENT EXEMPLAIRES

        A PONTOISE   le   15 FEVRIER 07
              Signature du Déposant :

(SPECIMEN ... / ou : formules imprimées en vente au Cercle de la Librairie / 117, boul. St-Germain, PARIS)

Figure 16: Deposit slip for Le Serpent Rouge at Bibliothèque nationale, Paris. (Image credit: Franck Marie, P.F...? le Serpent Rouge Preuves (Malakoff, Editions S.R.E.S. Vérités Anciennes, 1978) p.19.)

or explain every phrase or line or reference of the short work, but rather to assemble a frame of reference which will permit a coherent reading of the poem. I can now summarise the main questions from this chapter which are provoked by the poem:

The first is simply: what is it all about? We can recognise the broad outlines of the action as a journey undertaken by the unnamed narrator, but the details are deliberately obscured beneath the surrealist language. What is the nature of this walk? Why was it considered significant enough for the narrator to present us with his account of it?

The next set of questions relate to the identities of its subjects: who is it about? Who is the "grand voyager of the unknown"? Why is his number that of a "famous seal"? Who is the narrator? Who is the "queen of the lost kingdom"? And crucially: who or what is the *serpent rouge*?

Then there are questions of location: where does it take place? Can the path followed by the narrator be identified in the real world? What is the significance of the zodiac signs allocated to the stanzas and how do these relate to the location of the poem's setting?

There are also a number of specific questions posed within the text of the poem itself. What are the "64 scattered stones", and how are these to be reassembled? What is the secret of the Seal of Solomon? Why is the meridian of St Sulpice brought into the poem?

Beyond the contents of the poem itself, there are questions about how and why it came to be written. What was the motivation for its creation and the context in which it appeared? What are the sources and influences out of which it has been assembled? What are its literary antecedents? How is it related to other works?

Finally, and perhaps most crucially, there are questions surrounding its authorship. Who was the anonymous author? What was the reason for the mystification surrounding the three named authors on the front cover, and their untimely deaths? How does this relate to the overall strategy behind the poem's publication?

When I first encountered *Le Serpent Rouge*, it never crossed my mind that I might have anything to contribute to unravelling its mysteries. My focus and attention were firmly directed at the questions of lines on maps, and my interest in the poem remained purely incidental as part of familiarising myself with the broader topic of the Affair of Rennes.

But in an entirely unanticipated twist of fate, I stumbled on some seemingly insignificant scraps of information in unrelated locations which, as it turned out, were to have unexpected and far reaching consequences for my understanding of both the poem, the identity of the author and the geometry in the landscape.

Chapter Three
FIRST INKLINGS

A FORTUITOUS ACCIDENT of reading led me to the first clues. In my small, much-loved library, I had acquired several volumes of the writings of Carl Jung on alchemy. These included *Psychology & Alchemy* (1953) and his final book, the summation of his lifework, *Mysterium Coniunctionis* (1955).[24]

Both are rich and profound works that draw on a wide array of alchemical, esoteric, literary and historical sources. They are copiously illustrated with a remarkable selection of images collected from many old and often rare alchemical volumes. These texts are certainly challenging, and not to be read quickly, but they are also extremely rewarding and amply repay the effort required to digest them.

I dipped in and out of these two books when the mood took me. The images were a visual feast, and the text was equally fascinating. I began the long process of attempting to come to grips with Jung's complex understanding of alchemy and the true nature of the alchemist's work.

As I did so, something began to stir in my intuition in response to certain passages and themes. At first it was little more than a fleeting thought, to which I did not pay much attention. The deeper I immersed myself in Jung's dense difficult work, however, the more I began to take it seriously. I had the sense that whoever had composed *Le Serpent Rouge* must surely have also read these books – and certain crucial passages in particular.

It was not because there were any direct extended quotations from the Jung text appearing in the poem, though there were certainly

---

24   C.G.Jung, *Psychology and Alchemy* (London, Routledge, 1968) (originally published in German in 1944, and appearing in English in 1953).
C.G.Jung, *Mysterium Coniunctionis: An Inquiry into the Separation and Synthesis of Psychic Opposites in Alchemy*, (London, Routledge, 1970) (first published in German in 1955, and translated into English in 1963).

resonances between terms and concepts in both texts. (We will look at a striking cluster of parallels shortly.) Rather, it was a more subtle sense that the architecture, format and even the narrative shape of the poem seemed to be conscious expressions of the themes which underpinned Jung's reading of the alchemical texts.

I was not able to prove this, by any means, at this early stage, but neither could I shake the growing feeling that Jung's alchemical works were, in some sense, precursors of *Le Serpent Rouge*, and that the poem had been directly inspired by some of his specific and unique insights in these texts.

While the ideas I am referring to are found throughout Jung's works, there were two passages in particular that led me to this conclusion. The first was a relatively short discussion leading to page 368 in *Psychology and Alchemy* on certain medieval alchemical texts in which the work of the alchemist is presented in allegorical terms as a mythical journey. Jung introduces this notion of the alchemical opus as world voyage through several examples of various ancient texts which treat of the travels and wanderings of certain legendary heroes, including Hercules, Enoch and Osiris.

The second passage, in *Mysterium Coniunctionis,* is much longer, and is an expanded commentary on these same ideas as developed later in his career. It is based on a case study of one of these allegorical journeys, an account written by the alchemist Michael Maier in the sixteenth century. In this discussion, Jung explores in detail, using this text, why the central task of the alchemist was sometimes depicted as a journey of a heroic figure around an imagined landscape, symbolically constructed to represent the circle of the world. He arrives at some deeply fascinating insights.

For Jung, the apparent task of alchemy – the transformation of base metals into gold – was not the true goal of the sincere alchemist. Rather, this physical aspect of the work was merely the outer display of a psychological process that could be characterised as the struggle to balance the contradictory forces operating within the inner man, or woman. The genuine aim of the alchemist's efforts was to resolve these tensions in order to achieve a restoration of wholeness within the soul.

The physical aspects of the manipulations performed by the alchemist in their laboratory, with their furnaces and crucibles, could be understood as projections of these internal struggles. The "gold" which was sought was therefore, in truth, the psychological treasure of wholeness, rather than temporal riches.

Jung's reading of alchemical texts and treatises considered these works therefore as maps of psychological states and journeys towards the integration of soul forces within the practitioners, rather than recipes for the chemical manipulation of substances. This is the very briefest of summaries of his thought, which we will discuss in greater detail later in this book.

For now, we turn to one highly specific parallel which left me in little doubt that there must be a connection of some kind between Jung's alchemical works and *Le Serpent Rouge*.

The Salty and the Bitter

Jung devotes considerable space in *Mysterium Coniunctionis* to a discussion of the concepts of salt and certain related ideas. This extended passage, found in Chapter V of Part III, begins with this sentence:

> "In this section I shall discuss not only salt but other symbolisms that are closely connected with it, such as the 'bitterness' of the sea, sea-water and its baptismal quality, which in turn relates it to the 'Red Sea'".[25]

Remarkably, several of these terms in this passage occur within close proximity in the Sagittarius stanza of *Le Serpent Rouge*. The phrase *"salée et amère"*, which translates as "salty and bitter", is used to describe the *Red Serpent*, and there are references to water and to baptism immediately preceding.

Here is an excerpt from the stanza in English translation:

> "... close to me I felt a presence, feet in the water as one who has just received the mark of baptism...the enormous red serpent cited in the parchments, salty and bitter, the enormous beast..."

The tight cluster of related concepts – salt, bitterness, water, baptism, redness – occurring in both texts suggested to me that this passage of Jung's work might have provided direct inspiration for the Sagittarius stanza of *Le Serpent Rouge*

It certainly fuelled my growing suspicion that Jung's two books must have played a significant, specific role in the thinking life of the person who composed *Le Serpent Rouge* in late 1966/early 1967. The dates matched: the English edition of *Psychology and Alchemy* appeared in 1953, and *Mysterium Coniunctionis* in 1963. The themes matched:

---
25   C.G.Jung, *Mysterium Coniunctionis*. op. cit. p. 183.

whoever wrote *Le Serpent Rouge* was clearly fascinated by the notion of ritual journeys around symbolic landscapes, and the poem was even sprinkled with several alchemical references. Finally, there were direct textual resonances including the "salty and bitter" quotation. It seemed to me that a good case could be made for some kind of connection between the poem and Jung's work.

Nevertheless, I did not seriously consider that this slight insight would amount to much. After all, many people, certainly in European academic and literary circles, were reading Jung in the 1960s, so it can hardly have narrowed down the field much. But it was a tantalising sensation, nevertheless, to have found this thread from which the poem appeared to have been partially woven, and to be following in the author of *Le Serpent Rouge*'s footprints as I read and reread these intriguing passages of Jung.

Meanwhile, I was also pursuing possible connections between the landscape geometry at Rennes as claimed by Lincoln, and similar activity at other locations which might have been suggested by other observers, so I was actively searching for books that might provide some historical or cultural context into which this discovery of his might fit. Then, around this time, I first came across the research of Professor Jean Richer, the French author and academic whom we met in the prologue.

In 1994, SUNY Press (NY) published *Sacred Geography of the Ancient Greeks*, an English translation by Christine Rhone of Professor Richer's classic 1967 book *Géographie Sacrée du Monde Grec*.[26] Though he had authored a large collection of writings in his native France, this was the first time any substantial work of his had appeared in the English language.

Sacred Geography of the Ancient Greeks describes in detail Richer's discoveries of alignments in the landscape of Ancient Greece, and I read it with great interest. Although it deals broadly with the same topic, it is a very different book to *The Holy Place*, in every way.

Richer's work examines the relationship of the alignments he finds to Greek culture, religion and iconography, and positions it without hesitation at the centre of the ancient philosophical conception of the construction of the world and the organisation of space. As an academic, he brings a methodical approach to assembling his data and interpreting the results. My initial impression was that the evidence Richer had

---

26 Jean Richer, trans. Christine Rhone, *Sacred Geography of the Ancient Greeks: Astrological Symbolism in Art, Architecture and Landscape*, (New York, SUNY Press, 1994). Jean Richer, *Géographie Sacrée du Monde Grec*, (Paris, Hachette, 1967).

FIRST INKLINGS

Revenant alors à la blanche coline, le ciel ayant ouvert ses vannes, il me sembla près de moi sentir une présence, les pieds dans l'eau comme celui qui vient de recevoir la marque du baptême, me retournant vers l'est, face à moi je vis déroulant sans fin ses anneaux, l'énorme SERPENT ROUGE cité dans les parchemins, salée et amère, l'énorme bête déchainée devint au pied de ce mont blanc, rouge de colère.

5. SAL

a. Salt as the Arcane Substance

In this section I shall discuss not only salt but a number of Symbolisms that are closely connected with it, such as the "bitterness" of the sea, sea-water and its baptismal quality, which in turn relates it to the "Red Sea". I have included the latter in the scope of my observations but not the symbol of the sea

Figure 17: Comparison of Sagittarius stanza of *Le Serpent Rouge* with the opening of Jung's Part 5. Sal in *Mysterium Coniunctionis*. Notice the parallel terms in French and English: 'salée et amère' (salty and bitter), 'baptême' (baptism), and 'rouge' (red).

assembled was meticulous, detailed and overwhelming. The many striking examples of alignments and symbolic correspondences that he described were more than enough to convince me that his insights were based on a real historical tradition, and that Richer's discovery was genuine and tremendously important. In my view, he had indeed discovered a vital key for the understanding of Greek culture.

The descriptions of the alignments discovered in Greece and beyond were quite different from any of the books I had previously come across. They were very extensive and widespread. They appeared to be governed by a comprehensive overall system and combined to form grids and other forms. Traces of the alignments Richer had found were integrated into place names and local mythology.

There was something else that caught my attention. A short paragraph on the very first page of the first chapter made a curious impression on me. It read as follows:

"I was intrigued by a certain detail of the cult at Delphi: the sacred drama of the struggle of Apollo with the serpent

Python was performed at Delphi about every eight years. Now, the serpent is not only a symbol of the earth, it is also a representation of the path of the sun in the zodiac."

Something about this quotation reminded me of *Le Serpent Rouge*. Though it was by no means a direct reference, I was struck by the combination of identical elements in the paragraph and the poem – serpent, path, sun, zodiac – particularly given that Richer's book had been published in 1967, the same year as both the poem and Gérard de Sède's book, which launched the Rennes Affair.

I had set out to read Richer's book for potential insight into the landscape geometry of Rennes and I had not been disappointed. It seemed to hold significant promise for providing a historical and cultural framework within which the idea of ancient landscape intervention at Rennes might be understood; it was an uncanny twist to also find these resonances with *Le Serpent Rouge*. I started to think more closely about this strange poem, its origins and its connections to the rest of the mystery.

These questions hovered in my mind for years, but I never found any particular resolution or further breakthroughs. I began to suppose that I had taken things about as far as they would go. I moved on with other projects but these puzzles around the Rennes Affair never completely faded from my thoughts.

Then, in the mid-2000s, out of the blue, everything changed.

Chapter Four
SIGHTLINES

Rennes-le-Château Calling

ON THE evening of Saturday 24 June 2006, I was invited to dinner in a small outback town in the remote desert of northern South Australia. As we sat down to enjoy a delicious roast, the phone rang. Our host announced that it was her old friend, Michael, calling out of the blue from his home in Rennes-le-Château, France. Knowing of my interest in the Affair, she wondered if I would like to speak with him. My potatoes were cold by the time I returned to the table.

Michael was an Irishman, a jeweller, an author and a druid who had spent 25 years living alone on the West Coast of Ireland. He had moved to France several years earlier and taken a house in a small village in the Haute Vallée. He was also, as it happened, fascinated by the landscape mysteries of the area. As a practising druid with many years' experience, he brought a different perspective from the usual parade of treasure hunters, *chercheurs* and opportunists.

We began a correspondence and a friendship. Several months later, an unexpected business trip to Europe offered the opportunity to make my first visit to Rennes-le-Château. Michael very kindly offered to host me and show me around. He was a wonderful character and gave me a generous and very personal welcome to the Haute Vallée. He took me to many special locations, from churches to standing stones, and introduced me to the delights of the local food and wine, the cafés and the wonderful Sunday markets in Esperaza.

One day he announced we would take a drive to a place I had never heard of, Quéribus Château. This was one of the 'Cathar' châteaux, so-called because they had become places of refuge for Cathar men and women during the time of their persecution by the Catholic Church, culminating in the terrible events of the decades of the early fourteenth-century. These châteaux had mostly been built by the Templars, often on the sites of earlier fortifications and earthworks.

After driving for an hour towards Perpignan and the Mediterranean coast, we came around a bend in the road to see the château of Quéribus perched on an impossibly high ridge far above us. A wave of emotion swept over me as I glimpsed it for the first time, an awe-inspiring crumbling pile of ancient stone walls and turrets. I did not know at the time this was to be just the first of many visits to this extraordinary château replete with mysteries over the coming years.

Michael also shared with me many of his insights into the wonders and secrets of the landscape and the area. He was well-versed in the history and culture of the region and in the backstory of the Affair. He had discovered some curious items that seemed to have gone unremarked in any of the published accounts of the area. For example, he had found a special alignment between certain ancient sites in the landscape that coincided precisely with the direction of the Winter Solstice Sunrise, a bearing of 123° at this latitude. This was a significant observation, in my view, which I took as a further reinforcement of the suggestion of conscious intent on the part of some ancient landscape engineers.

I was completely captivated by the deep beauty and charm of the landscape and could only wonder why it had taken me so long to finally visit. After having been immersed in the literature, geography and history of this place for over a decade whilst living on the other side of the world, it was as if a light had been switched on. A world I had only known through a map suddenly became a living reality.

Over the next few years, I began spending extended periods of time in France, travelling back and forth from Australia. I met Judith on one of these trips. One thing led to another, as these things do, and we fell in love. Eventually, in late 2009, I moved to France and for the next few years, we lived in a wonderful narrow terraced house in the middle of the village, not too far from Rennes-le-Château itself. In 2010 we were married, by the mayor, in the *mairie* (town hall) of the village of Fa, in front of friends and family. We hung boughs of wild laurel gathered that morning from the surrounding hillsides throughout the house and had a lovely wedding party that night in celebration.

It was a joyful period, and an incredible turn of events to be living in the landscape I had contemplated from afar. I made many friends and learned so much that was new to me. There was a surreal aspect to all of this: the terraced house we called home was a tiny square on the very map over which I had obsessed all those years earlier.

I set out to visit as many of the places I knew so well from my study of the map as I could. We explored far and wide: the villages, the winding pathways, the mountain peaks, the rivers and streams, the churches, the châteaux. I walked the crests, climbed the ridges and followed paths through forests that had undoubtedly changed little for centuries, if not millennia. It was thrilling to be able to observe the territory first-hand, to visit and see with my own eyes the alignments in person at last.

Places I had known as impossibly far-off and exotic locations on the other side of the world were now in reach of short walks from the village. The names that had become part of my inner world – Rennes-les-Bains, Arques, Lavaldieu – became familiar territory, places to meet friends for coffee or lunch. The rivers and streams, the pathways through the forests, the peaks and lookouts that had once been mere words on the surface of the map, became part of my daily reality.

I had a huge empty wall in the attic of our house, on which I had mounted the original 1:25,000 scale map of the area immediately around the two Rennes. As the area of investigation grew beyond its boundaries, I added adjoining maps. I expected that I would soon find the edges of the geometry, but to my surprise and slight consternation, this was not the case.

Meanwhile, another map resource was proving useful. The resolution of the landscape in this area on Google Earth was quite coarse in the years before 2006, but it was steadily improving. By 2008, the changes were dramatic and, at last, I could see the meridians in high resolution, from any angle, on my computer screen. I began to build a detailed model within the software of the alignments I had found. It was thrilling to be able to explore the terrain in this manner.

The combination of real-world excursions into the landscape, the expanding map-wall and the new high-resolution Google imagery was extremely powerful. After every walk or excursion, I would review my personal experiences on both the map and the digital model.

Being able to visit these alignments for myself, to observe them with my own eyes, to walk them and to explore them with the map and Google Earth, any lingering doubts as to the reality of what I had begun to call The Complex, dissolved. Indeed, the impact of many of these only becomes fully apparent when viewed in person in the landscape from a suitable vantage point. Relationships between locations and alignments were revealed.

Amid this, there was also progress in other areas. It was easy to access books and other materials which would otherwise have been unobtainable. I met many incredible people and learned so much about the local history, culture, way of life and landscape.

I found myself drawn irresistibly into absorbing and reading as much as I could about this part of France, including naturally the questions surrounding the *Affaire de Rennes*. Between the landscape, the people, the maps, and the books there was always an adventure of some kind unfolding. There were walks around Rennes-les-Bains, trips to Rennes-le-Château and outings to all kinds of places near and far, well-known and remote, public and private.

The Cathar Châteaux and Sighting Lines

Judith and I also took up the challenge of visiting as many of the ancient châteaux as we could – and there are more than fifty just within the Languedoc region. Many are situated in spectacular locations, on the highest point of ridges, or rocky outcrops, or peaks. Each is unique in its design, siting and scale. Each has its own special sense of place, of deep antiquity. Some became favourites we would return to regularly. Some were in remote out-of-the-way corners of the Languedoc and were not easy to access.

Close to Fa, the village where we lived, stood an ancient tower, the crumbling remains of a larger château, occupying the summit of a conical hill. The age of the Tour de Fa, as it is known, was uncertain: the consensus of local historians was that it was built sometime between the sixth and the fourteenth centuries, which did not seem to narrow it down very much. A brisk ten-minute walk from our house, in time to catch sunrise, was a great way to start the day, before the van from the *boulangerie* pulled up at our front door, every day at 8 am, with fresh croissants and baguettes.

One morning, I noticed that it was just possible to see the bell tower of the Espéraza Church from the Tour at the far end of the valley, about two kilometres away, via a clear line of sight that managed to avoid all of the intervening hills. When I checked on the map, I found to my surprise that the church was also exactly due east of the Tour de Fa. To my even greater surprise, I then found that these two sites formed a near-perfect equilateral triangle with the church in Campagne-sur-Aude to the south, the local headquarters of the Knights Templar in the

Figure 18: The Fa Triangle. View from overhead in Google Earth. Esperaza church and the Tour de Fa are due east-west from each other and form an equilateral triangle with Campagne-sur-Aude. The summit of Mt Sec falls on the continuation of the Campagne-sur-Aude – Esperaza alignment on a 30° bearing.

region. Here was a regular triangle, aligned to the compass, with its upper side running east-west, along a line-of-sight, connecting three local major sites dating from the eleventh to the thirteenth century and possibly earlier.

Furthermore, the eastern side of the equilateral triangle, on a bearing of 30°, when extended further northwards terminated very exactly on the peak of Mt Sec, a very distinctly shaped, prominent landmark (as shown in Figure 18). This very neat set of results seemed highly significant to me and was to prove even more compelling later, in the light of subsequent discoveries. It demonstrated a convergence between physical sightlines in the landscape, prominent high points and geometrical formats that could be traced on the map. Here were traces of an ancient system of signalling between settlements separated by peaks, ridges and valleys, which was also geometrical.

There were many other line-of-sight connections to be observed at other locations as I began to pay attention to this feature of the landscape. At certain high positions marked by châteaux, especially, one could quite often see other châteaux on peaks in the far distance through fortuitous gaps in the intervening ridges and peaks. It slowly became apparent that this must have been one of the considerations by which these sites had been chosen, perhaps even the *main* consideration. Often, it was not easy to predict from the map whether two sites would be inter-visible with each other. There was no substitute for being there in person.

To this day, it is challenging territory to move around in rapidly. It began to dawn on me that the ability to communicate over long distances in this landscape would certainly have offered a strategic advantage for those living here. Sightlines offered a means to send a message between two châteaux, miles apart, located high on ridges or mountain peaks. A network of such sighting-lines would permit the flow of information over a large area in a short space of time. This would have been possible day or night when visibility permitted.

I soon learned that I was by no means the first person to notice this. Several articles had been published describing how the Romans had employed a system of fire-tower signalling between peaks and valleys throughout the Languedoc and Pyrenees, parts of which were already in place before they arrived. Georges Kiess in an article entitled *Les Tours Signaux au Pays de Rhedae* (or Signalling Towers of the Countryside of Rhedae) describes how later, during the Middle Ages, a network of towers, Templar châteaux and churches across the Haute Vallée itself, including at Bézu and other local sites, were used as a sophisticated fire-signalling system to send messages across large distances in short time. These were indispensable for military strategy and defence.[27]

I began to appreciate the importance of these connections between locations in the landscape: they provided a practical function which would have helped guide the choice of peaks on which to site the lookouts. It was obvious that the system was evidently known to the Templars and others in the eleventh to thirteenth centuries but that they did not originate it. Rather, they brought back to life an earlier system that had been used by the Romans, who in turn, had found it already in use when they arrived. These networks therefore date back to the bronze age and perhaps even earlier, to the megalithic era. Their arrangement was not haphazard. An order began to emerge slowly as I continued to explore and consider. Intriguingly, in many cases, the network of lines-of-sight coincided with the network of geometrical alignments.

---

27   Georges Kiess, *Les Tours Signaux au Pays de Rhedae*, Association Terre de Rhedae, Bulletin No. 9, December 1995.

If these suggested alignments really were the result of a conscious, intentional shaping of the landscape, then the amount of effort involved must have been significant. In turn, this implies that there must have been a strong reason for doing so, and that the alignments must perform some essential function that justified the expenditure of such effort. The practical necessity of communication would have been at least one powerful motivating force in every era.

Visiting the Meridians

After all the years of staring at the meridians on the map, it was a revelation to be able to experience them in the landscape at first hand, to see them with my own eyes and explore them for myself in person.

The 'sighting ridge', as I called it, was accessible via a steeply winding road from Alet-les-Bains to the tiny village of St Salvayre perched high on the upper slopes. The ancient stone church here dates back to the ninth century or earlier. Beyond the village, the road turns to the right, to run along the ridge from west to east. It is a short walk from the bend to the viewing point for the La Pique meridian, L'Homme Mort.

To my excitement and wonder, the location offered superb views to the south over the Pyrenees and the meridian was clearly visible: the symmetrical hill of Le Sarrat Rouge lay in perfect alignment with the distinctive form of La Pique, and beyond, further south, the highest points of the two ridges.

Planditou, the sighting point for the Pech Cardou meridian, lay a few minutes further away, a short drive through mature pine forests. Slight disappointment lay in store however, when I first arrived at the position marked on the map. For ten years I had waited patiently to be able to view the alignment, but when I arrived at the site, I found that all views to the south were effectively blocked by the dense cover of trees, and it was impossible to catch more than a glimpse of the peaks to the south from there.

There was abundant consolation and confirmation at other sites to make up for this. I was able to visit many of the various peaks and other points on the meridians and sight the alignments between them for myself. There was something deeply satisfying about this. No matter how clear the evidence on the map, it was another matter to witness it in the landscape. These explorations shook off any last vestiges of remaining doubt in my mind.

I climbed the small but perfectly shaped hill of Le Sarrat Rouge, the first peak on the La Pique meridian. This is a very curious landform. It is the highest point of the landscape to the west of the River Sals and south of the River Rialsesse until La Pique itself, further due south precisely. Le Sarrat Rouge is a small rocky mound that rises up from the surrounding plateau to form a natural viewing platform. It is not easy to scramble through the dense foliage which now overgrows it, but it is possible, and its summit offers an excellent vantage point to the south, over La Pique and the two ridges beyond.

We visited La Pique itself many times. This is also an unusual landscape feature, consisting of a large triangular limestone plateau that slopes gently upwards to one of its corners. This is the peak named La Pique, though as noted earlier, the French name does not mean "peak", so the homophony is an accident of English translation. Rather, *une pique* is a lance, or a long, thin stick of some kind. It is not difficult to imagine that a type of pole or mast could have been installed here as a meridian sighting aid in the ancient past, and to suggest that such a *pique* may have given rise to the name of the peak.

In any case, the peak is today easily accessible by a charming walking trail. It is marked by a gnarled, ancient tree and a modern geodetic stone, and offers true panoramic views of the Haute Vallée: to the east, Pech Cardou and Mt Bugarach; to the south, the two further ridge peaks on the meridian, and the peaks of the Pyrenees beyond, to the west, Rennes-le-Château and the river Aude.

I managed to visit the highest peak in the La Pique meridian, on the ridge named Les Crêtes d'al Pouil, by walking along a very old but seldom-used track that follows along the crest of the ridge. I had wondered if I would find any evidence of carving or removing rock when I arrived at the summit. The answer was a resounding yes: virtually the entire outcrop of rock along this ridge has been heavily mined and excavated. The signs of work were everywhere and obvious. I was amazed by the scale of effort that must have been expended in earlier times to remove and extract rock from these heights, presumably then taken elsewhere for use in construction.

I was more than satisfied from these explorations that the La Pique and Pech Cardou meridians were a genuine physical feature of the landscape. We climbed Pech Cardou many times. It's a great two-hour walk to the summit with spectacular views from the top in every direction. We even celebrated Judith's fiftieth birthday with a group hike with friends and a picnic at the top of the mountain.

Though I had been unable to sight along the meridian to the south from Planditou, I found that it could be viewed successfully from the

Figure 19: The La Pique Meridian: the view looking due south from the summit of Le Sarrat Rouge. It shows La Pique in the centre of the image, and the two ridges behind of Serre de Calmette and Les Crêtes d'al Pouil.

southern side of the summit of Pech Cardou, from where the accompanying photograph was taken (Figure 24). These clearly show the effect of the peaks "stacked" above each other from this vantage point on the meridian.

Col du Vent was another fascinating location that we visited. The road that passes over the ridge at this point had clearly been in the same location for a very long time, as there is no other way over the crest except to go around the long way in either direction. Here was one of the meridian locations literally carved into the landscape as a V-shaped groove where it crossed this prominent ridge.

Over time, I was able to visit most of the other locations on the meridians I had identified. The most memorable of these adventures

Figure 20: The side profile of Le Sarrat Rouge. The photo shows the view from the road that climbs the hill to Rennes-le-Château, looking to the east.

Figure 21: Summit of La Pique. View towards north-east, with Pech Cardou visible at left rear.

Figure 22: View looking south towards Les Crêtes d'al Pouil. The large rocky outcrop at centre marks the highest point of Les Crêtes d'al Pouil, where the La Pique meridian crosses this ridge.

Figure 23: Looking south towards Col du Vent. The ancient trackway passes over the ridge at the V-shaped notch, where the Pech Cardou meridian also crosses.

took place on the summit of Pech dels Escarabatets, the highest and most southerly of the mountains on the Pech Cardou meridian. At 1165 feet, this is a much higher peak in the Pyrenees than any of those in the Haute Vallée to the north.

Conveniently, it is possible to drive most of the way up the mountain on forestry roads, and to park not too far from the top. Judith and I set off from there one fine summer's day. I was hopeful I could get to the summit and see the view back to the north along the meridian. Judith decided to linger and pick wild strawberries in the forest, leaving the final summit push to me.

I had to make my way through some thick brambles and dense forest before bursting out into a clearing at the top of the mountain. The views in every direction were simply spectacular. I drank it in for several minutes, wandering around along a thin strip of clearing between a sheer, near-vertical drop down the side of the mountain to the west, and thick impenetrable cover of trees on the east flank. To the north, along the meridian, the view was uninterrupted. The air was crisp and clean, and the sky was the brightest shade of blue. I was standing on top of the world, taking it all in, when I heard, quite close to me, the growl of what sounded like nothing so much as a large bear.

For a moment, I froze, rooted to the spot in total silence. Had I imagined it? Was it really a bear? Time stood still. I considered my options. I looked down at the wooden stick I held in my hands. Then, there was a rustle in the bushes, only a matter of yards away, undoubtedly the sound of a large, heavy animal moving through the dense foliage. The bear let out a second growl, longer and louder than the first, and the sound of sniffing as he caught my scent in the mountain air.

I did not need any further time to think. I began to backtrack my way along the route, off the summit, away from whatever it was, as quickly and as quietly as I could, with my heart pounding out of my chest. I scrambled down through the forest, crashing through the undergrowth. Collecting Judith at the bottom, I still could not quite speak, nor was I ready to stop running. By the time we got back to the car, I had calmed down a bit.

I found out later that it was most likely one of the Slovenian brown bears that had been re-introduced in the Pyrenees, not without controversy, as a response to a decline in the local brown bear population. I was glad to learn that they are less aggressive than the original indigenous species and would rarely attack a human. It was just letting me know

Figure 24: The Pech Cardou meridian, looking due south from the upper slopes of Pech Cardou. Visible are: Col Doux in the foreground, followed by the ridges of Soula de la Carbonnière and La Garosse, the unnamed 802 m peak, the V-shaped groove of Col du Vent and Pech dels Escarabatets in the distance.

that I was in its territory. I had climbed the final peak on the meridian, turned to the north, and encountered a bear. It made my day, and I dined out on the story for weeks.

Another moment from one of those first expeditions made an impression on me. I had hiked along the sighting ridge from L'Homme Mort to the west, to visit a standing stone shown on the map in that area. It was late afternoon, and daylight was fading. I strained to make out various peaks and other landmarks in the distance, but it became increasingly difficult as visibility dwindled. Then, as I watched, a bright electric light was suddenly turned on in the distance. It came from the peak on which Rennes-le-Château itself stood. The summit of the hill had become a blazing beacon in the gathering gloom of approaching darkness.

In an instant, the realisation hit me: these landscape alignments were not observed, or tended, during the day, but at night! Visibility during the day was a function of sunlight, weather, clouds and rain, and it was not easy to make out distant details. But wait for night to fall, and light fires on the tops of the peaks and everything changes. Now it was possible to observe distant peaks with ease and accuracy. The meridians might be difficult to see clearly during the day, but once the sun had gone down, with fires on their peaks they would be like a string of fairy lights on a taut wire.

Three More Meridians

The mountainous region of higher peaks to the south of the Haute Vallée was interesting to me for many reasons. I wondered if any of the meridians I had discovered might continue further south into this more difficult terrain. Eventually, I added the map of the region to the south of the first map on my ever-expanding wall. With this and Google Earth, I could look methodically for further traces of possible meridians south of the area in which I had found the La Pique and Pech Cardou meridians. The search revealed an entirely new meridian, of excellent quality. This alignment passed through the tiny chapel at the eastern end of Château Bézu and then continued due south to pass exactly through three churches and three peaks, all located along the same deep mountain valley running north-south into the high country.

The remarkable Templar château of Albedunum, also known as Bézu, is situated on a high ridge overlooking Rennes-les-Bains and Pech Cardou to the north. Accessible by a sublime pathway that winds up, through ancient oak forest, the southern side of the ridge, the château had once occupied an extended section of the crest. Now it is in a glorious state of ruin, but many impressive stone walls remain as witness to its faded splendour. Its location is truly spectacular, with panoramic views of the entire region. From a strategic point of view, it offers unimpeded observation of all access routes in and out.

The observer looking to the north from Bézu can take in a wide vista of the peaks and valleys of the Rennes region, including Pech Cardou, Blanchefort, and Pech Bugarach, as shown in the panoramic photograph in Figure 25. The château offers an almost perfect vantage point from which to keep watch over this entire plateau to the north.

Two other prominent châteaux are also visible in the distance from this vantage point. To the east, beyond Pech Bugarach, the heights of Château Peyrepertuse can be seen poking out through the mountains in between, around twelve miles away. To the west, the tower of Château Puivert, famous centre of the Troubadour tradition of the thirteenth century, can be seen, perched on its ridgetop, thirteen miles distant from Bézu. These sightlines would have offered very convenient opportunities for communication.

To the south of Bézu, the landscape of the Pyrenees rises into a series of progressively higher peaks. As the mountains become more imposing to the south, so too do the valleys become deeper, carved out by the streams and rivers that drain the heavy winter snows each spring and summer. The landscape here is quite different already from that around Rennes, though it is not very far away geographically. The villages, often snowed in for months at a time, are built for alpine conditions and become further apart the deeper one moves into the mountains.

This meridian runs south of Bézu, directly along one of these alpine river valleys leading up into the high peaks. It corresponds to longitude 2°18'28"E, and it crosses the ridge at Bézu at the extreme eastern end of the château complex where the chapel, now in ruins, once stood. After Bézu, it passes over a peak called Tuc de Ramonet. Then it runs directly through the large church in the village of Lapradelles, at the beginning of the valley. The meridian continues close to the Château Puislaurens (on longitude 2°18'00"E) and even closer to the village church, (on longitude 2°18'22"E) but perhaps due to the steep topography of the valley here, does not directly run through these sites.

Further south, however, lies one of the most spectacular stretches of any of these meridians: it passes exactly through the church in the next village in the valley, Salvezines, then over two more peaks, the Col du Blaou, and a highpoint on a ridge on Sarrat de Babil. It then continues to the final church in the village of Montford-sur-Boulzane.

Figure 25: Panorama from Bézu, looking north-east. Pech Cardou can be seen at centre-left. The large peak at the right is Pech Bugarach.

This is the most southerly marked location on the meridian, deep in the mountains, some fifteen miles due south of Bézu, yet connected to it by this impeccable geometric thread. This new discovery of the Bézu meridian was especially impressive as it maintained its alignment even through these much higher peaks to the south.

Here is a list of the sites in the Bézu meridian, on longitude 2°18'28"E:

- Chapel at Château Bézu (747 m)
- Tuc de Ramonet (peak, 912 m)
- Church at Lapradelles (436 m)
- Church at Salvezines (537 m)
- Col du Blaou (peak 1,009 m)
- Sarrat de Babil, (ridge sighting point, 1,090 m)
- Church at Montfort-sur-Boulzane (754 m)

This brought the total of valid meridians to three, but the collection was not quite complete. The next meridian found, which passed through the church of Rennes-les-Bains itself, increased the total to four. To the south of the village, it runs through two significant ancient sites and another mountain peak. One of the sites is a ruined château in a small village called La Vialasse, whose physical setting is very

curious. It sits right at the bottom of a deep valley, by the River Blanque, accessible only by a steep winding road that descends from the main road south from Rennes-les-Bains. The château, which is marked on the map but consists of little more than a heap of ancient ruins, once stood on a prominent rocky outcrop in the midst of La Vialasse. The other marked position through which the meridian passed was the church in the village of St Louis, a little further to the south.

Whilst this meridian was not quite as spectacular as the other three I had found, in terms of the number of sites through which it passed, it was nevertheless perfectly accurate, and comprised a valid alignment in terms of my working guidelines.

There was, however, one noteworthy difference between the first two meridians I had found – through La Pique and Pech Cardou – and these next two, through Bézu and Rennes-le-Bains. The former pair consisted entirely of peaks, with no man-made structures involved, whilst the latter two were made up of a mix of peaks, churches and châteaux.

My catalogue of meridians marked in peaks, churches and châteaux had now doubled in size. It included four valid, exact meridians passing through this one small section of the map, or landscape: from west to

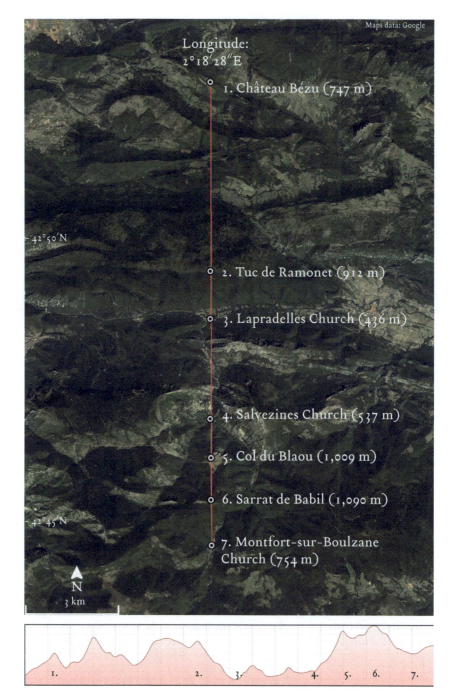

Figure 26: Bézu meridian. Shown from overhead and in elevation profile.

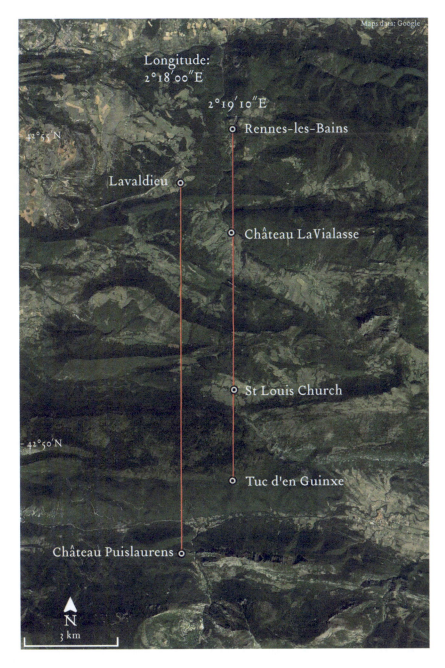

Figure 27: Meridians of Lavaldieu and Rennes-les-Bains. Shown from overhead.

east they were the La Pique meridian, the Bézu meridian, the Rennes-les-Bains meridian and the Pech Cardou meridian.

One further final meridian increased the total to five! It passed through Lavaldieu, one of the sites on the 45° alignment and also directly through Château Puislaurens located further south on longitude 2°18′00″E. Though it only passed through these two sites, its accuracy and proximity to the other four were sufficient for me to consider it another valid example of these extraordinary north-south lines engraved into the landscape.

Very remarkably, each of the five meridians terminated, to the north, on a local high point on the same sighting ridge running east-west across the landscape on which I'd already identified the high points of l'Homme Mort and Planditou, the sighting locations for the La Pique and Pech Cardou meridians respectively. This can be observed and confirmed in the accompanying Figure 28. From west to east the five sighting locations and their corresponding meridians are as follows:

Meridian Name	*Northernmost Peak on Sighting Ridge*	*Spot Height*	*Longitude*
La Pique	L'Homme Mort	734 m	2°17′50″E
Lavaldieu	Girbes de Bacou	758 m	2°18′00″E
Bézu	Pech d'al Bouich	769 m	2°18′28″E
Rennes-les-Bains	Pech de la Quartière	739 m	2°19′10″E
Pech Cardou	Planditou	802 m	2°19′40″E

These five meridians were real. They were physically observable.

They were all essentially perfect. They could be verified both on the map and first-hand in the landscape. I had now personally visited, viewed and explored many of the sites and seen the meridians with my own eyes. I had taken photographs of them. I had proven to my own satisfaction at least, that these meridians were genuine phenomena.

Meanwhile, other pieces of the puzzle were also slowly beginning to come together. A small but fortuitous discovery came one day when we were visiting a friend of Judith's, the owner of a yoga school and

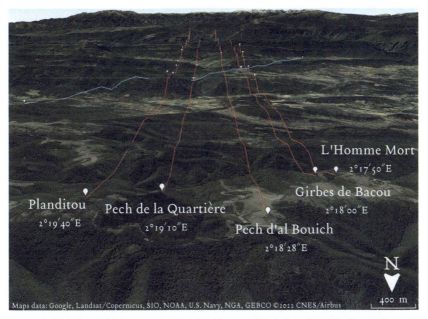

Figure 28: The viewing positions for the five meridians on the Sighting Ridge: Looking south over the field of meridians in Google Earth.

guesthouse which operates at Lavaldieu, one of the positions on the 45° alignment. Today it comprises a collection of buildings gathered around the old farmhouse. On a personal tour, our host showed us the ruins of what she described as a local Templar commanderie. Whilst Lavaldieu is marked on the map, there is no indication of any ruins here, Templar or otherwise, nor are they described in any book that I have seen, so it was a great privilege to be able to find out about it in this way, and to see it with my own eyes.

A side note to this was that I had begun to dream about the 45° alignment on many nights. I would wake up and realise I had spent the night, again, exploring its peaks and valleys in my sleep. Something was trying to get through to me, to tell me: *this is significant, keep going*.

Amidst these adventures, there was another real breakthrough that emerged when an element of the landscape geometry was considered in light of a fascinating item of local history and culture.

The "Sunrise Line" and the Pic de Saint-Barthélemy

The "Sunrise Line", as described in *The Holy Place*, and mentioned in Chapter One, is an alignment running west-to-east across the Haute Vallée at a bearing of approximately 74°. Lincoln traced it from

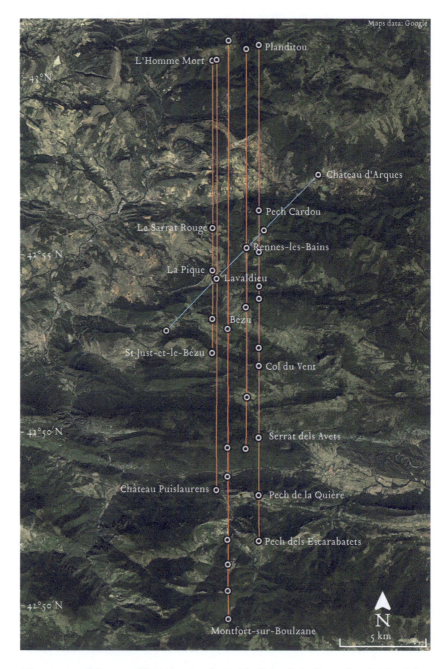

Figure 29: All five meridians on Google Earth, from directly overhead. The sighting ridge is shown at top of image. From east to west, the meridians of: La Pique, Lavaldieu, Bézu, Rennes-les-Bains and Pech Cardou.

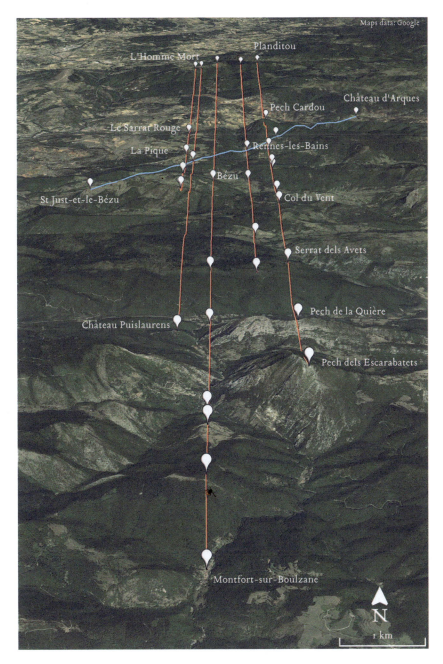

Figure 30: All five meridians on Google Earth, in perspective looking north. From east to west, the meridians of: La Pique, Lavaldieu, Bézu, Rennes-les-Bains and Pech Cardou.

Campagne-sur-Aude, (the local headquarters of the Knights Templar), through Rennes-le-Château and Château Blanchefort, ending at Arques Church. Note this is not the Château d'Arques, but the village church. Lincoln writes that according to local lore, the sun rose along this line as viewed from Rennes-le-Château, on July 22, Feast Day of Mary Magdalene, to whom of course the church in the village is dedicated. I set out to scrutinise these various claims, beginning with this one about the sunrise.

Using astronomy software and entering the correct latitude, I quickly discovered that there was a problem: the sunrise observed on July 22 from these latitudes takes place on a bearing of 62°, nowhere even remotely close to the Sunrise Line bearing of 74°! This was very odd. Lincoln's information was clearly incorrect. What was going on?

When I proceeded to check the line itself, I found that the alignment between the sites that Lincoln had named—Campagne-sur-Aude, Rennes-le-Château, Château Blanchefort, Arques Church—was accurate. Using Google Earth, I was then able to trace the line further in both directions to see if it passed through any other sites further afield.

I found that when extended to the east, the Sunrise Line continued to pass through Château de Villerouge-Termenès, an imposing fortress in the wild country of the Corbières, around twenty kilometres east of Arques. When continued further to the west, it passed quite close to Nébias Church and then continued very accurately to the summit of Pic de Saint-Barthélemy, (2,348 m) the highest, holiest, and certainly one of the most significant of the peaks in this entire region of the Pyrenees.

The full Sunrise Line therefore can be traced over 45 miles from Pic de Saint-Barthélemy in the west to Château de Villerouge-Termenès in the east, along a compass bearing of 74°.

Pic de Saint-Barthélemy is a mountain held in great reverence by the local population. It has been associated with the sun and sunrise since earliest times. The ruins of an ancient chapel were found at the summit. It has also been used as a major geodetic site for cartography and earth measure. *Histoire des comptes de Foix, Béarn et Navarre* by Pierre Olhagaray describes the mountain, under its earlier name:

> "It is necessary to know that Tabe, or Tabor is the highest mountain in the Pyrenees mountains, from where we see many notable secrets of nature, the rising of the sun with an incomprehensible grandeur & Majesty."[28]

---

28 Pierre Olhagaray, *Histoire des comptes de Foix*, Béarn et Navarre, (Paris, 1629), p. 704.

He also describes a ritual which had taken place on this mountain each year for many centuries. On the evening of August 23, the inhabitants of the local area would climb to the top and spend the night on the peak so that on the following morning, August 24, the Feast Day of Saint Barthélemy, they could observe the rising sun from the summit.

Curiously, the custom seems to have predated the name. From at least the tenth century AD, it was known as Mt Tabe, or Tabor, after the mountain in Palestine on which the Transfiguration of Christ took place. The name of Saint Barthélemy only became attached to the mountain in the seventeenth century.

The fact that Pic de Saint-Barthélemy was accurately aligned to the continuation of the Sunrise Line revealed its origin and significance. The line extended across the landscape from this most eminent and auspicious of peaks which has had an association with the sun from earliest times. The obvious next step was to find out which day of the year the sun rose along bearing 74°. Was it perhaps related to the ritual?

Using the software again, I was astonished to discover that it was indeed on August 24, the very day of the annual sunrise ceremony on the Pic de Saint-Barthélemy, the Feast Day of Saint Barthélemy.

At the ritual gathering on Pic de Saint-Barthélemy on August 24 each year, those that had spent the night on the summit watched the sun rising along the very same "Sunrise Line" alignment discussed by Lincoln, passing through Campagne-sur-Aude, Rennes-le-Château, Château Blanchefort and Arques Church. This was an electrifying discovery, to find that a major landscape alignment corresponded with an astronomical event which was memorialised in local cultural practise as a ritual annual observation. The historical record now stood witness to the reality of the geometry.

The annual ceremony of watching the rising of the sun from the summit of Pic de Saint-Barthélemy on August 24 each year along the Sunrise Line is an extraordinary confirmation that there is a genuine phenomenon present here. It testifies that this ancient complex of alignments had arisen from a conscious, intentional engagement with landscape in the broadest sense. They formed a system of observation posts on mountain peaks, and sightlines across landscape co-ordinated with rising and setting positions of the sun and stars.

The fact that Lincoln cited the incorrect date on which the sun rose along the Sunrise Line as viewed from Rennes-le-Château is intriguing. We can only assume that this "snippet of local lore" that had reached his ears must have become slightly garbled in transmission. It was not

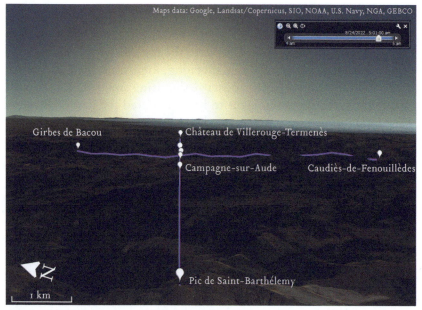

Figure 31: View from Pic de Saint Barthélemy, along the Sunrise Line at a bearing of 74°, at dawn on August 24. The Bugarach Baseline lies at right-angles.

the Feast Day of Mary Magdalene, but the Feast Day of St Barthélemy on which the rising of the sun took place along the Sunrise Line.

Nevertheless, this apparent error only serves to confirm that knowledge of this exquisite interaction between landscape, sun, calendar, ritual and myth has persisted into modern times, even if the details have faded or become obscure. The Sunrise Line was clearly not merely a figment of Wood and Lincoln's imagination. It was a genuine element in a wider complex of geometry. This hinted at the possibility of a connection between the Sunrise Line and Bugarach Baseline pair and the field of the meridians.

Meanwhile, I found that the Bugarach Baseline – the alignment at right angles to the Sunrise Line which we also met in Chapter 1 – also passed through two more locations when extended in both directions beyond the original extent identified by Lincoln and Wood. At the northern end it led directly to Girbes Bacou, the northern sighting position for the Lavaldieu meridian! To the south, beyond Bugarach Church, it passed very exactly over the church at Caudiès-de-Fenouillèdes. Both lines, therefore – the Sunrise Line and Bugarach Baseline – were even longer and more significant in the landscape than they had first appeared.

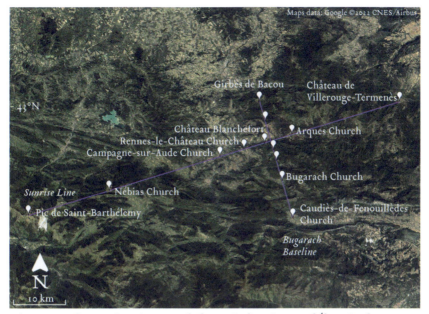

Figure 32: The Sunrise Line extends from Pic de Saint Barthélemy in the west to Château de Villerouge-Termenès in the east. The Bugarach Baseline also shown.

Many elements of the problem of the landscape geometry had become clearer after spending time living in the region itself. I had begun to appreciate that there was at least one practical reason for maintaining a system of alignments, namely for establishing line-of-sight communication, and that this practice had certainly left its traces in the siting of many of the ancient châteaux. I was also becoming more comfortable with the reality of the alignments and how they integrated into the local lay of the land. There seemed to be an association with rivers, particularly when they passed through the many spectacular gorges with sheer vertical limestone walls. There was also evidently a connection between certain lines and the rising and setting of the sun on certain dates.

Moreover, it was also becoming clear that these various schemes overlapped to a degree. The same alignments which served as communication sightlines also adhered to certain simple geometrical relationships. The meridians included sites that were also part of the same sighting networks.

My Google Earth model continued to grow, and now stretched from beyond Montségur in the west, to the Mediterranean in the east. I chose to colour-code the alignments, depending on their direction, and found many groups of lines in parallel clusters. Evidence began to accumulate to suggest that these might be related to regular compass angles, the winter solstice sunrise and other rising and setting positions on the horizon, perhaps of major stars.

Meanwhile, I continued to explore the writings of Professor Richer. It was much easier to track down his other (untranslated) works in France, and over time I managed to acquire a small collection of his books. Slowly, I worked my way through these. They are not easy texts to digest, but I found them to be deeply engrossing and full of fascinating insights. It is time now to look more closely at some of the remarkable conceptions about landscape to be found in the works of Professor Jean Richer.

Chapter Five
SACRED GEOGRAPHY

D ESPITE HIS initial enthusiasm, it was to be a further two years from the initial intuitive insight prompted by his dream in Athens before Professor Richer commenced a systematic search for other examples of such relationships between ancient sites. Once he had begun, however, it did not take him long to find further instances, encouraging him to persist in his efforts. Slowly, a picture began to emerge of a vast network, or forest, of these alignments, which encompassed the territories of ancient Greece and even extended to lands beyond.

Jean Richer was born on 4 February 1915 in Paris. He studied at the Faculté des lettres de Paris, where he earned his degree, and doctorate. He went on to become a professor in arts faculties in Athens, Algeria and Nice. His field of expertise was symbolist poetry and, in particular, the work and life of the French writer Gérard de Nerval. He wrote many books and monographs on Nerval, in addition to studies on Cazotte, Nodier, Gautier, Verlaine, Shakespeare and others. He was the co-editor of the definitive Pleiades edition of the *Oeuvres completes* (Complete Works) of Nerval. He passed away in 1992 and was honoured by a special issue of the *Cahiers de la Société Gérard de Nerval* dedicated to his life, work and career.[29]

Richer's dream in 1958 of the Apollo line in Athens would eventually lead him to forge an entirely new understanding of the concept of cosmic or sacred landscape in ancient Greece. He went on to publish a sequence of four books, in France, documenting his research findings, the first of which was *Sacred Geography of the Ancient Greeks*. In these works, only the first of which has been translated into English, Richer proposed a radical new blueprint for the understanding of ancient

29 *Cahiers de la Société Gérard de Nerval*, No. 15, 1992.

Greek culture which opened a rich and valuable window into certain remarkable aspects of the ancient world that have been all but forgotten.

Sacred Geography of the Ancient Greeks

As introduced in the Prologue, the French edition of Richer's *Sacred Geography of the Ancient Greeks* was first published in 1967 by Hachette in Paris as part of their popular *Guides Bleus* series of travel guides. With their familiar blue cloth covers and embossed gold lettering the *Guides Bleus* have become a staple item of French life since they first appeared in 1919. With each title devoted to a different destination, they have an emphasis on historical and cultural depth.

Though it might seem slightly unusual that a work such as this would be published under a series of guidebooks, during the late 1960s several titles were released which extended the scope of the *Guides Bleus*. These included volumes focussing on cultural themes of more general interest, including, for example, the *Guide Religieux de la France* (1967), and the *Guide Artistique de la France* (1998), so the inclusion of Richer's book amongst such titles must have been considered an appropriate fit.

Sacred Geography of the Ancient Greeks explored Richer's discovery of landscape geometry in the ancient world, from the Far East to the Mediterranean and Europe. It was followed in 1970 by *Delphes, Délos, et Cumes: Les Grecs et le Zodiaque*, which Richer considered to be the continuation and complement to the first volume.

Later, he also wrote *Géographie Sacrée dans le Monde Romain*, which extended the discoveries to the lands making up the Roman empire. Finally, in *Iconologie et Iconographie* he showed how the same symbolic language persisted in European and Christian architecture and art through to the Middle Ages and beyond.[30]

In these seminal works, Richer presented his extraordinary rediscovery of a network of alignments which encompassed the entire landscape of Greece and extended to other lands around the Mediterranean, connecting natural mountain peaks and other features, with cult centres, cities, temples and other significant locations.

The system appeared to be the result of a highly co-ordinated and comprehensive programme of geographic intervention with origins in deep antiquity. The reason the ancients had undertaken the huge task of reshaping the landscape, in Richer's view, was to create a reflection on the earth's surface of the layout of the heavens in the skies above.

30   Jean Richer, *Delphes, Délos, et Cumes : Les Grecs et le Zodiaque*, (Paris, Julliard, 1970) ; *Géographie Sacrée dans le Monde Romain*, (Paris, Guy Trédaniel, 1975) ; *Iconologie et Iconographie*, (Paris, Guy Trédaniel, 1980).

As he was to describe it, years later:

> "The evidence of the monuments shows in an undeniable way, but not yet clearly perceived, that during more than two thousand years, the Phoenicians, the Hittites, the ancient Greeks, and then the Etruscans, the Carthaginians, and the Romans, had patiently woven a fabric of correspondences between the sky, especially the apparent course of the sun through the zodiac, the inhabited earth, and the cities built by humanity."[31]

The ultimate purpose of this careful arrangement of locations and landmarks within the landscape was as he suggested:

> "... to create unity and solidity in a cosmos that was conceived as the harmonious equilibrium of the three planes of the underworld, the earth, and the celestial spheres."[32]

The notion that landmarks on earth might have been arranged to mirror the position of stars in the night sky can be found across many ancient cultures. A raft of examples is cited in a 1975 essay by Lewis M. Greenberg and Walter B. Sizemore, which draws on work of M. Eliade.[33] These collectively demonstrate that "astrogeography", as they term it, was the universal practice and determined the layout and orientation of all of the great cities of early humanity. Here is a selection of instances they discuss:

> "In the Sumerian Creation myth, a cosmic, celestial Nippur, antedating the creation of the earth, was the model of the terrestrial Nippur. In Mesopotamia, 'all the Babylonian cities had their archetypes in the constellations: Sippara in Cancer, Nineveh in Ursa Major, Assur in Arcturus, etc.'
> Sennacherib has Nineveh built according to the form 'delineated from distant ages by the writing of the heaven-of-stars'. (...) Even 'the Tigris has its model in the star Anunit and the Euphrates in the star of the Swallow'. In India: all the Indian royal cities, even the modern ones, are

---

31 Jean Richer, *Sacred Geography of the Ancient Greeks*, op. cit. p. xxv.
32 Ibid, p. 62.
33 Lewis M. Greenberg and Walter B. Sizemore, "Cosmology and Psychology", Kronos, Vol 0101, April 1975.

built after the mythical model of the celestial city where, in the age of gold (*in illo tempore*), the Universal Sovereign dwelt. (...) In Iranian cosmology of the Zarvanitic tradition, 'every terrestrial phenomenon, whether abstract or concrete, corresponds to a celestial, transcendent invisible term, to an "idea" in the Platonic sense.'"[34]

Richer's ideas therefore, far from being novel, were consistent with a widespread conception of ancient cosmology in which cities and sacred places on earth were understood as reflections of corresponding forms in the heavens or stars. These conceptions were related to the known ancient practice – mentioned at the beginning of Aeschylus' *Agamemnon*, for example – of communicating news over long distances by means of fire signals on high places.

The play, dating from the fifth-century BC, opens with a speech by a guard who describes waiting for a beacon on a distant peak that will signal news transmitted by relay, from mountain-top to mountain-top, all the way from Troy.

"*Guard*: The gods relieve my watch: that's all I ask.
Year-long I've haunched here on this palace roof,
Year-long been the all-fours watchdog of the Atreids,
Learning by rote the slow dance of the stars,
Spectator of the brilliance in black skies
That brings to men their winters and their suns:
The stately light-lords' settings and their rise.
I'm here still. Still watching for the fire,
The relayed beacon that will bring the word
That Troy is taken."[35]

This lone sentinel watching for the light from a distant beacon cannot have been operating in isolation. His presence implies the existence of an entire infrastructure.

Aeschylus' guard is only the visible sign of what must have been a widely distributed communication network, involving considerable planning, construction and co-ordination, stretching over vast distances. Notice also the connection being made in the same breath

---

34   Quotations in single quotation marks within the main excerpt from Greenberg and Sizemore are from M. Eliade, *The Myth of the Eternal Return: Cosmos & History*, Volume 4 of Bollingen Series Mythos Series, Princeton University Press, 1971, p. 6–9.
35   Aeschylus, *Agamemnon*, quoted in Richer *Sacred Geography of the Ancient Greeks* op. cit.

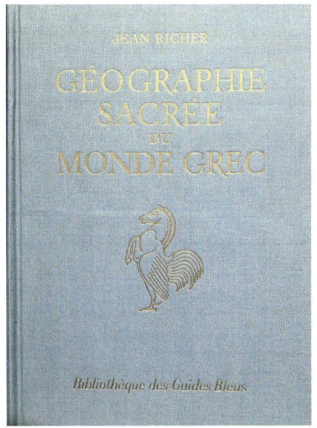

Figure 33: Front cover of first edition of Jean Richer's Géographie Sacrée du Monde Grec.(Paris, Hachette, 1967).

between observing the heavens ("learning by rote the slow dance of the stars") and watching for the distant beacon fires. We see here the subtle traces of an ancient system which seamlessly integrated cosmology, knowledge of the motions of the stars, and the use of sightlines for communication. In this way, a web of relations is built between the heavens, the earth and humanity. Fires on peaks and stars in motion inscribe themselves in the landscape as a geography of myth, a poetic geometry woven into territory.

Landscape Grids and the Zodiac of Delphi

Richer's unique contribution to the recovery of this ancient science in modern times was to grasp the wider significance of these fragments, and to recover certain specific technical details that enabled the ancient system to be described. He found that several major alignments established a geometrical framework that encompassed the entire territory of Ancient Greece. One was a meridian that ran like a spine down the central axis of the land, through multiple key sites, from Mt Olympus in the north and through Delphi to Cythera in the south. Another was the latitude line through Delphi, which passed through several significant locations, including the ancient temple of Sardis in present-day Turkey, to the east. Together, these and other alignments at significant compass bearings formed a grid of east-west and north-south lines, creating a vast network linking temples, cities and cult centres.

He also found that many of the major ancient, sacred sites of Greece, beginning with Delphi, lay at the focus of what he referred to as "landscape zodiacs", formed by the intersection of up to six alignments, converging on a city or sacred centre and dividing the territory surrounding the focal point into twelve segments.

These formats projected symbolic star-maps over the landscape of Greece with profound consequences for the understanding of ancient Greek culture and myth. He wrote:

> "It is as if, the whole territory of historical Greece having been considered as a vast zodiacal clock, with its centre at Delphi, a distribution of functions between the cities had been carried out."[36]

We have so far met three of the alignments through Delphi out of the full complement of six which make up the zodiac centred on this most sacred of all ancient Greek sites. The first two are the primary axes through Delphi, which crossed north-south and east-west.

The third is the Delphi-Athens-Delos-Camiros alignment described earlier in the Prologue, whose discovery had been inspired by Richer's dream in Athens. This was of course the very first line that Richer found and it passed through Delphi at a bearing of 120°.

The fourth is an alignment on a bearing of 150° that led to Mt Ida, the highest mountain in Crete, lying some 800 kilometres to the southeast of Delphi. This peak was sacred to the Goddess Rhea and was the legendary birthplace of Zeus. Two further alignments through Delphi, at bearings of 30° and 60°, complete the division into twelve 30° segments.

These twelve-fold divisions of landscape around sacred centres, including Delphi and others, were conceived as terrestrial reflections

---

36 Jean Richer, *Sacred Geography of the Ancient Greeks,* op. cit. p. 59.

of the zodiac, as if the earth itself acted as a mirror of the heavens. Each segment was associated with a sign of the zodiac via an elaborate system of symbolic correspondences reflected in place names and local iconography. For example, Richer showed that coins from a particular city would often display the emblem of the corresponding zodiac sector in which they were located.

The same scheme applied to designs on the pediments of temples, which engaged with both the landscape in which they were set and the sky above, via detailed displays of relevant symbols and appropriate gods and goddesses.

Richer also showed that the decorations on Greek classical vases could be interpreted in an entirely new light with this new-found zodiac key. Configurations of animals and figures could be read as astronomical statements concerning solstices, equinoxes, planetary positions, stars and the zodiac.

> "Detailed study of the iconography of towns, cities and regions located in the various segments revealed that these elements of zodiac symbolism were pervasive throughout the ancient culture. The system offered profound opportunities for the reinterpretation of the entirety of Greek mythology."[37]

Unexpected as this discovery was, it found firm foundation in the works of Plato. In *The Laws*, for example, he gives directions for the implementation of such a system of twelve-fold division of landscape. He remarks:

> "... the founder must see that his city is placed as nearly as possible in the centre of the territory... Then he must divide his city into twelve parts: Next, he will be at pains to assign the twelve divisions to twelve gods, naming each section after the god to whom it has been allotted and consecrated, and calling it a tribe."[38]

Richer shows that certain modern writers had discussed the existence of these ancient techniques in general terms. Saintyves for example wrote:

> "The influence of the stars in the sky, and especially of the twelve signs that mark the path of the moon and the

---
37 Jean Richer, *Sacred Geography of the Ancient Greeks*, op. cit.
38 Plato, *The Laws*, V, trans. A. E. Taylor, 1934. p. 745-746. Richer op.cit. p.24.

sun, could be enhanced and made entirely benevolent if the earth itself was arranged as a reflection of the heavens. The division of a state, population and territory, into twelve sectors was an expression of reverence for the gods of the zodiac, the twelve great celestial deities and it would surely attract their blessings." [39]

Much of Richer's first book is dedicated to cataloguing a wealth of evidence of such correspondences. We will cite just a few selected examples relating to the zodiac of Delphi to demonstrate the flavour of his research.

Aries: Richer states that the beginning of the cycle, the Aries segment, symbolising spring and corresponding to the spring equinox, was located on the coast of the Ionian Sea at a place called Cap Leucat, also known as the cliffs of Leucas, or the White Rock, due west of Delphi. There is an intriguing connection here between the dawning sun and the concept of "whiteness" which we will explore in more detail in a later chapter.

The Island of Kefalonia, which means "head", is entirely within the sign of Aries. Whilst it is said the name is derived from the shape of the island, Richer notes that it is not very "head-shaped". The traditional association of Aries with the head, however, he suggests, offers an alternative explanation for its origin. Coins from the town of Cranii on the island show a ram's head on the front and a ram's hoofprint on the reverse.

Gemini: The city of Sparta was located in this sign. Richer explains that "texts, coins and a great many steles confirm the prominence of the cult of the Dioscuri at Sparta." The Dioscuri are of course the twin half-brothers Castor and Pollux in Greek and Roman mythology. The constellation of Gemini, which is Latin for twins, comprises the stars Castor and Pollux. Hence the Spartan coins displayed the symbol of Gemini.

Leo: The island of Ceos (Kea) falls in Leo, and a monolithic sculpture of a lion has stood there since earliest times. The island of Hydra is also found in this sign. This is very suggestive as the constellation of Hydra lies within the sign of Leo. Here we have an example of a geographical place name deriving directly from a matching astronomical feature in the correct zodiac location.

---

39   Pierre Saintyves, *Deux Mythes évangéliques : les douzes apôtres et les soixante-douze disciples*, (Paris, Librairie Emile Nourry J. Thiébaud, 1938), p. 159.

Figure 34: The main axes of Greece through Delphi. Alignments and text are as depicted in Figure 10 of Jean Richer's Aspects ésotériques de l'œuvre littéraire (Paris, Dervy-Livres, 1980), p. 145.

Delphi as Navel or Sacred Centre

Sacred to Apollo, Delphi was the home of the most revered oracle in the ancient world. It sits in a cleft of the earth, as the meaning of its name denotes, at the foot of Mt Parnassus, ancient home of the muses and site of the Castalian Spring. Physically, the central point of the temple precinct is marked by the omphalos, or "navel", an egg-shaped stone carved with decorations of a network of skeins of thread. Symbolically, it represents the axis around which the heavens and the earth move. This stone was the navel of Greece, the cosmic axis, the spindle of the rotation of the sky. It may be taken, therefore, as an image, or reflection, of the polestar.

The omphalos, and by extension Delphi itself, represents the immutable central point around which the universe turns. The surrounding landscape radiating out from Delphi then naturally stands in relation to the omphalos as the night sky, or zodiac, to the polestar. Delphi is ideally suited to act as the ritual centre and focal point of such an "amphictyony", a word derived from the ancient Greek term for a twelve-fold division of the landscape.

Figure 35: The Grand Alignments. All alignments and text are as as depicted in Carte 1 of Jean Richer's *Géographie Sacrée du Monde Grec* (Paris, Hachette, 1967), p. 26.

Richer was able to demonstrate beyond doubt, therefore, that a system like the one Plato described was embedded in the territory surrounding Delphi, testifying to the reality of the zodiac system as a template for city, temple and landscape design amongst the ancient Greeks.

For Richer, this inspired discovery of landscape zodiacs became the organising principle which unlocked virtually every aspect of ancient Greek culture and thought, including architecture, art, sculpture and more. It was also to become the master key and *leitmotif* for much of his subsequent writings and thought on many different topics. The revelation opened many new possible avenues for investigation. Before we explore this further, however, armed now with the insight of the landscape zodiac, we can revisit and resolve the question of Richer's dream in Athens, which began his entire journey in this material and ours in this book.

The Riddle in the Dream Resolved

With the discovery of the zodiac of Delphi, Professor Richer was at last able to deduce the answer to his original question about the statue of Athena Pronaia there. As we have mentioned, the alignment leading from Delphi to Athens, which had been revealed to him in his dream, turned out to be one of the six major division lines converging on Delphi to create its zodiac. Specifically, in the Delphi zodiac, the Delphi-Athens line marked the beginning of the segment dedicated to Virgo; this sign was associated with the virgin goddess, Athena.

Now, at last, it became clear to Richer why at the entrance to the sanctuary of Delphi stood a statue of the virgin goddess of wisdom, Athena Pronaia. It is because the route by which the traveller from Athens arrives in Delphi coincides with the Virgo alignment in the landscape zodiac of Delphi. Therefore, a statue of Athena, representative of the sign of Virgo, could not be more appropriate to offer a welcome to the Athenian pilgrim. Indeed, perhaps this may even be the original reason why the people of Athens held the goddess in such reverence that they named the city after her: it marked the Virgo alignment on the Zodiac of Delphi.

In any case, Richer's question about the enigma of the statue of Athena Pronaia at Delphi had been answered. The dream had inspired him to rediscover the Apollo line and later, the zodiac around Delphi and other centres. A specific question was posed, and a precise response came in a dream.

Yet, there is a twist to this story. In preparing this book, I decided to recreate Richer's original alignment, from Delphi to Athens, Delos and Camiros on the island of Rhodes, on Google Earth. After some internet sleuthing, I was able to locate the ancient site of Camiros on the northwest coast of Rhodes. Connecting this site to Delphi, I noticed, to my slight concern, that the line narrowly missed the island of Delos on the northern side. No doubt on the smaller-scale map Richer had to hand in the middle of the night in 1958, the alignment seemed good, but it was clear to me now that if the alignment passed through Delos, then it must arrive at the coast of Rhodes some distance to the west of Camiros.

On further research, I learned that the tallest peak on Rhodes was Mount Atabyris, considered the most holy mountain on the island in ancient times.[40] A temple dedicated to Zeus was located on the summit, founded according to legend by Althaemenes, son of Catreus, King of Crete, who chose the site as the only position on the island from which he could see his homeland. The connection of a sighting peak on Crete with one of the gods was encouraging, but Mount Atabyris was much

---

40   Strabo, *Geography*, 14.2.12: "Atabyris, the highest of the mountains there on the island of Rhodes, which is sacred to Zeus Atabyrios."

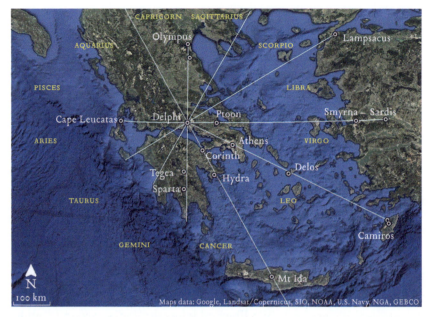

Figure 36: The Zodiac of Delphi. The major alignments and the allocation of zodiac signs to the twelve landscape segments according to Prof. Jean Richer.

too far west to be a suitable candidate. I wondered if Richer's alignment perhaps terminated on another peak on Rhodes. From a practical perspective, this would make more sense than a location at sea-level on the coast. And, indeed, I found that the line from Delphi, when extended over Athens and Delos, did in fact arrive at another peak on Rhodes – a mountain called Profitis Ilias, 798 metres in height, and named after the biblical prophet Elijah. The alignment to the summit was accurate, but was there any connection to Apollo?

I had been exploring Mount Profitis Ilias and its surroundings on Google Earth with the "placenames" feature turned off for clarity; now I turned it back on to see if there were any helpful clues. I was astonished to see, immediately, that there was a village close to the summit named Apollonia! A quick search revealed that it was so named because it stood on the site of an ancient temple to Apollo, erected very near to the top of the mountain.

Incredibly, it appeared Richer had been correct in his intuition about this alignment, and yet slightly off in the detail of its termination on Rhodes at Camiros. The Apollo line from Delphi through Athens and Delos is aligned to Mount Profitis Ilias on Rhodes, site of an ancient temple to Apollo himself.

I learned that there are in fact many mountains in Greece and surrounding regions which bear this name. It has been suggested that one possible reason for this is that the peaks were in earlier times dedicated to Helios, the sun god.

When certain landmarks were renamed in the Christian era, substituting Christian figures for the Greek deities, Elijah was associated with Helios because he did not experience death but was taken up into the heavens in a solar chariot.[41] Later, Apollo took over from Helios as the sun god. So, by symbolic equivalence, Profitis Ilias represents Helios, who represents Apollo.

The Path of the Hero

One implication for this renewed vision of the relationship between myth, astronomy and landscape was the possibility for a re-interpretation of the symbolic meaning of the journeys of Greek mythic heroes.

And now we arrive at Richer's decisive insight for our unfolding story: if the landscape was a reflection of the heavens, then stories involving the travels of gods and men around the Greek mainland and the islands could be read allegorically as trajectories of heavenly bodies across spaces mapped as zodiacs and other representations of the sky.

Major alignments, like the vertical axis of the Delphi zodiac, which connects the sacred oracular temple to Mt Olympus in the north, become gateways or passages by which the great figures of Greek mythology move between heavens and the earth.

Richer observed:

> "The zodiac and the gods of the zodiac seem more ancient than the gods of the planets. Zodiacal diagrams will act as keys to decoding or deciphering the hidden meaning of certain ancient texts, which have never before been interpreted from this point of view."[42]

By re-imagining the land as a mirror of the heavens, the grand journeys of the heroes of myth on earth can be re-interpreted as the motions of celestial bodies in the sky.

> "For example, the epic travels of various heroes of Greek mythology when mapped as paths onto these landscape zodiacs allows them to be read as a reflection of the journey of the sun through the planetary houses."

---

41   2 Kings chapter 2 verse 11.
42   Jean Richer, *Sacred Geography of the Ancient Greeks*, op. cit. p. 79.

In *Sacred Geography of the Ancient Greeks,* Richer analysed several ancient narratives in light of these new insights, to conclude that:

> "It is now possible to show that a great many mythological stories concerning heroes, demigods, or gods take on their full significance only when they are seen in context of an astral religion that places major emphasis on the figures of the zodiac and the constellations. Here again, as far as geographical locations are concerned, the zodiacal wheels act as keys to decoding, which shows the extreme antiquity of the system."[43]

He found that many puzzling references and problematic details from these stories are clarified and resolved when they are examined under the zodiac paradigm. As I became more familiar with his ideas, it gradually dawned on me that this motif of the journey of gods and heroes around a landscape configured as a zodiac, could just as easily be describing the underlying narrative format of *Le Serpent Rouge*.

Was it possible that Richer himself knew about *Le Serpent Rouge*? Had the author of the poem perhaps discussed these ideas with him directly? He certainly seemed to have been on the same wavelength as the author(s) of *Le Serpent Rouge*, but was it anything more than that? Had there been any kind of direct contact between the professor and the poets? Or were these ideas just "in the air" in Paris, in 1967?

My reading of Richer's book had been prompted, at least in part, by a search for a historic context in which to frame Lincoln's alignments in the south of France. I was not disappointed. *Sacred Geography of the Ancient Greeks* provides more than ample food for thought.

Without taking it too far, there are certain broad similarities between the landscape interventions that Richer described from his discoveries in Greece, and those that Lincoln was suggesting in *The Holy Place*. They both extended over wide areas. They both formed interconnecting networks. The alignments involved in each scheme connected temples, or churches, together with landscape features such as mountain peaks. Yet there was a marked difference between their approaches. Richer was able to identify a historical and cultural context on which he could build a deep and systematic insight into the nature and purpose of the alignments.

It was clear that Lincoln had never read Richer, which would only have been possible in the original French edition of 1967 prior to the

---

43 Ibid, p. 109.

publication of *The Holy Place* in 1991. But it is interesting to speculate on how he might have approached his topic if he had, as Richer certainly has some valuable insights and observations that are relevant to the questions Lincoln poses, even if the geographical location is obviously very different. It offered the possibility of a way forward in understanding the landscape geometry of *The Holy Place*, which was what I was looking for, but surprisingly it also seemed to suggest a connection to *Le Serpent Rouge*, which I had not been expecting at all.

So now I had found two unexpected textual pathways leading to *Le Serpent Rouge*, one through the writings of Jung and the other through those of Richer. Moreover, both authors had formulated a particular view about the symbolic nature of certain voyages.

For Jung, the voyages of alchemists described in their texts could often be read as allegorical journeys around landscapes constructed to represent the archetypal world circle.

Meanwhile, Richer had interpreted the voyages of the heroes of ancient Greece as allegorical journeys around landscapes constructed to reflect the heavens, or zodiac.

In both conceptions, we have the idea of heroic figures undertaking symbolically charged voyages around landscape configured as world-space.

Expressed in these terms, the resonance between the thought of the two authors is so suggestive that it provokes an obvious question: were Jean Richer's ideas on sacred geography consciously influenced by the alchemical writings of Carl Jung?

In this case we do not need to speculate. The answer is a resounding yes. We can be certain that Richer had been reading Jung on alchemy because he tells us so explicitly in many places in the text and footnotes throughout his books. We will explore some of these references later in Part IV, but for now a single example is sufficient to prove the point.

In *Sacred Geography of the Ancient Greeks*, in the context of a discussion on the notion of the journey of the hero as representative of the sun's passage around the earth, Richer writes:

> "A psychologist of the stature of Jung saw the hero as an image of the 'integrated man' who contains in himself the four elements and the four cardinal points."[44]

This quotation is then footnoted with a reference to the very same crucial passages on page 368 of Jung's *Psychology and Alchemy* discussed

---

44 Ibid, p. 103.

earlier in Chapter Three of this book in which the work of the alchemist is described metaphorically as a voyage of the hero around a solar path in a ritual or symbolic landscape. Evidently, Richer was indeed familiar with these same passages of Jung on alchemical journeys that I had identified years earlier as being a source of certain ideas that the author of *Le Serpent Rouge* had drawn on in composing the poem. Whilst the exact nature of these connections was by no means yet obvious, far less proven, it was becoming increasingly clear to me that there were close links between *Le Serpent Rouge* on the one hand, and both Richer and Jung on the other.

This was perhaps not altogether surprising regarding the works of Jung; Richer, as a professor of literature in the 1960s in Europe would certainly be expected to be familiar with the work of the Swiss psychoanalyst. But the suggestion of a link between Richer and *Le Serpent Rouge* was a different matter. Was it possible that there were any links between him and Plantard's team? This, of course, raises in turn another obvious question that will have by now also occurred to the reader: could Jean Richer himself have been directly involved in the creation of *Le Serpent Rouge*?

Chapter Six
CONVERGING CIRCLES

WERE THERE any discussions or interaction between the professor and the Priory of Sion pranksters? I set myself the task of trying to locate any evidence that might indicate if Richer had known de Sède, Plantard or de Chérisey.

In order to do so, I realised I needed to learn as much I could about Jean Richer. I found there was little information available on his life and work in English, and even in France he had remained a relatively obscure figure. As of this writing, he does not even have a French Wikipedia entry!

The only book of his I had read up until this point was his *Sacred Geography of the Ancient Greeks*, in translation. I realised I now needed to tackle this and his other untranslated works in the original French.

All Richer's books are now, of course, long out of print, but second-hand copies are available. I began to search for his various works in local bookshops in France and online, and to slowly order them as I tracked them down. One of the first titles I managed to obtain was his follow-up volume to *Géographie Sacrée du Monde Grec*, entitled *Delphes, Délos et Cumes*.[45]

Published in 1970, this work extended his discussion of landscape zodiacs beyond Greece into Italy and the surrounding Mediterranean. In fact, *Delphes, Délos et Cumes* was more than just the next work written and published by Richer; the author states plainly that it comprised the continuation and the complement to his *Géographie Sacrée du Monde Grec*, as if they comprise a single work in two volumes.

When my copy of *Delphes, Délos et Cumes* arrived in the post, I slipped the book out of its envelope and there it was: the vital clue I had been

---

45 Jean Richer, *Delphes, Délos et Cumes: Les Grecs et le Zodiaque*, (Paris, Julliard, 1970).

hoping and looking for. In large white letters, beneath the photo, the cover proclaimed that this book was part of a series called *Les Lieux et les Dieux*, (or "Places and Gods"), and that the series was "*dirigée par*" or "directed by": Gerard de Sède.

I had found the first definite connection between Richer and the Team. Beyond doubt, by 1970, they not only knew each other but were collaborating professionally. Moreover, the book on which they worked together was the sequel to Richer's 1967 publication, with its resonances to *Le Serpent Rouge*.

It was also intriguing to note that the publisher of Richer's 1970 follow-up was Julliard, who had been responsible for de Sède's book on Rennes-le-Château and a previous work on the Cathars, rather than Hachette who had published his *Géographie Sacrée du Monde Grec*.

Gérard de Sède was a fascinating character, author of *L'Or de Rennes*, published in 1967, the book which had kick-started the modern episode of the Affair. [46] An accomplished Surrealist poet, and author of many historical books, he was also an associate of Pierre Plantard, and played a prominent role in the Priory of Sion events.

Six years before *L'Or de Rennes* was published, de Sède had interviewed Plantard at length for his 1961 book *Les Templiers Sont Parmi Nous*, about Gisors, another French village with an esoteric secret. The published interview is a very strange *mélange* of esoteric lore and subtle nonsense and shows Plantard and de Sède working together in an early attempt to conjure a mysterious narrative at the intersection of history, mythology and complete fantasy.

In 1966, he wrote *Le Trésor Cathare*, his last published work prior to *L'Or de Rennes*. Though it has not been translated into English, this work deserves wider reading. It provides an intriguing background account of many historical and esoteric aspects of Cathar life and culture in the Languedoc. It is a serious work, devoid of the kind of loopy mystification which is at play in the 1961 book on Gisors and, to a lesser extent, in the 1967 book on Rennes-le-Château. It displays a deep knowledge of the culture of the Languedoc and shows de Sède as a very capable writer and researcher.

Le Trésor Cathare includes some fascinating material on the ancient conception of landscape and its relationship to the heavens, and clearly such considerations hold deep interest for de Sède. He discusses at length the pivotal role of Toulouse as both cultural, historic and

---

46 Gérard de Sède, *L'Or de Rennes ou La Vie Insolite de Bérenger Saunière*, (Paris, Julliard, 1967), op.cit.

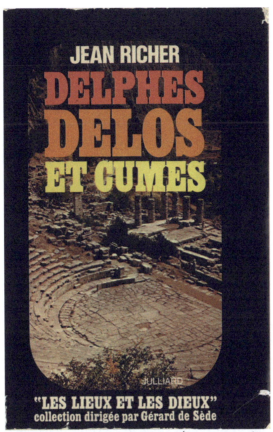

Figure 37: The cover of Jean Richer's *Delphes, Délos et Cumes (Paris, Julliard, 1970)*. This was the sequel volume to his earlier book *Géographie Sacrée du Monde Grec (Paris, Hachette, 1967)*.

geographic centre of the south of France. The title of *Le Trésor Cathare* is somewhat of a misnomer. In fact, the book refers only briefly to the idea of a Cathar treasure. Perhaps a more fitting title would have been "The Treasure of Toulouse".

It comprises three long chapters and several appendices. The first chapter is an account of the more than two thousand years of history of Toulouse, emphasising the curious links of destiny between the city and ancient Greece, particularly Delphi. It recounts in detail the critical story of the Tectosages, Brennus and the cursed "gold of Toulouse", and discusses its deep role in shaping the historic and cultural legacy of the city. The second chapter discusses the history of the Cathars, while the third and last chapter focusses on the art of the Troubadours. Finally, three lengthy epilogues discuss, in considerable detail and with genuine insight, many fascinating items relating to the sacred geography of Toulouse.

The constant and recurring theme of *Le Trésor Cathare* is the landscape of Occitania itself and how it has forged the destiny of the peoples who have inhabited it. The opening chapter lays the groundwork and establishes the important themes: the cycle of origins and returns between Asia Minor and the Languedoc.

The understanding of these events which de Sède is at pains to emphasise here is that the invasion of Delphi by the Celtic tribes in 279 BC should be regarded not so much as a conquest but rather as a *return to source*. According to the historian M. Markale, quoted by de Sède, it was in this sense a quest for purity, an attempt by the Celts to rediscover and rekindle the source of religion as it was practised in this most sacred of Greek sanctuaries. If gold and silver were taken, these were only representative of a much more valuable and profound exchange which had taken place between Delphi and Occitania.

> "'Now', writes Mr Markale, 'if the Sun is the most perfect image of divinity, gold is the symbol of the Sun. The gold of Delphi is therefore the image of the god, a perfectly valid image for a Celt who refuses to accept anthropomorphism. This could explain the attraction exerted by Delphi on Brennus. The attitude of Brennus bursting out laughing in the temple takes on new meaning: it was a question, in the mind of the Gallic chief, of scorning idols and restoring to the solar cult its former simplicity.'"[47]

Ultimately, according to de Sède, this is the reason why the 'gold of Toulouse' carries a curse: not only because it is ill-gotten, but also because it has been mistaken for the genuine treasure of Delphi, the inheritance of the solar cult and its influence on the landscape itself, through the zodiacs and other geometric forms that help to bring about a unity between the earth, its inhabitants and the heavens.

To the extent that the legend of cursed gold has also attached itself to the region around the two Rennes villages in the Haute Vallée, this suggests that its origins might also be traced to the same archetypes of mythic history. The connections and similarities between *Le Trésor Cathare* and *L'Or de Rennes* do not by any means end there. The two books are so closely related that one could almost consider the 1967 work to be the sequel to his book on the Cathars written in 1966. In

---

47 Gérard de Sède, *Le Trésor Cathare*, (Paris, Julliard, 1966). p. 45.

fact, it was being completed and printed during the same time that he was working on the materials for the Rennes book. The poem of *Le Serpent Rouge* is dated October 1966, and the front page of the work itself bears the date 17 January 1967. The contracts for *L'Or de Rennes* were signed on 15 January 1967. The printing of *Le Trésor Cathare* (according to the final page, as is customary of French books) was completed on 18 January 1967!

Although de Sède's book does not include any reference to Jean Richer's work in this field, as it had not yet been published, one can easily imagine that the two of them would have had much to talk about. Indeed, many passages in *Le Trésor Cathare* almost seem to anticipate themes in *Géographie Sacrée du Monde Grec*.

All of this made for a highly intriguing situation: four separate texts, *Le Serpent Rouge*, de Sède's *Le Trésor Cathare* and *L'Or de Rennes*, and Richer's *Géographie Sacrée du Monde Grec*, all appeared in 1967, in Paris. The four texts are very curiously interlinked and seem to hint at some kind of common origin, at least in their underlying ideas, if not in exact content.

The Python of Delphi

Delphes, Délos et Cumes showed that Jean Richer and Gérard de Sède were working together in a professional collaboration that had led to publication in 1970, thus demonstrating that they had known each other from some time prior to that date at least. And it also contained something more, another unexpected and very welcome surprise: the first tangible evidence of a direct connection between Jean Richer and *Le Serpent Rouge*.

Delphes, Délos et Cumes includes an extended passage which reveals the source for some of the cluster of references surrounding the "*serpent rouge*" itself, as it is described in the Sagittarius stanza of the poem, reproduced in full here below:

> "Then returning to the white hill, the sky having opened its floodgates, it seemed to me that nearby I could feel a presence, its feet in the water like one who has just received the mark of baptism; turning to the east, facing me I saw, unrolling endlessly his coils, the huge RED SERPENT cited in the parchments, salty and bitter, the huge unbridled beast became at the foot of this white mountain, red with anger."[48]

---

48   *Le Serpent Rouge.* p. 5.

This was, of course, the stanza with the "salty and bitter" reference that I had stumbled across in Jung years earlier. But now, in Richer, I had found a passage taken from an ancient text which provided the source for several more of the distinctive terms it contained. It occurs near the end of the first chapter which is devoted to Delphi and is entitled: *"Delphes, Centre du Monde et Site Oraculaire"*. The text he quotes comes from an ancient Greek manuscript known as the *Homeric Hymn to Apollo* and recounts the mythic history of Delphi, including the building of the temple by Apollo, and his encounter there with Pythia, the giant serpent, or dragon, who was the original tutelary spirit of the sacred location.

The excerpts Richer cites begin with a description of the location where the *"Seigneur Phobos Apollon"*, that is, the Solar Lord Apollo, decided to erect a magnificent temple. It was to be built:

> "... at the foot of the snowy Mt Parnassus ... it is there that Lord Apollo resolved to build a beautiful temple."[49]

It then describes his vision for the temple, and how it was constructed. At the end of the passage, the serpent Pythia is introduced. Here is the key sentence:

> "Close by is the spring of beautiful ripples where the Lord, son of Zeus, killed the female Dragon with his powerful bow – the enormous and huge Beast, the savage monster who, on earth, made such trouble for men, and such trouble also for their sheep on slender legs: it was a bloody scourge."[50]

If we compare these two texts, the excerpt above from the *Homeric Hymn to Apollo* and the Sagittarius stanza of *Le Serpent Rouge*, we can see that they share the exact same imagery, even to the extent of employing the same words. First, compare the descriptions of the serpent. Here is the key passage for comparison from the Sagittarius stanza of the poem:

> "... the enormous red serpent cited in the parchments, salty and bitter, the enormous beast, unchained at the foot of the white mountain, red with anger."[51]

---

49   Jean Richer, *Delphes, Délos et Cumes*, op. cit. p. 32. quoting Vers 280-304 *"Suite pythique" de l'Hymne homérique d'Apollon* (vers 214 and following.). Translated by Jean Humbert, Belles-Lettres éd.
50   Ibid, p. 32.
51   *Le Serpent Rouge*, p. 5.

Reading the two versions side-by-side, especially in the original French, reveals the identity between the excerpts. The serpent Pythia, the guardian spirit of Delphi, is described in the *Homeric Hymn* as an "enormous beast" (*"la Bête énorme"*). In *Le Serpent Rouge* she is also called an "enormous beast" (*"l'énorme bête"*).

Pythia is also said in the *Homeric Hymn* to be a "bloody scourge" and "a savage monster making trouble for man and sheep". In the modern poem, *"le serpent rouge"* is said to be "red" and "angry". The clear implication is that this red angry serpent, *"le serpent rouge"*, of the modern poem, is based on, or identical to, Pythia, the "bloody", "savage" ancient spirit-serpent of Delphi, as she is described in the *Homeric Hymn*.

Now let's compare the location. The action in the *Hymn* takes place at the "foot of the snowy Mt Parnassus" which overlooks Delphi, whilst in *"Le Serpent Rouge"*, the setting is described as "the foot of the white mountain". Again, they are essentially identical in description. From this one small quotation from the Sagittarius stanza, therefore, a cluster of words can be traced and connected to the passage from the *Homeric Hymn to Apollo* cited by Richer in *Delphes, Délos et Cumes* in 1970. I felt confident enough to conclude that whoever wrote *Le Serpent Rouge* must have derived this set of descriptions from the same passage from the *Homeric Hymn to Apollo*, as quoted by Jean Richer in this book.

However, the author of *Le Serpent Rouge* could not have found these quotations in *Delphes, Délos et Cumes*, because the latter was not published until three years after the former. Of course, it is possible that he or she sourced these concepts from the *Homeric Hymn to Apollo* independently from Richer's interest in the same passage, but it was becoming more difficult to resist the conclusion that the author of *Le Serpent Rouge* was, in fact, Jean Richer himself.

It is not just a question of a word here or a phrase there. Richer's writings are filled with such material. In light of passages such as the following, from his *Delphes, Délos et Cumes*, which is entirely typical of his work, all the signs seemed to be pointing in the same direction:

> "The oracular texts or the texts relating to Delphi enlighten us on certain categories of images that must have appeared to the seer. They are, as is normal, of a dreamlike nature, but at the same time they are 'coded'. There is a predominant animal symbolism with multiple meanings, but it has an astrological basis which is superimposed on the patterns we have indicated. So much so that

the zodiacal geography inherent in the Homeric Hymns acts as a deciphering grid."[52]

This paragraph could almost qualify equally as a description of *Le Serpent Rouge*!

Now I was all but convinced that my intuition of a possible connection between the authorship of *Le Serpent Rouge* and the writings of Jean Richer was on the right track. Richer's work on landscape zodiacs, and the sacred journeys of the heroes of Greek myth around these forms, were echoed in the landscape zodiac of *Le Serpent Rouge* and the pilgrimage of the unnamed protagonist. I could now demonstrate that Richer and de Sède had known each other by 1970 at the latest. In addition, there were quotations from both his 1967 and 1970 works that seemed to indicate, at the very least, that the author of *Le Serpent Rouge* exhibited a familiarity with the same material.

Nevertheless, it still did not quite add up to a positive identification of Richer as the author of *Le Serpent Rouge*. All I could say for certain at that stage was that it appeared the research and writings of Jean Richer and Jung must have played a role in the thinking of whoever it was had composed the poem. To positively identify this person as Jean Richer himself would require something much more specific, some unique, identifying trait, or fingerprint, which would leave no doubt as to who had been responsible for the poem.

Then, on another memorable day, the crucial clue appeared.

---

52   Jean Richer, *Delphes, Délos et Cumes*, op. cit. p. 49.

Chapter Seven
THE NUMBER OF THE FAMOUS SEAL

The Identity of "Cet Ami", "the Friend"

B Y 2008, on the evidence of *Delphes, Délos et Cumes*, I could now say that Richer had certainly been in contact with members of the Rennes team by 1970. More than this, key quotations from the book referring to the serpent seem to underlie key passages of *Le Serpent Rouge*. The question of Richer's influence, even involvement, in the creation of *Le Serpent Rouge* seemed entirely possible, perhaps even plausible. Nevertheless, I knew I would need stronger proof than this to establish his authorship beyond doubt, so I continued to learn as much as I could about him.

I was now becoming more familiar with his work on sacred geography, but at this stage, I still knew next to nothing about his main career focus, the work of the nineteenth-century poet and writer Gérard de Nerval, on whom he came to be considered the pre-eminent authority. Richer makes the claim that it was expressly due to his long immersion in the thought and writings of Nerval that he was able to arrive at the insights that led to his discoveries in the sacred geography of the ancient world.

> "The attentive and sustained study of the work of the poet Gérard de Nerval was a discipline that accustomed me to taking a comprehensive view of complex systems."[53]

Nerval produced a body of work that consisted mostly of short stories, reportage and travel writing, but he was also responsible for a small yet very highly esteemed collection of poetry. Proust considered him amongst the greatest French writers of the nineteenth century,

---
53 Jean Richer, *Sacred Geography of the Ancient Greeks*, op. cit. p. 261.

and his novella *Sylvie* an inspiration for his own work. Baudelaire called him one of the few authors of his age who managed to remain forever lucid, even in death. André Breton considered that he was a precursor of the Surrealists. Yet, for all the high praise, Nerval somehow still eludes the spotlight. As his friend Théophile Gautier said of him, "Gérard seemed to take pleasure in disappearing from himself, in vanishing from his work, in leading his readers astray."

If Nerval's place in the history of French literature is assured, he is still relatively unknown in the English-speaking world. His writings remained almost entirely untranslated until Penguin issued a *Collected Works* in the mid 1990s, accompanied by an excellent essay on the poet by Richard Sieburth, which offers perhaps the best introduction to his work for English readers. There is another fine recent extended chapter on Nerval by the author Richard Holmes in his 2008 book *Footprints* which offers an incisive and deeply personal insight into Nerval's world. He describes becoming interested in Gérard's story as a young literary student living in Paris, and his fascination with the poet, which nearly led to a full-blown obsession. Such can be the effect of coming into the orbit of Nerval's life!

Born in 1808, as Gérard Labrunie, Nerval achieved early literary fame at the age of just nineteen, through his translation into French of Goethe's *Faust*. Goethe himself highly praised this translation as the finest produced and remarked that it captured the spirit of the work as well as the original.

Nerval's life and career as a writer were beset by challenges. He was constantly in and out of debt and suffered from various mental episodes and crises that required hospitalisation during his adult years. Yet he vehemently rejected any diagnosis of madness. Richard Holmes observed, "what is called Nerval's madness seems to have come from a difficulty in distinguishing between, on the one hand, reminiscence, introverted feeling and, on the other hand, revelations of intuition or inner guide."[54]

The ever-present theme which runs through Nerval's work is the interplay between the dreaming and waking lives. His final short story, *Aurélia*, begins *"Le rêve est une double vie"* or "Dream is a second life". In both his life and his writing, there is a certain blurring of the boundary between the waking and sleeping worlds. He inhabits a liminal realm where consciousness and the dream state are intermingled, and

---

54 Richard Holmes, *Footprints : Adventures of a Romantic Biographer*, (London, Harper Perennial, 1985). p. 143.

Figure 38: Gérard de Nerval (22 May 1808 – 26 January 1855).
Photograph by Félix Nadar.

myth mixes freely with the mundane. In the Nervalien universe, every object, person and event takes on an archetypal meaning and is linked via chains of correspondence to infinite realms beyond.

The finest fruit of his labours was a small perfectly formed body of poetry, the distillation of a life lived immersed in this world of sympathetic resonances. Richer expanded on these ideas in a 1957 monograph:

> "Gérard was convinced that sleep puts man in communication with the spirit world and he made no distinction between sleep and reverie (see Aurélia I-1). However, he also conceived of poetry and all literature as an exploration of the mystery. Even in the waking state, in each being he saw the archetype."[55]

---

[55] Jean Richer, *Poètes d'Aujourd'hui: Gérard de Nerval*, (Paris, Hachette, 1970), p.98.

His work is now regarded as one of the early forerunners and inspiration for the Surrealist movement, which was to emerge in the first decades of the twentieth century amongst a circle of writers, and later artists, in Paris and beyond.

He was also obsessed with dates, numbers and visions. In his writings, the ancient pagan gods mingled freely with the visions of Christianity. He read very widely, studied the esoteric sciences and travelled to Egypt and the Near East. The anecdote that is always recounted about Nerval's life is that he would take a pet lobster for walks in the Jardins des Tuileries at the end of a ribbon. It sounds almost too surrealist to be true, but it was apparently not apocryphal. Richard Holmes tracked down the origin of the story to Gautier, who, he said:

> "... told the lobster story, not as an example of Nerval's exhibitionism or fashionable flamboyance—it soon became clear that Nerval was the most retiring and secretive of men—but as an example of his friend's obsession with symbols, and the extraordinary power of his inner imaginative life. The whole point, said Gautier, was that Nerval thought it was a perfectly reasonable thing to do."[56]

He died tragically at the age of just 46, ending his own life on a cold winter's night in Paris in 1855. He was found hanging from a grating in the rue de la Vieille Lanterne, a small alley by the Seine. Much loved in life, and in death, his funeral was attended by a huge gathering of mourners. He was buried in Père Lachaise cemetery after his friends managed to convince the authorities that there was sufficient doubt about his state of mind and intentions to permit him to be buried on consecrated ground.

Jean Richer perhaps more than anyone else was responsible in the twentieth century for establishing Nerval's reputation, through a series of books and texts in which he explored in great depth the esoteric basis of the poet's writings. Yet, Richer did not always treat Nerval's work from a strictly objective point of view. In his own way, the professor was as much of an esotericist as his subject of study. His books on the poet do not shrink from using analyses based in astrology, tarot, hermeticism and numerology to analyse his writings.

It must be said that Richer's writings on Nerval go well beyond the usual boundaries of literary criticism; indeed, they are esoteric works in their own right. He breaks the cardinal rules of modern scholarship:

---

56 Richard Holmes, *Footprints: Adventures of a Romantic Biographer*, op. cit.

he does not hesitate to inhabit Nerval's ideas, to bring them to life, to work within their world, rather than to dispassionately observe them from a distance, or resolutely interpret his work in terms of historical and cultural context. Richer goes as far as to claim that he brings to the surface the deep motivations of Nerval.

For these and other perceived sins against the norms of contemporary literary scholarship, Richer was relieved of the editorship of the venerable Pleiades edition of Nerval's work. A slightly acrimonious war with some of his critical interventions was carried out in the footnotes in later editions by new editors. Nevertheless, while he encountered some controversy, he was much-loved amongst peers and colleagues. The special memorial edition, published in 1992, of the journal of the *Société Gérard de Nerval*, which Richer had founded, was devoted to his life and career and carries many fond tributes to his work and memory from colleagues.

Now that I had begun to familiarise myself with the scope of Richer's work, I could see that there were two topics which permeated his writing and to which he constantly returned: Gérard de Nerval and the zodiac. Considered in light of the question of whether he was the author of *Le Serpent Rouge*, a tantalising possibility presented itself.

If Richer had indeed been involved in the writing of the poem then there was an obvious candidate for the identity of the mysterious *"cet ami"*, or "this friend", the "grand voyager of the unknown", namely: Gérard de Nerval.

Even stronger than that, it seemed to me that this suggestion offered a litmus test to resolve the question. If Jean Richer was indeed the true author of the poem, then *"cet ami"* could only be Gérard de Nerval. On the other hand, if it could be shown that *"cet ami"* in the poem was certainly intended to be Gérard de Nerval, then only Jean Richer could be the author.

Were there any hints or clues which might help? The identity of the mysterious person introduced at the beginning of the poem turned on the riddle in the crucial passage in the second (Pisces) stanza:

> "This Friend, how would you introduce him? His name remains a mystery, but his number is that of a famous seal."

The identity of *"Cet Ami"* was the lynchpin. If I could identify him, I could go on to understand the mystery. If I could not, then the poem would remain an enigma. So, who was this friend? And how could his number be that of a famous seal? I set out to see if I could solve the riddle again, now that I had glimpsed the tantalising possibility that the answer might be Nerval.

I was familiar with the idea that a name could be turned into a number, by the process known as gematria in English, or *arithmosophie* in French. Both Richer and Nerval were fascinated by these kinds of games, and the works of both are replete with examples of converting names into numbers by adding up the values of the letters, using the familiar A=1, B=2, C=3, etc., cipher. I could conceive of turning Nerval's, or anyone's, name into a number, but I could not find the breakthrough to understand what the number of the famous seal might denote.

Indeed, what was this "number" of the "famous seal"? What might this mean? I wondered, like many others before me, no doubt, if the famous seal might be the six-pointed star known as the Seal of Solomon. But if so, what number might be said to be associated with it? I was unable to find any discussion anywhere about a number representing the Seal of Solomon – or any other seal. I gave this considerable thought, and searched through a lot of material, but I could not make any progress. It simply did not seem to be valid, this notion of a number associated with a famous seal.

There seemed to be no way forward.

Then, one day in June 2008, in the golden light of a perfect summer afternoon, I was taking a walk in the countryside around Fa, turning the problem over in my mind, when a startling possibility presented itself. It suddenly occurred to me that perhaps the answer to the riddle was that the number of the "famous seal" was not some number associated with a seal, but the number of the words themselves, *"un sceau célèbre"*.

What would happen if I added up the numbers of the letters making up that phrase? What would the total be? And could it compare to the total of the letters in Nerval's name?

I could hardly wait to get home to test it out with paper and pencil.

The first step was to lay out the alphabet and the simple number code known to every schoolboy.

A B C D E F G H I J K L M N O P Q R S T U V W X Y Z

1 2 3 4 5 6 7 8 9 10 11 12 13 14 15 16 17 18 19 20 21 22 23 24 25 26

Then, I had to calculate the number of "Gérard de Nerval". This was a simple matter of finding the numerical equivalent of each letter and then adding them up to find the total.

I laid out the letters of his name, and the corresponding numbers:

G E R A R D   D E   N E R V A L

7 + 5 + 18 + 1 + 18 + 4 + 4 + 5 + 14 + 5 + 18 + 22 + 1 + 12 = 134

Finally, I added up the total of *"un sceau célèbre"*

U N   S C E A U   C E L E B R E

21 + 14 + 19 + 3 + 5 + 1 + 21 + 3 + 5 + 12 + 5 + 2 + 18 + 5 = 134

It worked. *"Un sceau célèbre"* summed to 134, as did Gérard de Nerval! His number, 134, *"son nombre"*, was equal to the number of *"un sceau célèbre"*, 134. I had solved the riddle. *"Cet Ami"*, the Grand Voyager of the Unknown, was *Gérard de Nerval*.

There are several aspects to this result which make it profoundly satisfying. The first is the sheer elegance of the riddle itself. It is a beautifully constructed puzzle that has withstood fifty years of determined attempts to resolve it by the ingenious, simple trick of concealing the solution in plain sight within the words *"un sceau célèbre" themselves*, rather than in the *meaning* of the words. But there is more to it than this because there is a specific literary allusion here to the works of both Nerval and Richer. As I have noted, both authors were hopelessly addicted to the charms of *arithmosophie*.

Nerval's notebooks are filled with complex calculations connecting names, numbers and dates into a web of semantic relationships which allowed endless scope for his imagination to play. Richer does not hesitate to discuss this aspect of Nerval's work, treating it perfectly seriously, but in addition, he too presents his own arithmosophic speculations amongst his textual analyses. Whilst this approach to an arithmetic interplay between numbers and names is well known in esoteric writings, it is not usually encountered in the serious works of twentieth-century literature professors. It is perhaps little wonder that Richer was eventually relieved of the editorship of the Complete Works of Nerval!

Considered in the context of the *"un sceau célèbre"* riddle, however, the use of gematria to conceal the solution is an exquisite reference to the persistent habit of both Nerval and Richer and is itself an unmistakeable fingerprint which confirmed that I had indeed identified both the protagonist and the author of the poem.

So, I could be quite certain that the solution was valid. The number of *"un sceau célèbre"* was 134, which was also the number of Gérard de Nerval, and moreover the method of obtaining these numbers was itself a watermark that validated that the approach was the correct one. Further, given that Nerval had been revealed as *"cet ami"*, I was confident that the driving force behind the creation of the poem must be Jean Richer. Could I cite any further evidence from the poem that would confirm the identification of *"cet ami"* as Nerval?

There are two occasions in *Le Serpent Rouge* where the number fourteen is introduced, although in both instances there is no obvious reason why this number should have been chosen. The two instances occur in the fifth (Gemini) and ninth (Libra) stanzas:

> "Position yourself in front of the fourteen stones marked with a cross"
> "The cross on the crest stood out under the midday sun; it was the fourteenth…"

The fact that it is mentioned twice is a signal from the puzzle-maker to be alert. Why would the number fourteen be highlighted in this way? The answer is found in one of Richer's works on Nerval, his 1987 book, *Gérard de Nerval Expérience vécue et création ésotérique*.

This occurs in the context of a discussion of Nerval's short story, *Sylvie*, to which we will return in later chapters. For now, we will just note that on page 267 Richer is discussing the significance of the date 1832 which occurs in the very last line of the novella. He observes of the number that the individual digits add up to 14. Then he remarks:

> "…which carries a numerical allusion to the number XIV, (1 + 8 + 3 + 2 = 14) and recall that N is the initial of Nerval."[57]

So, here we have Richer in his own words declaring that the allusion to the number fourteen in a Nervalien text can be taken as a potential reference to his name, because 'Nerval' begins with the fourteenth letter, N. Thus, we are at liberty to interpret the two fourteens in *Le Serpent Rouge* as confirmation clues by Richer that the name which is hidden is indeed Nerval's.

In the previous chapters, I began to make some inroads into the tangle of various problems connected to *Le Serpent Rouge*. I showed that Jean Richer played a hitherto unnoticed role in the Affair of Rennes and was actively collaborating with de Sède by 1970, at the latest. I have

---

57   Jean Richer, *Gérard de Nerval : Expérience vécue et création ésotérique*, (Paris, Guy Trédaniel, 1987), p. 267.

also found that the solution to one of the key riddles of *Le Serpent Rouge* is the name of the poet, Gérard de Nerval, and that he is therefore the "grand voyager of the unknown". This identification is entirely consistent, and to be expected, if Richer was the narrator and author.

This proof of Richer's involvement changed everything. He was a respected academic professor of literature with, until then, no known active connection to the Affair of Rennes-le-Château. Identifying him as the likely author of *Le Serpent Rouge* in mid 2008 was a deep breakthrough. His background in the study of French symbolist poetry and his discoveries in ancient landscape offered a rich and fertile foundation for understanding the depths of the poem – and as the poem is a microcosm of the entire Rennes Affair, it also offered a window into other aspects of the story.

Solving this puzzle broke the seal. Over the next two years there were to be surprising resolutions to major questions I had been asking for a long time. The poem began to open and reveal its secrets.

PART TWO

Orientation

Chapter Eight
THE ZODIAC OF RENNES-LES-BAINS

NOW THAT some of the issues of identity which hover over *Le Serpent Rouge* have been resolved, the next set of questions which I will address relate to location and orientation. Throughout Part II we will explore certain clues and cues which will allow us to identify the setting of the poem, and to find our bearings.

The action of *Le Serpent Rouge* unfolds around several different spaces. The surface layer is clearly recognisable as the church of St Sulpice in Paris – as suggested by the full title of both the poem and the pamphlet. Two of the stanzas (Virgo and Scorpio) are devoted to imagery relating to the architecture, art and history of the church. These references create a frame that acts as a starting point and provides a certain context to the poem. Meanwhile however, an entirely different world unfolds beneath the surface in the other eleven stanzas.

In these layers, the protagonist, the unnamed narrator I have now identified with the poetic voice of Jean Richer, undertakes a journey within a dream around an unidentified landscape of wooded forests, streams and mountain peaks.

Where is this place depicted in the poem? What is the route that he takes, and does it follow a path which can be identified in the real world? And perhaps the most interesting question of all: how does the progress of the journey relate to the zodiac structure of the poem?

Given that the poem appeared amongst the Priory of Sion publications known as the *Dossiers Secrets*, which are largely concerned with the Affair of Rennes, we might expect that it, too, is set in this region. And indeed, although the specific location is not explicitly named in the poem, it is possible to infer, from several oblique references to local landmarks and other items, that *Le Serpent Rouge* is set in the landscape surrounding the village of Rennes-les-Bains. One such example occurs in the Leo stanza, which refers to a spring of healing waters named Madeleine. This is clearly identifiable as the "Source de la Madeleine", a much-beloved local site found a short distance south of Rennes-les-Bains.

Another hint is the phrase "*Par ce signe tu le vaincras*", which occurs in the sixth stanza (Cancer) of the poem. This is a French translation of the Latin "*In hoc signo vinces*", traditionally associated with the famous vision of Emperor Constantine, and usually rendered in English as "By this sign you shall conquer". As quoted in the poem, however, the phrase includes the additional word "*le*", so that it reads: "By this sign you shall conquer him". This same variant form appears as an inscription in the church of Rennes-le-Château, and nowhere else other than the poem, whilst the Latin version may be found in the foyer of the church of Rennes-les-Bains, on a memorial to the priest Abbé Boudet.

We can be confident, therefore, that the poem is set somewhere in the vicinity of the two villages. In order to delve deeper, we need to investigate the early history of Rennes-les-Bains.

The Roman Cardo and Decumanus Axes

The oldest foundations of Rennes-les-Bains predate the Roman era. The village was laid out according to the well-defined rules of spatial organisation of towns and cities of the ancient world. The central meeting place, the *agora*, is located at the intersection of the crossed axes of the *cardo* (usually the north-south line) and the *decumanus* (the east-west line). In the Greek expression of this system of civic architecture, the crossing point of *cardo* and *decumanus* was known as the *omphalos*, which means "navel". This was considered the focal point of the conception of landscape, the meeting place of earth and sky.

Nominally, the *cardo* of a town or city was defined as the meridian passing through the centre, but this was not a rigid rule and consideration was always given to the topography of the site, to take into account any strong natural axis in the landscape. Thus, the precise orientation of the *cardo* for any given location was always informed by the local geography. The massive compendium of lore and learning that is Godfrey Higgins' *Anacalypsis*, published in 1893, discusses the subject of the *cardo* at length. Higgins traces its use to the Etruscans.

> "*Cor* was the Latin name for both heart and wisdom. From this came the word *Cardo*. (...) It gave name to the line drawn from North to South, the pole or axis of the earth, used by the Etruscan Agrimensores to make their

squares for the collection of the sacred tenths or tithes. (...) This line regulated all others. Where the *Decumanus*, the line, crossed it from East to West, the point of intersection at which a cross was set up, was called *cor*, or *cardo*.

He also discussed its role as the symbolic centre:

> "It was in each district the centre, the heart, of all their operations. From this point of intersection two roads always branched off, which is the reason why we have a cross or merestone in the centre of every village, which arose by houses collecting round the sacred X: for this was, for many evident reasons, declared most sacred and holy, and in suitable places the temples arose around or over these crosses."[58]

The layout of the village of Rennes-les-Bains was established in its earliest era of settlement and is described in the monograph *Histoire de Rennes-les-Bains* by Jacques Rivière and Claude Boumendil. They describe the *cardo* and *decumanus* axes:

> "Orienting North-South, this essential axis of Roman town planning (*Cardo*), leads to the heart of the village, to the south, towards the old castrum and to the north, to the sumptuous thermal baths of the Arena. A second track cuts, at a right angle, the Place du Forum, now the Place du Marché, and provides the East-West or *Decumanus* axis."[59]

At Rennes-les-Bains, due to the narrowness of the valley in which the village is set, the urban axis was established alongside and parallel to the bank of the River Sals. This happens to run at a bearing of 15° east of north and defines the orientation of the main street to this day.

> "Urbanism has been imposed by the narrowness of the Sals Valley. The axis of the main street or *Cardo* follows the river on the left bank."

These two alignments are depicted and labelled on a map in Rivière and Boumendil's book, the details of which have been reproduced in Figure 39. The existence and orientation of the ancient Roman axes

---

58   Godfrey Higgins, *Anacalypsis an Attempt to Draw Aside the Veil of the Saitic Isis or an Inquiry into the Origin of Languages, Nations and Religions*, (London, Longman, 1836). Vol. II. p. 413.

59   Jacques Rivière and Claude Boumendil, *Histoire de Rennes-les-Bains*, (Cazilhac, Bélisane, 2006). p. 59.

of Rennes-les-Bains are therefore a matter of historical record. The reader will also have noticed already no doubt that the *cardo* has even given its name to the mountain through which it passes to the north of the village. Rivière and Boumendil comment on this:

> "The very name of the (Pech) Cardou, which is precisely in line with the main street evokes, without ambiguity, the reality of the Roman city."

The authors could not have been aware, however, as we are, of the meridian which also passes through Pech Cardou on a bearing of exactly 180°, as detailed in Part I. This is really an extraordinary situation: two distinct *cardos* pass through its summit, namely, the historic *cardo* axis of Rennes-les-Bains at a bearing of 15° and the *cardo* meridian through ten peaks, which has been revealed earlier in this book. Was ever a mountain more aptly named?

Hence the traces of the ancient *cardo/decumanus* system of Rennes-les-Bains may be found inscribed into the local culture, history and landscape itself, even to the very names of local features.

It is fascinating to observe that if we trace the continuation of this *cardo* further to the south of Rennes-les-Bains, the alignment leads directly to the Templar Château of Albedunem, or Bézu. This fact is not mentioned in *Histoire de Rennes-les-Bains,* but it provides additional confirmation of the historical reality of these axes, and links them to the wider system of structures built by humans in earlier times in the region.

The Château at Bézu is the remarkable site discussed earlier that served as a major stronghold of the Templars and from which they were able to keep watch over the entire valley. Rebuilt in the eleventh century on foundations dating from the sixth century, and possibly much earlier, it stretches along the crest of a mountain ridge with spectacular panoramic views in every direction. Today it lies in majestic ruins, but the sights from its craggy summit remain as magnificent as ever.

For the viewer standing at Château Bézu, looking northwards along the *cardo* on its 15° bearing, the gaze is directed through the centre of Rennes-les-Bains, and onward to the peaks of Bazel and Pech Cardou. (The exact 15° bearing passes precisely through the marked summit of Bazel and grazes the highest point of Pech Cardou very slightly to the west). This *cardo* alignment is strongly marked in the landscape, linking man-made structures and natural landforms: from the Château

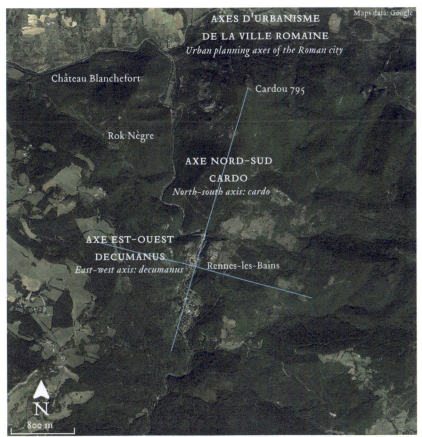

Figure 39: The ancient Roman cardo and decumanus axes of Rennes-les-Bains. The labels and alignments shown are as depicted in the map on page 59 of Jacques Rivière and Claude Boumendil's Histoire de Rennes-les-Bains, (Cazilhac, Bélisane, 2006). (Original French all caps, translation to English in italics).

at Bézu, through the church and the central square of Rennes-les-Bains, parallel to the bank of the River Sals and the main street, and onward to the peaks of Bazel and Cardou, the imposing mountain overlooking the village.

At right-angles to the *cardo* is the *decumanus*, and their crossing point is traditionally the location of the *agora* or public square.

The *cardo* of Rennes-les-Bains lies on a bearing of 15°, and therefore the *decumanus* in this case lies on nominal bearing of 105° (eastwards) and 285° (westwards). The alignment to the church of Rennes-le-Château, approximately two miles away, is within 2 degrees of this ideal figure, on bearing 107°/283°. Again, due allowance is always made for local topography and lore. Today, the central square of Rennes-les-Bains, where the axes cross, features a sign which reads *La Place des Deux Rennes*. This is silent witness to the fact that the alignment to Rennes-les-Bains that connects these two villages which share the name of Rennes was considered to be the axis of the *decumanus*.

For the sake of consistency, I will continue to refer to this axis as a bearing of 105°/285°, with the understanding that the alignment between the Two Rennes differs very slightly from the ideal theoretical value.

A Discovery in the Landscape Around Rennes-les-Bains

It was on my first visit to the Rennes area in 2006 when I came across a copy of *Histoire de Rennes-les-Bains* in a local bookshop and learned of the historical evidence for the *cardo-decumanus* axes. After returning to Australia, filled with renewed enthusiasm, I began to look again at all my previous results in light of my visits to France, my experiences there and the various fascinating new books I had acquired, including the Rivière and Boumendil volume. After lapsing into hibernation for several years, my project had burst back to life. Then, one memorable day in early 2007, I had another decisive breakthrough which was to take the journey to an entirely new level.

I wanted to explore whether these ancient axes of Rennes-les-Bains might interact with any other alignments I had found. Beginning with a clean copy of the 1:25,000 scale map, I ruled them in. First, the cardo on its 15° bearing, from Château Bézu to the church of Rennes-les-Bains, and on through the summit of Bazel. Next, the *decumanus* nominally at right-angles, leading to the church of Rennes-le-Château to the west, and to the peak of a mountain named Pech de Brens to the east.

On a whim, I then decided to add to the map the 45° alignment, discussed in detail earlier, which passes through the church of St-Just-et-le-Bézu, Lavaldieu, the church of Rennes-les-Bains, the château of Montferrand and the château of Arques. Now the map had just three lines ruled on it, all passing through the church of Rennes-les-Bains. Proceeding clockwise from north, with their respective bearings, these were the *cardo* at 15°, the 45° alignment, and the *decumanus* at 105°. I sat back and contemplated the result, as shown in Figure 41.

Up until that moment, I had probably spent hundreds of hours poring over that map during the preceding dozen years or more. I had tested countless theories and traced a myriad of complicated geometric proposals of my own and of others. I had wrestled with a growing

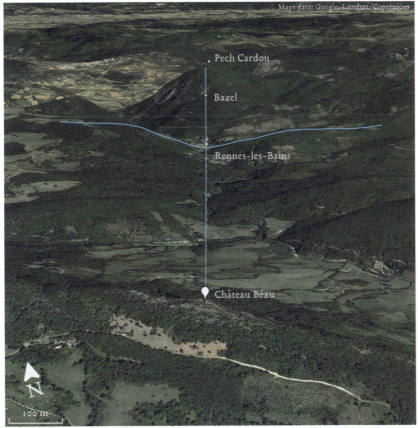

Figure 40: The ancient Roman cardo and decumanus axes of Rennes-les-Bains shown on Google Earth. Perspective view from south of Château Bézu looking slightly east of north, along bearing 15°.

complexity of lines that had often left me baffled, confused and frustrated. But that day, in a sudden flash of intuition, the chaos suddenly dissolved before my eyes and was replaced by a single blindingly simple observation. I realised with a jolt that the first two of those lines through Rennes-les-Bains – the *cardo* at 15° and the 45° alignment – formed an angle of 30° with each other. Together they created a 30° segment centred on Rennes-les-Bains.

The implications of this insight were immediately obvious, against the background of all the reading and consideration I had given to Jean Richer's discoveries of zodiac forms marked around sacred and civil centres throughout the ancient Middle East and the Mediterranean. That first 30° segment between the *cardo* and the 45° alignment

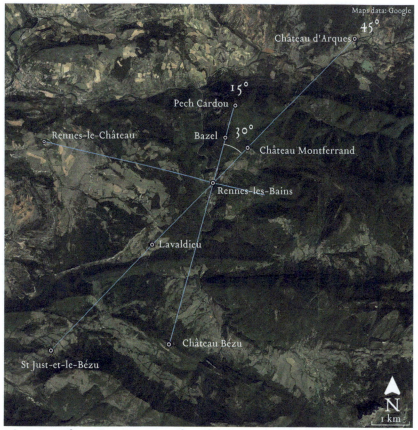

Figure 41: The ancient Roman cardo and decumanus axes of Rennes-les-Bains, on Google Earth, together with the 45° alignment through St-Just-et-le-Bézu, Rennes-les-Bains and Château d'Arques.

hinted at the possibility of a full division of twelve 30° segments around Rennes-les-Bains with the church at the focal point. Had I stumbled on an instance of one of Richer's zodiac divisions of landscape in this location, I wondered?

The next step was to check for any further alignments that might complete such a twelve-fold division. It only took a few moments to find the next crucial line. It passed from Rennes-les-Bains church, running slightly west of north, over the rocky outcrop named Rok Nègre and continuing to Château Blanchefort. This was the ruined structure perched high on the peak on the west bank of the River Sals, providing an ideal vantage point to watch over the sole road from the north by which visitors make their approach to Rennes-les-Bains.

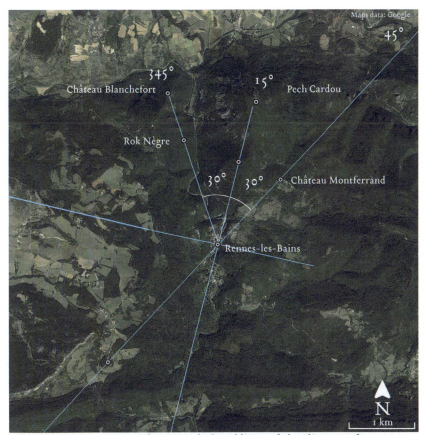

Figure 42: As per previous figure, with the addition of the alignment from Rennes-les-Bains, over Rok Nègre, to Château Blanchefort. This line, the cardo and the 45° alignment create two 30° segments centred on Rennes-les-Bains.

This alignment was of very good quality, with the line connecting the two end points passing exactly through the site in between, namely Rok Nègre. I reached for my protractor to check the angle that it made to the meridian. It measured approximately 14° and thus lay on a bearing of close to 346°, measured from the Rennes-les-Bains church. Again, I will refer to this as "345°", representing the nominal value, which in this case differs by around 1° from the actual physical value, keeping in mind that we are dealing with acceptable variations on the ideal format due to the particular local topography. While this was, again, very slightly off from a theoretical perfect 345° bearing, it was clearly another exemplary case of the strong local topography itself defining the line on a slight variation from the ideal theoretical value. Thus, it too formed an angle of (close to) 30° with the *cardo* of Rennes-les-Bains at 15° bearing, and thereby created a second segment of a twelve-fold division.

Now there were four ruled lines on the map, all passing through the church of Rennes-les-Bains: the "345°" bearing to Blanchefort, the *cardo* at 15°, the 45° alignment, and the *decumanus* at "105°". These are shown in Figure 42.

I soon completed the full complement of six lines required to create twelve 30° segments by adding two more to the four already marked. The fifth, at right-angles to the 45° alignment, I ruled in on a bearing of 135°. I noted that it passed through the prominent peak of Pech de Rodes, due south-east of Rennes-les-Bains. The sixth and final alignment was the bearing at 75°, at right angles to the Château Blanchefort alignment. I then took a step back to survey the results.

I now had six alignments converging on Rennes-les-Bains ruled on the map, of which five passed through at least two significant landmarks. Taken together they marked out the territory around Rennes-les-Bains into a circle of twelve equal segments.

The remains of an ancient zodiac division of the landscape centred on Rennes-les-Bains were emerging into view before my eyes: a pristine example of exactly the same format that Jean Richer had identified around Delphi and other ancient sites. This was the breakthrough I had been looking for in over a decade of wrestling with the map. I sat staring at it, elated, slightly stunned, incredulous.

I took in the simplicity of what was now visible: an accurate twelve-fold division physically marked in the landscape around Rennes-les-Bains, centred on the church, which correlated to and grew out of an acknowledged, valid historical form, the ancient Roman *cardo/decumanus* axes. Here, at last, was something beyond dispute on which I could stand with complete confidence.

The evidence for this geometrical format was buttressed by multiple independent sources. It could not simply be a figment of my own imagination, or a hopeful theory, because it was consistent both with the historical record of the original layout of Rennes-les-Bains, and with Richer's discoveries of similar landscape divisions in the ancient world, in the very location where I had already concluded that he had written about in *Le Serpent Rouge*. I had no doubt: the twelve-fold division, as I called it, was real.

This discovery marked a major milestone in my journey and changed everything. The mists were beginning to dissipate, but I still was not

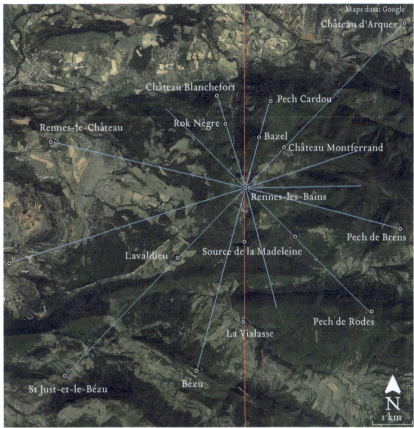

Figure 43: Completing the alignments at right-angles to the Blanchefort and 45° lines. The result is a twelve-fold division of landscape around Rennes-les-Bains, an ancient tradition embedded in the geometric layout of the structures and peaks.

quite there yet. At least several further steps remained before I would be able to positively identify the twelve-fold division with the zodiac format that underpinned the poem. One was to determine a suitable allocation of the zodiac signs to the landscape segments in some order that might make sense of the unfolding action of *Le Serpent Rouge*. Another small but significant problem was to resolve the apparent contradiction between a twelve-fold division based on 30° segments, and the 13-sign zodiac of the poem.

But before I could make progress on these issues, there was a more fundamental problem to be resolved. It was reasonably clear that the path followed in the poem must be somewhere in the vicinity of Rennes-les-Bains, but what was the exact route, and why was it chosen? In order to explore these questions, I need to introduce another source where the answers are to be found.

Chapter Nine
THE CROMLECH OF RENNES-LES-BAINS

IN THIS chapter, I will explore an enigmatic book written in the late nineteenth century by the priest of Rennes-les-Bains. His name was Abbé Henri Boudet, and his work was called *La Vraie Langue Celtique et le Cromleck de Rennes-les-Bains* or, to give the English translation, *The True Celtic Language and the Cromlech of Rennes-les-Bains*.[60]

Boudet's book, which bears the date 1886 on its title page, is a very strange volume that has perplexed readers since it first appeared. It came to wider public prominence when it was republished in 1978, in two separate editions by two different Parisian publishing houses.

The first of these, from Belfond, included an introduction by Pierre Plantard himself. The second, from Bélisane, featured an essay by Gérard de Sède. Both included a facsimile reproduction of Boudet's book, including the full text and a map of the region that had accompanied the original. This fascinating item will be discussed in detail in a later chapter.

The central premise of *La Vraie Langue Celtique* is a patent absurdity: Boudet proposes, with a perfectly straight face, that modern English was the original language of the Celtic peoples and the root from which all other tongues had been derived. This delightfully bonkers linguistic theory is interwoven with discussions of the history of Rennes-les-Bains and the surrounding region and encompasses many unusual items from a wide range of obscure sources.

Boudet sets out to prove his bizarre hypothesis by a convoluted method that involves homonyms between different languages. To select one example at random to illustrate: on page 124 he explains the origin of the Basque word *etchôla*, meaning "cabin", as deriving from the combination of two English words "head" and "shoal". Keep in mind

60 The word cromlech usually means a stone circle.

that he performs these feats of linguistic acrobatics in French, which only adds to the strangeness.

> "A cabin, *etchôla*: A crowd of heads under the same roof – *head*, head – *shoal*, a crowd or troop."[61]

He maintains this apparent madness without flinching for more than three hundred pages. In his 1978 introduction, de Sède does not mince his words in declaring the book "crazy":

> "When it is read carelessly, this work appears, it is true, like a tissue of absurdities. The theory that the author develops is that the primitive language of humanity is that of the Celts, that it is maintained intact in English. (…) If we consider it as a linguistic treatise, the work is evidently crazy."

And yet, perhaps things are not quite as foolish as they might seem:

> "We must not be fooled by the apparent stupidity of *La Vraie Langue Celtique*. In certain cases, what is foolish is sealed. It is therefore necessary to reread this work with a wary eye which looks for things other than those which seem at first apparent."

In the midst of the "apparent stupidity" of the crazy linguistic theories, Boudet introduces another theme, an extended description of what he calls the "cromlech", or stone circle of Rennes-les-Bains. He claims to have identified in the landscape a large collection of standing stones and boulders that he asserts were put into position by an unknown race who occupied the land before the arrival of the Celts. The purpose, according to Boudet, was to form a huge enclosure for large gatherings of people.

This description of the cromlech is difficult to reconcile with the reality of the landscape. Whilst the rocky outcrops and stones that he identifies as making up this "stone circle" certainly exist, they do not by any means form a coherent enclosure in any meaningful way, nor is it remotely possible that they were positioned intentionally. They are, clearly, naturally occurring geological forms which may be found scattered throughout the entire region.

---

61 Henri Boudet, *The True Celtic Language and the Stone Circle of Rennes-les-Bains*, (Paris, Les Editions de l'œil du Sphinx, 2008), pp. 124, x, xi. English translation of *La Vraie Langue Celtique et le Cromleck de Rennes-les-Bains*, (Cazilhac, Bélisane, 1986).

LA VRAIE
LANGUE CELTIQUE

ET

Le Cromleck de Rennes-les-Bains

PAR

l'Abbé H. BOUDET

CURÉ DE RENNES-LES-BAINS (AUDE)

CARCASSONNE

Imprimerie François POMIÈS, rue de la Mairie, 50.

Droits de traduction et de reproduction réservés.

Figure 44: The title page of La Vraie Langue Celtique by Abbé Henri Boudet, priest of Rennes-les-Bains. Notice the date of publication, 1886, prominently displayed in the centre within an elaborate arabesque design.

To this day, nobody has ever arrived at any sensible interpretation of what Boudet had in mind when he presented his cromlech to the world. Many have written off *La Vraie Langue Celtique* as little more than the deranged ravings of a lunatic.

On the other hand, Boudet was also known as a serious, scholarly priest, and this notion of his book as crazed nonsense does not seem consistent with the known facts of his life and personality.

Boudet clearly had an agenda when he wrote *La Vraie Langue Celtique*, but what this might have been has never been satisfactorily explained. His book remains as much of a mystery as the day it was published.

La Vraie Langue Celtique and Le Serpent Rouge

As noted in an earlier chapter, there are a number of clues scattered throughout *Le Serpent Rouge* that suggest that the action is set somewhere in the landscape around Rennes-les-Bains.

Intriguingly, the poem also seems to contain several lightly veiled references to Boudet's book. For example, the Pisces stanza of *Le Serpent Rouge* includes the following phrase (in translation):

> "... sitting on the white rock, looking beyond the black rock to the south".

These terms echo a passage in *La Vraie Langue Celtique* in which Boudet is discussing the environs of the ruined Château Blanchefort:

> "The natural peak of this rock was removed in the Middle Ages to allow the construction of a small fort to serve as an observation post. There remain some vestiges of masonry testifying to the existence of this fort. The white rock, which is the first to strike one visually, is followed by a stratum of blackish rock stretching to Roko Nègre. This special feature gave this white rock, placed on top of the black rocks, the name Blanchefort – *blank,* white – *forth*, in front."[62]

Here Boudet has identified Château Blanchefort, perched high on a rocky outcrop, as the "white rock" and Rok Nègre, an outcrop or pillar of dark rock visible a short distance to the south, as the "black rock". Hence, when *Le Serpent Rouge* refers to the "white rock" and the "black rock", it is clearly calling attention both to these prominent local landmarks, and to Boudet's description of them in his book.

In the Libra stanza of *Le Serpent Rouge,* we find:

> "The cross of the crest stood out under the midday sun, it was the fourteenth and largest of all at 35 centimetres!"[63]

This line echoes several reports by Boudet in *La Vraie Langue Celtique* of various Greek crosses that he had found carved into the stones in the area. Regarding these he writes:

> "One discovers on the neighbouring rocks some Greek crosses deeply engraved with a chisel and measuring from

---

62   Ibid, p. 231.
63   *Le Serpent Rouge.* p. 4.

twenty to thirty to thirty-five centimetres. (...) A rock of the crest carries a Greek cross engraved in the stone. It is the largest of all of these that we have encountered." [64]

The "largest" "cross" of the "crest" having a dimension of "thirty-five centimetres" in the stanza of *Le Serpent Rouge* is an unmistakable reference to these passages in *La Vraie Langue Celtique*. Note, however, that there is no mention of any "fourteenth" cross in this corresponding passage in Boudet. I showed earlier that Richer considered the number fourteen to be a symbolic reference to Nerval, because N is the fourteenth letter of the alphabet.

It is evident from these examples that *Le Serpent Rouge* is set in the landscape around Rennes-les-Bains, and that it contains elements taken from *La Vraie Langue Celtique*. But why has Richer brought Boudet's work into his poem? What is the precise nature of the connection between the two texts? In this chapter, I will investigate their relationship. I begin with this question: what exactly did Boudet have in mind when he wrote of the cromlech, or stone-circle, of Rennes-les-Bains?

What is the Cromlech?

The cromlech of Rennes-les-Bains is obviously intended to be a major theme of Boudet's book, as it is included in the title itself, yet his discussion on the topic takes up only a relatively small number of pages, compared to the endless chapters devoted to his bizarre linguistic theories. Most of the description of the cromlech is found in Chapter VII.

Boudet informs the reader that ancient peoples who predated the Celts had built stone circles, known as cromlechs, or *drunemotons,* in which they would hold large gatherings, and that examples of these may be found in various locations in France. Quoting an earlier writer on this topic, he writes:

> "When the menhirs are arranged in a circle, singly or in multiples, they are called cromlechs. These are vast enclosures of stones, usually positioned around a dolmen. (...) These stones monuments as we have already said, are no more Celtic than Druidic. The Celts ... found them already made when they arrived, and no doubt regarded them with as much astonishment as ourselves."[65]

---

64 Henri Boudet, *The True Celtic Language and the Stone Circle of Rennes-les-Bains*, (Paris, Les Editions de l'œil du Sphinx, 2008). pp. 235, 245.
65 Ibid, p. 163.

As I have already noted, Boudet's suggestion is an absurdity: there is no man-made cromlech, as he has described it, in the landscape around Rennes-les-Bains. Why then does Boudet assert the existence of a cromlech when according to his own definition, no such thing can be identified? What was he up to? Was he simply mistaken? Or was he pursuing some more subtle literary strategy?

De Sède comments in his 1978 introduction to Boudet's work:

> "While several menhirs exist around Rennes-les-Bains, on the other hand a cromlech cannot be found. As it is not possible to write about something that does not exist, even its title indicates that *it necessarily discusses something other than that which it pretends to deal with*."[66] [emphasis added.]

This is an intriguing idea. If under the guise of writing about this cromlech, Boudet was in fact discussing something else, what might this be? A suggestion springs to mind, considering what I have described in the previous chapter. We have found historical, cultural and geometric evidence for a twelve-fold division of landscape centred on Rennes-les-Bains. Could this be related in some manner to the cromlech? Was de Sède aware of the format we have discovered, and could this be what he is hinting at?

There are other clues scattered throughout *La Vraie Langue Celtique* which offer more detail on the nature of Boudet's enigmatic "stone circle". At the outset of Chapter VII, for example, he offers a surprisingly exact estimate of its size:

> "Indeed, its mountains crowned with rocks form an immense Cromlech of sixteen or eighteen kilometres circumference."[67]

There is something rather curious about this statement, especially for something that does not physically exist! It seems slightly more precise than one might reasonably expect under the circumstances. In the absence of a well-defined boundary, or perimeter, or indeed any discernible circular shape amongst the rocks Boudet identifies as making up his cromlech, why did he give it such an exact description as "sixteen or eighteen kilometres" in circumference? If he wanted to indicate its approximate size, why not just say, for example, "around twenty kilometres" or, if a range was more appropriate, "fifteen to

---

66 Ibid, p. 124.
67 Ibid, p. 225.

twenty kilometres"? He does not give a range at all, but rather two discrete values, sixteen *or* eighteen kilometres. It occurred to me that the dimensions he cited were somewhat *over-specified*, for some reason. At the very least, he seemed to be drawing attention to the circular nature of this cromlech.

Slowly, an alternative interpretation of what Boudet intended began to dawn on me. Was he covertly signalling to the alert reader through these clues, I wondered, that his cromlech should be understood in some sense as a *geometric* concept, rather than a physical artefact?

If so, it is amusing that Boudet states that the centre of the cromlech "is found in the place named by the Celts themselves, Le Cercle".[68] This is a tiny hamlet, still named Le Cercle, located close to Rennes-les-Bains, less than a few hundred metres approximately south of the central square and church of the village. "Le Cercle" translates, of course, to "The Circle". Was Boudet making a little joke here and in doing so drawing a veil over the true centre?

This question led to others. When he described it as "sixteen or eighteen kilometres" in circumference, was he perhaps implying that it could be thought of as a pair of "virtual" circles with respective circumferences of sixteen and eighteen kilometres? And, if so, might these circles share the same centre as the twelve-fold division which I had identified, namely the church of Rennes-les-Bains?

It seemed at least worth considering that the centre of Boudet's cromlech might be in Rennes-les-Bains, rather than a tiny cluster of a few houses a short distance away. After all, he did call it the cromlech of Rennes-les-Bains. Furthermore, the *omphalos*, or "navel", the focal point of the local conception of landscape in the ancient world, was historically always situated at the intersection of the *cardo* and *decumanus*, and as we have learned, at Rennes-les-Bains this coincided with what is now the town square, close to the church. We have also seen that this was the centre of the twelve-fold division of landscape.

These seemed to me like promising suggestions, worth exploring to see where they led, so I drew the two suggested circles in Google Earth. Here they are in Figure 45: two circles with circumferences of sixteen and eighteen kilometres, centred on the church of Rennes-les-Bains.

I examined the result. The circles neatly enclosed the area around Rennes-les-Bains and its surrounding ring of mountains, including the peaks of Pech Cardou, Château Blanchefort and La Pique. The 16-kilometre circumference circle fell exactly on the junctions of the

---

68   Ibid, p. 246.

rivers Sals and Rialsesse in the north. If the dimensions of the cromlech were intended to designate the local territory around the valley of the River Sals and Rennes-les-Bains, then it was a good physical fit to the landscape.

It still did not fully account for the remarkably specific values for the circumference that Boudet had cited, and I certainly could not prove that this was what he intended the reader to understand by his enigmatic description, but it felt like a promising development.

The notion that the cromlech was in some sense a circular geometric form, rather than a ring of physical stones making an enclosure, at least answered to de Sède's penetrating observation about its description quoted above:

> ".. it necessarily discusses something other than that which it pretends to deal with".

Now that we've briefly surveyed Boudet's book and begun to delve into the nature of his cromlech, we are ready to take a closer look at the relationship between *La Vraie Langue Celtique* and *Le Serpent Rouge*.

A Tour Around the Cromlech

The account of the cromlech in Chapter VII of Boudet's book includes a guided walking tour, in which the author, with the reader in tow, follows a path in the countryside around Rennes-les-Bains, visiting and describing its various landmarks in turn.

Many of the place names that Boudet mentions have disappeared from the contemporary map of the area, but using the map that he included in his book, we can identify the locations to which he is referring and recreate the route that he has taken.

For the most part, the path that he traces out corresponds with trails and tracks that still exist today. Things change very slowly in this part of the world. So, it is possible to plot, with a fair degree of confidence, the route of Boudet's walk on a modern map of the region. This result is shown in Figure 46 below. I have superimposed his map on to Google Earth at correct scale, placed markers at the locations that he mentions in Chapter VII, with their page numbers, and then mapped the path in red which he would have followed to visit them in order.

The walk begins in the north-east quadrant, nominally at the summit of Pech Cardou, and proceeds in a counter-clockwise direction, making a full circuit around the village. A brief summary of his itinerary is as follows: starting from Pech Cardou (accessible today

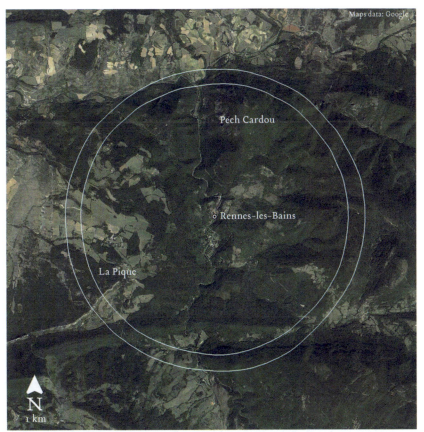

Figure 45: Boudet's cromlech conceived as a pair of circles of 16 and 18 kilometres in circumference, centred on Rennes-les-Bains Church.

via a well-maintained track up the north side which departs from the village of Serres), the walk descends from the summit, southwards, via the peak of Bazel, then down the south-west flank of the mountain to the east bank of the Sals.

After crossing the river, the path heads north on the west bank along a trail through the dense oak forest which leads to Rok Nègre and Château Blanchefort.

Boudet then retraces his steps back through the forest, before the path climbs the steep ridge to the west of Rennes-les-Bains, with commanding views over the valley. He then proceeds south. Along the way, he describes various sites on the route, including Cugulhou, Cap de l'Homme and Pla du Coste, arriving eventually at the stream of Trinque Bouteille.

After coming down from the ridge, he reaches Haum Moor and other sites located to the south of the village. Then, the path turns back to the north to continue the circuit. He climbs the hills which are east of the village, to reach La Fajole. From there, after a short loop through the outskirts of Rennes-les-Bains, the path leads north-east and climbs back up into the mountains. It passes Château Montferrand, and then finally arrives back at Pech Cardou to conclude the tour.

It is easy enough to follow this walk today on paths that have remained essentially unchanged for centuries, perhaps millennia. The entire journey covers around twenty kilometres and can be accomplished in a single day, as Judith and I did on 1 August 2008. There are some slight variations which are possible on some sections, but for the most part this circuit walk around the village, from Cardou to Cardou, follows a single easily identifiable route.

With the path of Boudet's tour of the cromlech mapped out, we now turn our attention to recreating the itinerary of the action of *Le Serpent Rouge*. There are a sprinkling of clues throughout the poem which refer to locations. From these, our aim will be to identify the path taken by the narrator.

At the outset, the location of the beginning of the walk does not seem to be obviously indicated, as there is no explicit landmark mentioned in the first stanza. However, in the penultimate stanza the narrator describes "returning again to the white hill". We saw in Chapter Six, in the discussion of Richer's *Delphes, Délos et Cumes*, that this may be understood as a reference to Mount Parnassus, from the Homeric Hymn to Apollo, and that it represents Pech Cardou in the poetic geography of *Le Serpent Rouge*. Hence, we may reasonably infer that the walk begins and ends on that mountain, the "white hill", Pech Cardou.

The first identifiable landmark mentioned in the poem occurs in the second stanza, which contains the references to the white and black rocks. As discussed, these stand for Château Blanchefort and Rok Nègre. So, we can deduce that the narrator has descended from Cardou, crossed the River Sals, and hiked through the woods to the ruins of Blanchefort on the west bank. From here he looks back, over the "black rock" towards the church of Rennes-les-Bains, as we saw in the last chapter.

The narrator then declares that his goal is to reach the place of the "sleeping beauty", who "certain poets saw as the Queen of a lost kingdom". We know that the poem was written by Jean Richer, at least in part as homage to Gérard de Nerval, so it would seem reasonable to

THE CROMLECH OF RENNES-LES-BAINS

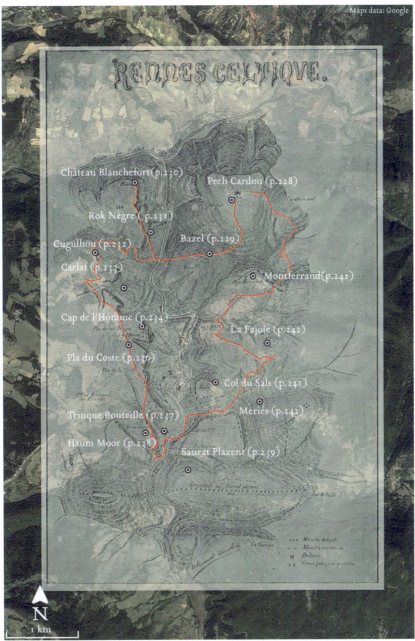

Figure 46: The route described in Chapter VII of Boudet's La Vraie Langue Celtique. It depicts the walking path around Rennes-les-Bains making a circuit of the "cromlech". Places mentioned in the text, with page numbers, are shown superimposed on the map from the book and Google Earth.

identify him as our "certain poet". The question then becomes: did Nerval ever write about a Queen of a lost kingdom? The answer is that he certainly did, and she is easily identifiable.

Nerval's works display a near-constant pre-occupation with the archetype of the divine feminine. She is referred to under many different guises, all of whom may be taken as images of the goddess Isis. He writes frequently in this vein of the Queen of "Saba", which is the term he uses for the land of Sheba, the kingdom whose precise identity and location has been lost to history. Richer had an entire chapter in his first 1947 book on Nerval's references to the *Reine de Saba*. Unsurprisingly, this obsession had its roots in his relationship with his mother, who died when he was less than two years old. This loss affected him profoundly and was at the heart of a certain longing that permeated his existence. The absence of his mother was transformed into a yearning for the divine feminine. As Richard Holmes noted:

> "Every woman and goddess in Nerval's story became a personification of his lost mother; every animal—the lobster in Paris, the parrot in the Valois, the scarab in Egypt—became messengers from the supernatural world."[69]

The woman was frequently identified as the Reine de Saba, and the goddess, of course, Isis. Richard Sieburth remarks that Nerval experienced a "mystical hierogamy with his anima, (when he) weds his Queen of Sheba".

If Nerval is our "certain poet", as he must surely be, then the "Queen of a lost kingdom" is the Queen of Sheba. By symbolic equivalence in Nerval's poetic universe, the "sleeping beauty" may be understood as representing Isis herself. This interpretation is confirmed in the Leo stanza, where this queen is explicitly named as Isis. She is also the Magdalene, or the Madeleine, as she is sometimes referred to in France.

The clues converge unmistakably on the Source de la Madeleine, a holy ancient spring which bubbles from a rock alcove by the banks of the Sals, south of Rennes-les-Bains. The poet states his goal is to reach the place of sleeping beauty. That is to say, his destination is the Source de la Madeleine, which as we have seen lies on the Rennes-les-Bains meridian.

The route from Château Blanchefort to the spring follows the same path along the crest of the ridge to the west of Rennes-les-Bains which Boudet traversed past Cugulhou and the other named locations.

---

69 Richard Holmes, *Footprints: Adventures of a Romantic Biographer*, location 4917

Figure 47: The path of the walk in *Le Serpent Rouge* in Google Earth. Beginning from the first named location Château Blanchefort at upper left, the tour proceeds counter-clockwise, visiting la Source de la Madeleine (bottom) and Château Montferrand (right) before returning to Pech Cardou in the upper right.

After reaching the Source de la Madeleine, the narrator must now turn and head back north so as to complete the journey back to the "white hill", Pech Cardou. While there are no further obvious named locations that would narrow down the exact route on the return leg northward, it must pass to the east of Rennes-les-Bains, then up to Montferrand before finally returning to the summit of Cardou.

The narrator has now completed a large circumambulation around the village. The path of the walk described in *Le Serpent Rouge* in the landscape around Rennes-les-Bains is shown in Figure 47. By comparison with the preceding Figure 46, and as the reader will have noticed by now, the path that he followed is the same as that described by Boudet in chapter VII of *La Vraie Langue Celtique*. The poem is modelled on Boudet's tour of the cromlech.

This is a tantalising result, as it offers a basis for comparison between the zodiac and the cromlech, with the twelve-fold division as the bridging concept between the two. If the twelve-fold division I had found was indeed the basis for the zodiac of *Le Serpent Rouge*, then I realised it should be possible to find a unique allocation of the zodiac signs to the landscape format which correctly matches the action of the poem in the various stanzas to their corresponding segments. How then should the zodiac be arranged on the twelve-fold division?

To begin: where does it start?

Allocating the Zodiac Signs to the Twelve-fold Division in Landscape

Observe firstly that the action of *Le Serpent Rouge* opens at Pech Cardou, and proceeds west, across the Sals river. This is the same area that Boudet identifies as the entrance to his cromlech, at the confluence of the Sals and Rialsesse rivers, north of Rennes-les-Bains. Indeed, it is the natural entrance to the Rennes-les-Bains area to this day. This dictates that the beginning of the *Le Serpent Rouge* walk and the entrance to the cromlech lie in the segment north of Rennes-les-Bains, between the *cardo* axis (at a bearing of $15°$), and the Rennes-les-Bains-to-Château-Blanchefort alignment (at a bearing of $345°$).

We know that the zodiac of *Le Serpent Rouge* begins with the sign of Aquarius. Hence there is only one possible solution to the problem of integrating the *Le Serpent Rouge* walk, Boudet's cromlech, the twelve-fold division and the zodiac: Aquarius must be allocated to the segment lying due north of Rennes-les-Bains, at the confluence of the Sals and Rialsesse rivers, between Pech Cardou and Château Blanchefort.

After placing Aquarius in that northern segment, the remainder of the signs of the zodiac can then be allocated to the twelve-fold landscape division, proceeding anti-clockwise.

There is no room for adjustment. If this zodiac does not merge the poem, the landscape and the cromlech into a single cohesive format, then something has gone wrong, and this hypothesis must be abandoned. We can now proceed to find out if the action within the poem falls within the correct zodiac signs according to the scheme suggested.

But wait: one thorny problem remains. We have not yet addressed the issue of the thirteen stanzas of *Le Serpent Rouge*. How are we to distribute thirteen zodiac signs to the twelve-fold division of the landscape? How does thirteen divide into twelve? This baffled me for a long time until one day, browsing in a popular reference work, I came across an article on astrology that included a short discussion about Ophiucus, the additional sign included in the zodiac of *Le Serpent Rouge*. It noted that whilst the sign is rarely encountered in English astrological writings, it is not uncommon in the European tradition, where it is often considered to be *the second half of Scorpio*. This was the vital clue I had been looking for!

If we allocate a normal 12-sign zodiac to the twelve-fold division of landscape around Rennes-les-Bains, beginning with Aquarius in the north as described above, Scorpio will occupy the segment due east of the village. Hence, it is readily divided into two halves by the east-west line through the centre.

This provides an elegant solution to the problem of how to assimilate the full thirteen signs of the poem to the twelve-fold division: we allocate Ophiucus to the upper half of that eastern segment, which leaves Scorpio occupying the lower half. In this manner, the 13-sign zodiac of *Le Serpent Rouge* can be distributed neatly and naturally around the twelve-fold division, with the simple, helpful addition of the east-west line, in a manner that is entirely consistent with the tradition of European astrological usage.

At last, we can superimpose the path of the walk of the narrator in *Le Serpent Rouge,* the thirteen-sign zodiac of the poem and the twelve-fold division of landscape around Rennes-les-Bains on Google Earth. (See Figure 48.) With that done, we can proceed around the zodiac and compare the contents of each stanza of the poem with the corresponding segment of landscape.

The next sign of the zodiac in order after Aquarius is of course Pisces, so it has been allocated to the second segment, proceeding anticlockwise, which lies between the 345° bearing and the 315° line.

Thus, the 345° line, the alignment from Rennes-les-Bains church, over Rok Nègre, to Château Blanchefort, is the line that marks the beginning of the Pisces segment.

Immediate and spectacular confirmation that this allocation is indeed the correct one is provided by the Pisces verse of *Le Serpent Rouge*, which as we have seen, includes the phrase which describes the viewer standing atop Château Blanchefort, looking southward directly over Rok Nègre, the black rock, and on to the church of Rennes-les-Bains.

> "... sitting on the white rock, looking beyond the black rock to the south".

Our twelve-fold division and the zodiac distribution have stipulated that the Pisces segment commences on the bearing of 345° from Rennes-les-Bains church. This line is, precisely, the Rennes-les-Bains–Rok Nègre–Château Blanchefort alignment. The allocation is a perfect match! Thus, the Pisces verse specifies accurately the segment to which it has been assigned.

Now we can continue around the rest of the zodiac testing whether the signs correspond with the locations identified in the poem. The next one is the ancient spring, *La Source de la Madeleine* south of Rennes-les-Bains, referenced in the Leo stanza of the poem. And indeed, on the map, it falls squarely within the Leo segment on the zodiac allocation to the twelve-fold division. Another direct hit!

There is more evidence to come, but already these results are sufficient to confirm the interpretation to which we have been slowly building over the course of the preceding chapters: the zodiac on which *Le Serpent Rouge* has been constructed is based on the twelve-fold division around Rennes-les-Bains which I have identified.

If this is true, then the author of the poem, Jean Richer, recognised in the landscape around Rennes-les-Bains the traces of a zodiac of the same type and form as those he had discovered at Delphi and many other locations.

Furthermore, as he modelled the walk described in the poem on the tour of the cromlech outlined in chapter VII of *La Vraie Langue Celtique*, we are justified in concluding that Richer considered that the zodiac and Boudet's cromlech were, in some sense, closely related forms, or different facets of the same underlying historic physical reality.

Or perhaps he might have put it more simply: Boudet's cromlech is a landscape zodiac. Either way, it is now apparent that in *Le Serpent Rouge*, Richer has brought the twelve-fold division, the zodiac and Boudet's cromlech together into a single unified format.

The third match as we proceed around the zodiac in order, is one that I did not recognise until later. It is of a slightly different flavour. In the Virgo stanza, we read (in English translation):

> "... in the jumps of the four horsemen, the hooves of a horse had left four imprints on the stone."

If we inspect the Virgo segment on the IGN map, shown in Figure 51, we can see that there are four rocky outcrops marked that do indeed offer the appearance of the hoof-prints of a horse. I hasten to add that I am by no means the first to notice this. David Wood in his book

THE CROMLECH OF RENNES-LES-BAINS 159

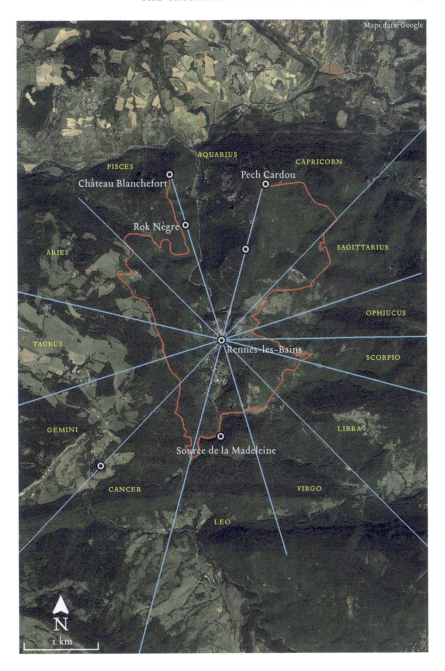

Figure 48: The Le Serpent Rouge path on the zodiac of Rennes-les-Bains.

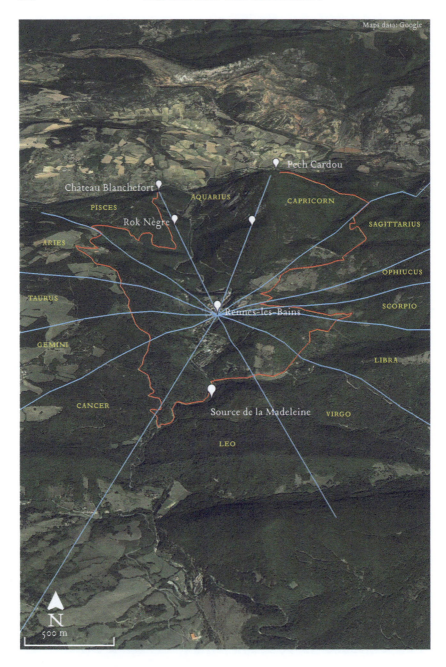

Figure 49: The Le Serpent Rouge path and the zodiac of Rennes-les-Bains. Perspective view, from the south looking north.

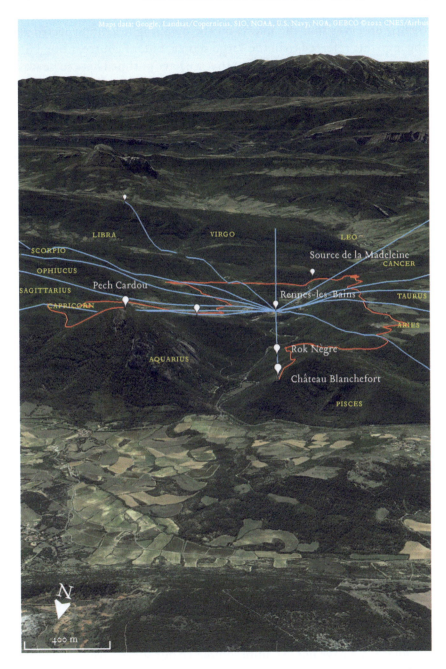

Figure 50: View south along the Pisces alignment of the Le Serpent Rouge zodiac in the landscape, from Château Blanchefort, over Rok Nègre, to Rennes-les-Bains church.

Genisis observes that the marks on the map might be interpreted as the imprints referred to in the Virgo stanza, but, of course, he had no knowledge of the zodiac solution that we have outlined here so he was unable to proceed any further along the correct path with this insight.

This is an intriguing result, because it can only have been discerned by reference to the IGN cartography. We can conclude from this that Richer had ruled the lines of the zodiac on an earlier edition of the map and noticed the same "hoof-prints" depicted in the iconography there representing the rocky outcrops.

The next opportunity to check the allocation is in Scorpio. This stanza in *Le Serpent Rouge* is primarily concerned with St Sulpice in Paris. It includes these words:

> "A heavenly sight for he who remembers the four works of Em. Signol around the Meridian line, in the very choir of the sanctuary from which this source of love for one another radiates, I turn around looking from the rose of the P to that of the S, then from the S to the P . . . ".

This is a reference to the brass meridian inlaid into the floor of the church, marked by a "P" and an "S", and made famous by Dan Brown's *The Da Vinci Code*. It lies on meridian 2°20'5.7"E which happens to pass directly through the zodiac. If we rule it in, as shown on the accompanying Figure 52, we notice that the path approaches very close, within a few hundred feet or less, of the St Sulpice meridian in the Scorpio sector, for the first time in the circuit.

I suggest that Richer ruled the St Sulpice meridian on the map out of curiosity and then noticed that it met the path in the Scorpio segment of his zodiac, so he wove it into the poem. It is another incremental piece of evidence that testifies to the reality of the zodiac of Rennes-les-Bains, and its use as the underlying template for the action of *Le Serpent Rouge*. This is not the last time we will encounter the St Sulpice meridian before the end of our journey.

With these examples, we have enough clues already to conclude definitively that the zodiac plan of the poem is consistent with the twelve-fold division centred on Rennes-les-Bains and the allocation of signs determined by the placement of Aquarius at the opening of the cromlech. We have not yet quite completed our survey around the full zodiac circle, but we can be quietly confident that the solution is correct.

THE CROMLECH OF RENNES-LES-BAINS 163

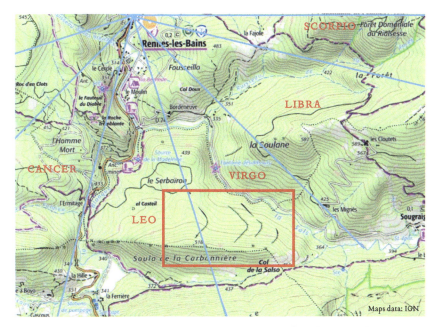

Figure 51: The Virgo sector, with the four hoof prints of the horse on the stone, depicted on the IGN map.

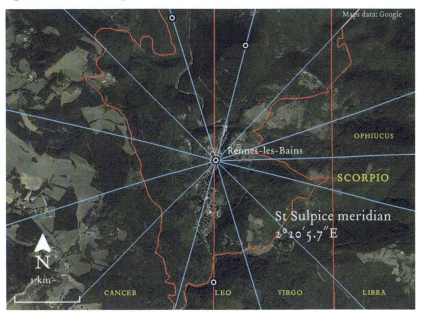

Figure 52: The St Sulpice meridian and the Scorpio segment.

Summary

Putting the pieces together, we are able at last to answer the question about the route traced by the path of the walk in *Le Serpent Rouge*.

Our protagonist, who is Richer himself, is recreating the tour described by Henri Boudet in Chapter VII of his book, *La Vraie Langue Celtique*. The path commences on Pech Cardou and proceeds in a counter-clockwise direction to perform a complete circuit in the landscape around Rennes-les-Bains.

This landscape is ordered around an ancient format, a twelve-fold division centred on the church that Richer discerned and to which he added zodiac signs, following the examples of Delphi and elsewhere from his own research.

By superimposing the zodiac structure on Boudet's walk and using this as the template or framework for the geography of *Le Serpent Rouge*, Richer was asserting that Boudet's cromlech and the Zodiac of Rennes-les-Bains, with its origins in the pre-Roman *cardo* and *decumanus* axes, are one and the same.

This is the message hidden in *Le Serpent Rouge*: Boudet's cromlech of Rennes-les-Bains and the zodiac are alternative descriptions of the same underlying physical reality, an ancient geometric construction that has been imprinted on this landscape in the same cohesive manner that Richer had identified at ancient sites around the Mediterranean. It has left its traces in the spatial relationships between man-made and natural structures, and in place names. It can be mapped, and walked, and sighted. It's invisible, but it is real. It is very old, but it has been recognised in modern times by Boudet and Richer.

A plausible scenario is not difficult to imagine. In October 1966, Professor Richer had made a visit to Rennes-les-Bains in the company of certain members of the Team, perhaps including Gérard de Sède or Pierre Plantard himself.

It was perfect late autumn weather, when the Haute Vallée is at its finest and ideal for rambling around in the countryside. There was much discussion of Boudet's book, and local lore, and maps. At some point during their stay, Richer and his colleagues had an idea. They decided to retrace Boudet's itinerary of Chapter VII for themselves and spend a memorable day walking a grand tour around Rennes-les-Bains. Setting out after no doubt a fine breakfast, they hiked the path which took them around the village in a long circuit. On their return, Richer was inspired to create a short work of poetry based around the experience, and to use it as a vehicle to encapsulate certain technical details he had discovered. He cast himself as the narrator and the result was *Le Serpent Rouge*.

The mystery-poem of *Le Serpent Rouge* has begun to reveal its secrets, but we are not quite finished yet. Now, it is time to come face to face with the *serpent rouge* itself.

Chapter Ten
DELPHI, APOLLO AND THE PYTHON

The Identity of Le Serpent Rouge

WE HAVE seen in Chapter Six how the sources for some of the language in the Sagittarius stanza which describe the *serpent rouge* can be traced to references in the *Homeric Hymn to Apollo*, as cited in Richer's *Delphes, Délos et Cumes*.

These texts provide the terminology, but what is it exactly that is being described in the poem? What is the nature and identity of this red serpent? Is it a real serpent, or a symbolic one, or perhaps both? And why is it ascribed to the Sagittarius segment of the zodiac? In this chapter we will explore these questions to see if we can precisely identify and locate this mysterious *serpent rouge*.

Recall the exact moment in *Le Serpent Rouge* when the encounter with the serpent takes place. According to the poem, on arriving at the Sagittarius segment along the path around the circuit of the zodiac/cromlech, the traveller should be approaching a "white hill" from which they earlier departed.

Then, as they make a turn to the east, at that exact moment, directly in front of the narrator, a giant "red serpent" should appear.

> "Then, coming back to the white hill, the sky having opened its floodgates, it seemed to me nearby I felt a presence, feet in the water like one who has just received the mark of the baptism, turning myself to the east, in front of me I saw, endlessly unrolling its coils, the enormous RED SERPENT quoted in the parchments, salty and bitter, the enormous, unchained beast at the foot of this white mountain became red with anger."[70]

---
70 *Le Serpent Rouge*, p. 5.

Notice that the narrator explicitly claims to be able to *see* the *serpent rouge*. We have earlier identified the "white mountain" as Pech Cardou, the local equivalent of Mount Parnassus. What, then, is this *serpent rouge* that the walker sees in front of him at the foot of Pech Cardou?

This question offers the definitive test of the proposed zodiac of Rennes-les-Bains format. If our interpretation of the relationship between the poem and the landscape holds, then we should be able to identify the location and the direction of the gaze of the narrator. If everything is correct, according to the poem, and we position ourselves in the Sagittarius segment, we should then expect to see *something* directly ahead which corresponds to a red serpent.

Around 2008, I had begun to create a detailed model of the landscape geometry I had found in Google Earth. After relying on the 1:25,000 IGN map alone for all those years, it was a revelation to be able to zoom in and out, fly around in 3D, and to examine the alignments from every angle and orientation as if I were hovering above the landscape itself.

I carefully laid out the zodiac of Rennes-les-Bains, including the *cardo*, the *decumanus*, the 45° line and the other alignments, and began to explore them and their relationship to the wider landscape with Google's amazing mapping tool. Late one night, I was roaming around looking down from above on the rivers, valleys, mountains, tracks and roads of the *Haute Vallée de l'Aude*, gazing in wonder at this model of the geometry, when something caught my attention.

It was a forestry track on the south flank of Pech Cardou that traversed the steep terrain in a zigzag pattern, crossing back and forth across the steep side of the mountain as it climbed. And as it did so, it traced a path in red soil which stood out boldly against the green background of the pine plantations through which it had been carved.

I suddenly realised with a jolt of recognition: the track looked exactly like a *red serpent* climbing up the mountainside of Pech Cardou. In another of the sudden moments of intuition which have punctuated this journey, the answer to the riddle had arrived in a flash. This was the *serpent rouge*! The shape of the track was so serpent-like, the colour of the soil so red, and the location so perfectly apt, that there could surely be no doubt. This humble forestry track, however surprising it might sound at first, was revealed as the mysterious *serpent rouge*.

I was so elated by this observation that I took note of the exact time and date. It was precisely midnight on the evening of 3 February 2008. Only later did I discover that Jean Richer's birthday was 4 February. This also happens to be the day of the year when the sun passes through the zodiac position of 15° Aquarius, the precise degree of the landscape zodiac which falls on the meridian of Rennes-les-Bains!

I had found the *serpent rouge* on the stroke of midnight on the day marking Jean Richer's birthday, as the sun was crossing the midline meridian of the zodiac of Rennes-les-Bains. It would be impossible to overstate my elation at this synchronicity, and I can only imagine that Richer himself would have been equally as overjoyed.

As confident as I was, there remained the small matter of checking to see whether this interpretation did in fact meet all the criteria for a correct solution. I still had to determine if the forestry path was even visible from the Sagittarius segment, or whether the route of the path conformed to the description in the poem.

Let's now compare the Google Earth vision with the reality in the landscape on the path we have identified around Rennes-les-Bains by revisiting the itinerary of the walk. To recap: having departed from Pech Cardou in the north, and descended the south-west flank of Bazel, the narrator then crosses the River Sals from east to west. After a visit to Château Blanchefort, he turns south, and follows the track along the plateau and rocky heights that lie to the west of Rennes-les-Bains. His goal is to visit the ancient spring, the *Source de la Madeleine*, which is found south of the village by the River Blanque in the Leo segment. Having reached the spring and rested by its sacred waters, the return journey back towards Pech Cardou commences. The traveller crosses the River Sals again, and now heads north on the return leg of the circuit.

If the route follows Boudet's itinerary in Chapter VII and indeed there is really only one walking path that is possible to complete the circuit around the village, then the path climbs the hills to the south-east of Rennes-les-Bains as it traverses the Libra and Scorpio segments of the zodiac.

It then turns back towards the village, crosses over into the Ophiucus segment and then rises along a gentle slope until it reaches the alignment that marks the line of 0° Sagittarius. Here, the traveller clears the brow of the hill.

At this point, the road makes a turn to the right, or the east, as can be seen in Figure 54. The traveller has now crossed over into the Sagittarius segment. Directly ahead, across a slight valley, lies the impressive sight of the imposing bulk of Pech Cardou. And there it is. On the flank of the mountain, the trace of the forestry track is visible as a line snaking through the trees.

At the present time, the plantation growth of pine is quite mature, so from the vantage point on the road today it is not possible to see the soil of the track itself. It is clearly visible however from above, or close-up, or from the other side of the valley, and from these perspectives its distinctive feature is immediately obvious: it is deep red, the colour of the local soil. Indeed, the hue is so distinctive that the local hiking route around the surrounding area is even named after it. It is called *Le Sentier Rouge*, the Red Trail.

The track would have been highly visible in the early years after the plantation was seeded, when the pine trees had not yet grown sufficiently to obscure the view of the access road from the Sagittarius location.

I have been unable to confirm precisely when the forestry track was first carved into the mountainside, but I have been able to narrow down the range of possible dates. Historic aerial photos of the region are now available on the IGN website which show the southern flank of Pech Cardou covered in its natural vegetation before the plantation and forestry road were formed. The images are dated to the period 1950 to 1965 but the exact year is not recorded. The pine plantation and the track were laid down sometime after these photographs were taken, so it could not have been before 1950 at the earliest.

If it was between that date and October 1966, when *Le Serpent Rouge* was written, then the young trees would have been no older than 15 years. From the vantage point of the Sagittarius segment, the forestry track would have stood out as a distinct red scar against the green shoots of the growing pines.

The only remaining alternative, that the plantation was not laid out until after 1966, would be a coincidence beyond belief. If the rest of our evidence has shown that we have definitely identified the path of the walk of *Le Serpent Rouge*, then the idea that this dramatic, red serpentine path should appear sometime later, in the exact correct place to correspond with the dictates of the poem, yet was not visible at the time, would be, well, some kind of miracle. I confidently predict that the date of construction of the forestry track winding up the southern flank of Pech Cardou, when it is confirmed, will be between 1950 and 1966.

To summarise: the traveller arriving at the Sagittarius segment comes over the hill, turns to the east, and sees before him a *red serpentine path* climbing the side of an imposing mountain. Here it is: *le serpent rouge*, with the correct shape, the correct colour and in the correct position.

Figure 53: The serpentine path of Le Serpent Rouge visible on the south flank of Pech Cardou, as seen from above in Google Earth.

The *serpent rouge* has a precise physical identity: it is the forestry path that winds up the side of Pech Cardou. This is an unexpected result, to say the least, but it is surely confirmed by the fact that the description occurs in the Sagittarius segment, at the very location uniquely identified on the path from which the landscape serpent emerges into view for the walker.

The poem describes a tour around the landscape of Rennes-les-Bains, in which the progress is calibrated by reference to a landscape zodiac based around the ancient *cardo* axis. Its path imitates the tour of the cromlech described in Boudet's Chapter VII, but it is also clearly based on a real walk undertaken by the author. The view of the serpentine path from the Sagittarius segment is a detail that was observed first-hand, an authentic recollection from experience.

Now we can answer the question as to what Richer was describing in the Sagittarius stanza of *Le Serpent Rouge* when he invoked the language of the *Homeric Hymn to Apollo* in its account of the serpent/dragon of Delphi. He was describing *the path around Rennes-les-Bains itself*.

Thus, in *Le Serpent Rouge*, Richer was assimilating the route of Boudet's walk around the cromlech with the sun's annual passage around the sky through the zodiac, and representing this by the symbolism

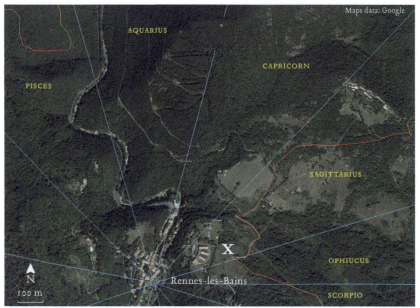

Figure 54: The path of Le Serpent Rouge is shown in red. The traveller arriving at the position marked x has crossed into the Sagittarius segment. Directly ahead on the south-west flank of Pech Cardou, the serpentine forestry track may be seen.

of Pythia. Now we can see why the progress of the narrator as he followed the path through the landscape was calibrated by the zodiac. Let's now explore Richer's other acknowledged works at the time to explore the extent to which such a reading is consistent with what he wrote elsewhere.

Delphi and the Serpent Python

The home of the most sacred oracle in the ancient world, Delphi is situated on the south-west slopes of Mount Parnassus, in the valley of Phocis. The name derives from the Greek word *delphus* meaning womb, hollow or cleft. It was also known by the name of *Pytho*, after the name of the original tutelary spirit, the giant serpent Pythia.

Delphi was considered the navel, or *omphalos*, of the world. Its location was said to have been determined when Zeus released a pair of eagles, one from the eastern-most extension of the earth and the other from the west. The point at which they crossed defined therefore the centre of the earth. The earliest myths recount that Pythia was the guardian of the oracle in pre-classical times. It was rededicated to Apollo sometime around the eighth century BC and subsequently

Figure 55: The serpentine path of Le Serpent Rouge just visible on the south flank of Pech Cardou, snaking through the green pine plantation, now grown to maturity. This photograph was taken in 2008 from the location marked x in Figure 54.

became the most important site for the worship of the sun god in the ancient world. The myths of its founding recount that Apollo engaged in a battle with the serpent Pythia and slew her, taking over the guardianship of the oracle from that time.

Curiously enough, as I have noted, Richer mentions the serpent of Delphi prominently on the very first page of *Sacred Geography of the Ancient Greeks* (1967). The book begins with his first visit to Delphi in 1957, and how he was filled with inspiration at the majesty of the site. He describes its setting in a cleft in the landscape, close by the sacred mountain of Parnassus and the Castalian springs from which the Priestesses would drink before delivering their prophecies. Then, as noted earlier in Chapter Five, he wrote:

> "I was intrigued by a certain detail of the cult at Delphi: the sacred drama of the struggle of Apollo with the serpent Python was performed at Delphi about every eight years. Now, the serpent is not only a symbol of the earth, it is also a representation of the path of the sun in the zodiac"[71]

---

71   Jean Richer *Sacred Geography of the Ancient Greeks*, op. cit. p. 1.

Figure 56: Photograph of "le serpent rouge" from a vantage point on the west side of the valley of the River Sals.

There it is: "the serpent is a representation of the path of the sun in the zodiac". Richer could hardly have been plainer. In this book published in the same year as *Le Serpent Rouge*, he has left a clear explanation of his symbolism: the serpent Python at Delphi is a representation of the path of the sun in the zodiac. He does not cite a reference to support such a reading, but later in these pages as we delve deeper into his sources, we will discover exactly where he found it.

Meanwhile, in the poem, he uses language applied to the serpent Python at Delphi in the *Homeric Hymns to Apollo*, to describe the *serpent rouge* which is revealed as the literal path which snakes up the southern flank of Pech Cardou. This path runs, of course, through the zodiac which marks the thirteen stanzas. Thus, in both Richer's book and poem of 1967, we have the serpent Python of Delphi assimilated to the zodiac path of the sun.

If we now turn to his follow-up volume, *Delphes, Délos et Cumes*, we find that Richer has repeated and underscored the same message.

The first chapter is entitled *Delphes, Centre du Monde et Site Oraculaire*, or Delphi, Centre of the World and Site of the Oracle. Again, the book opens with a description of the striking physical setting of Delphi and its sacred character. Richer remarks:

37. Bas-relief de Délos : Omphalos au serpent, entre deux palmiers.

Figure 57: Plate 37 from Jean Richer's Delphes, Délos et Cumes, *(Paris, Julliard, 1970), p.64, showing a stone carving at Delos of a serpent coiled around an omphalos between two palm trees, a symbolic representation of the founding myth of Delphi and the Python.*

> "From very early times, the place is said to be the completion of a hierogamy of the Earth and Heaven."[72]

It is only a few pages later that Richer brings in the extracts from the *Homeric Hymn*, which we quoted in Chapter Five, and which contain the key references to the serpent in the Sagittarius stanza of *Le Serpent Rouge*. Then, immediately following this passage, Richer mentions again the festival held each eight years in Delphi:

> "At Delphi, the festival of the Septerion was a representation of the murder of Python by Apollo. It was celebrated at the end of each period of eight years."[73]

This is a restatement of his comment on the first page of *Sacred Geography of the Ancient Greeks* which we have already noted. He has omitted this time the next sentence about the path of the sun being represented by the serpent, but the intent is obvious: he was tying the two texts together, as clearly and neatly as possible. Richer has ensured

---

72 Jean Richer *Delphes, Délos, Cumes*. op. cit. p.28.
73 Ibid. p. 32.

that the precise key for deciphering the riddle of the *serpent rouge* has been provided, in print, in plain sight, in two separate locations, in order to provide unimpeachable verification.

This firm identification of the *serpent rouge* of the poem with the serpent Python carries another profound implication. It means that Richer is also associating Rennes-les-Bains *itself* with Delphi.

It is not hard to see why he might have been led to formulate this impression, as there are certainly some strong physical similarities between the two sites. Both are situated in clefts of the earth, or valleys, and are close to underground springs of healing or what might even be considered sacred waters. Both also have an imposing and significant mountain peak standing to the north: Parnassus at Delphi and Pech Cardou at Rennes-les-Bains. And we have now also established that both Delphi and Rennes-les-Bains lie at the centre of landscape zodiacs.

For Professor Jean Richer therefore, Rennes-les-Bains was in some sense patterned after the model of Delphi. The importance of this identification to understanding his contribution to the mystery cannot be overstated.

In an earlier chapter, we noted parallels between the *Homeric Hymn to Apollo* as reproduced in Richer's *Delphes, Délos et Cumes* and *Le Serpent Rouge*, including the suggestion that the snow-covered Mount Parnassus near Delphi mentioned in the ancient Hymn had its counterpart in the "white mountain" in the modern poem. If this is the case, then we ought to be able to identify a mountain that corresponds to Mt Parnassus and is close to Rennes-les-Bains. It is immediately obvious that there is only one possible candidate, and that is Pech Cardou. Why, though, should this peak be described as "white"?

Of course, in winter there are many occasions when the mountain is covered in snow, but in October 1966, when the poem was said to have been written, this would have been unusual so it could not have been intended as a physical description. On what basis then could Richer have associated the colour white with the mountain in order to justify linking it to Parnassus? Is there any warrant or authority for a symbolic association of whiteness with Pech Cardou?

The answer to this question may be found in Robert Graves' seminal mid-twentieth century book, *The White Goddess*. This compendium of rare and valuable knowledge is an indispensable guidebook for the poetic imagination in navigating the waters of ancient and classical myth. Graves has some fascinating passages that are directly relevant to our story. Could Richer have been aware of these?

In fact, there are multiple footnoted references to *The White Goddess* from the final chapter of *Delphes, Délos et Cumes*, and indeed the pages noted lead us to the very passages that contain Graves' discussions of the etymology of certain terms for whiteness. Let's now look closer at some of these observations, knowing that Richer had the pages of this extraordinary book at his elbow as he wrote.

Pech Cardou and Mount Parnassus

The name of the mountain Pech Cardou is an obvious reference to the function of the peak within ancient landscape practice. In fact, as we have previously noted, it marks the intersection of two separate *cardos*: first, the 15° bearing that is the historic *cardo* axis of Rennes-les-Bains, and second, the *cardo* that passes through ten peaks on the meridian of longitude 2°19′40″E.

Robert Graves sheds fascinating light on this word *cardo* in the context of exploring the origins of certain words for the concept of white in ancient languages. For example, he states that the earliest name for Britain, Albion, as it was known to Pliny, is derived from Albina, ("the White Goddess").[74] He then extends this connected linguistic chain through various related terms, including the River Elbe, (*Albis* in Latin), and words such as *alphos*, *albus*, *alphiton* and *Alphito*.

Graves' far-reaching etymological musings on this notion of whiteness encompass a dizzying array of interconnected concepts. His aim is to flesh out the rich associations inherent in the name given broadly across cultures and history to the Goddess of Ten-Thousand Names, or who Graves prefers to call: The White Goddess. One of the names, as he explains, under which she was worshipped was Cardea.

> "The Latins worshipped the White Goddess as Cardea, and Ovid tells a muddled story about her in his Fasti, connecting her with the word *cardo*, a hinge. He says that she was the mistress of Janus, the two-headed god of doors and of the first month of the year and had charge over door-hinges. (...) He says that she first exercised this power at Alba ('the white city')."

Even if Ovid's account is "muddled", it is enough to establish the chain of associations as they have been reported by Graves, and undoubtedly read by Richer: the goddess Cardea, manifestation of the White Goddess and mistress of Janus, is assimilated with this word

---

74 Robert Graves, *The White Goddess*, (London, Faber & Faber, 1997) p. 62.

cardo. This is not merely some passing reference, either. Graves devotes pages of his book to exploring the rich web of associations that cluster around the connections between the word-stems *cardo*/Cardea and *albus/alba/albino*. Richer footnotes directly in the final chapter of *Delphes, Délos and Cumes* to one of these very pages.[75]

We can conclude from this that Richer found rich textual support in Graves for the association of Pech Cardou with the quality of whiteness, and perhaps even the source for such ideas. In any case, Richer would not have had to look far in the local landscape around Pech Cardou to find further traces of "white".

On the other side of the River Sals (or "salt"), one finds Château Blanchefort, (or "white-fort"). The river that meets the Sals just south of Rennes-les-Bains is called the Blanque, the "White River" Finally, the Château at Bézu, which is sited on the southern end of the *cardo* through Rennes-les-Bains, was known as Albedunum. Here again we encounter the word-stem "albe", which denotes "white". Albedunum is the "white fortress".

So now we can read the opening phrase of the Sagittarius stanza of *Le Serpent Rouge* with renewed insight. When the poet writes *"Revenant alors à la blanche coline"* or "returning then to the white hill", he is referring to Pech Cardou in the full context of the associations mapped out by Robert Graves in which the word Cardou/*cardo*/Cardea is assimilated to whiteness, as well as the numerous local placenames which derive from the same concept.

For Jean Richer, the sacred geography of Rennes-les-Bains was modelled after the original pattern of Delphi, and he has left a trail of clues confirming this throughout *Delphes, Délos et Cumes*. This has been established by tracing the identities of the *serpent rouge* with the serpent Python, and the symbolic equivalence of Pech Cardou with Mount Parnassus, via the linked concepts of "whiteness" and the Goddess of Hinges, Cardea.

One further major revelation awaits in the last chapter of Richer's *Delphes, Délos et Cumes* that seals the identification of Rennes-les-Bains with Delphi in a remarkable manner. It begins with a discussion of various place names derived from the stem "leuk-" and related terms, all of which signify "whiteness", deriving from the Latin root *"lux"*, or light. Again, Robert Graves provides background discussion from where Richer could easily have either learned or refreshed his

---

75 Jean Richer, *Delphes, Délos et Cumes*, op. cit. p. 210. Richer cites page 434, corresponding to page 425 in the edition I have used and cited.

understanding in this area. He opens Chapter IX of *Delphes, Délos et Cumes* with the following observation:

> "As if the idea of radiant whiteness, evoking what had to be the purity of the candidate for initiation, being indissociable from the beginning of the zodiacal cycle, all the places symbolically linked to the vernal point bear a name in which the radical Leuce appears."[76]

Richer begins this chain of thought with the idea of whiteness, and associates it with the vernal equinox, or spring, as the place of the rising sun at the commencement of the annual solar cycle. He then suggests that this association carries over into landscape. He notes that, not infrequently, the location marking the vernal equinox, typically due east or due west, in many of these landscape zodiacs is given a place name based on the root-stem "leuké". Of these "white stones of the threshold", as he calls them, he then selects and presents several examples:

- Cape Leucade in Greece
- Cape Leuké or Leuca in the south of Italy
- Leukas on the coast of ancient Ionia, in modern-day Turkey, at the latitude of Sardis
- The island of Leuké at the mouth of the Danube

All these places named with variations of the leuké term, are located at a point due east or west, equivalent to the vernal equinox in respect to one of the landscape zodiacs identified by Richer. He notes, after Graves, that the term leuké also denotes the white poplar, and that this tree is associated with the autumn equinox in the ancient Celtic tree calendar related to Ogham writing. In this case, the substitution of autumn for the spring equinox is simply an acceptable variation in the same underlying symbolism.

The first placename on the list, Cape Leucade, is found on the coast of the Greek mainland, directly due west from Delphi. Richer devotes considerable discussion to this arrangement in *Géographie Sacrée du Monde Grec*, where he describes an ancient celebration which took place here, in which young men would perform ritual dives into the sea from the rocky cliffs. He cites this occurrence of the name Cape Leucade at the point on the map where the east-west line through Delphi meets the Peloponnesian coast as one of the principal confirmations of the presence of the landscape zodiac at Delphi.

---

76 Jean Richer, *Delphes, Délos et Cumes*, op. cit. p. 209.

There is a reason why I have devoted some space to describing in detail this insight of Richer's, namely that landscape zodiacs sometimes exhibit a place called leuké, or a variation, at the location of the vernal equinox or equivalent. Given that I have proposed that he considered Rennes-les-Bains to be patterned after the sacred geography of Delphi, we might conjecture whether he turned to a map of the Languedoc to see if the symbolic leuké arrangement might also be repeated there.

If he did, he was in for a pleasant surprise. A line drawn exactly due east from Rennes-les-Bains to the Mediterranean, as shown in Figure 58, intersects the coast at a place called Cap Leucate! After Graves and Richer, we might translate this as the White Cape. One can observe the young men diving into the sea at this location also today but in this case, it is the modern ritual of a day at the beach in the August summer holidays. Nevertheless, the ancient symbolism persists. The point due east of Rennes-les-Bains, the direction of the vernal equinox, is marked by the exact same distinctive place name that Richer has associated with Delphi and other landscape zodiacs!

This amounts to an extraordinary independent confirmation of my hypothesis. Now we can say with some confidence that it is not merely in the poetic imagination of Jean Richer that Rennes-les-Bains may be assimilated to Delphi, but the traces of a genuine historical resonance may be found inscribed into the very place names of the region itself.

I now turn to the second section of the final chapter of Richer's *Delphes, Délos et Cumes*, entitled *Le Réseau des Centres Zodiacaux*, or The Network of Zodiac Centres. In these pages, Richer reveals that he has identified the presence of landscape zodiacs in many other locations, across the Near East, and Europe, and even including the British Isles. He discusses these at some length and even includes a list, which runs to twenty examples, including zodiacs centred on such locations as Jerusalem, the Isle of Man, Rome and Toledo. This material would later be expanded to another book-length treatment in his 1985 work, *Géographie Sacrée dans le Monde Romain* (Sacred Geography of the Roman World). Buried in these passages in the final pages of *Delphes, Délos et Cumes*, is a nugget of gold. Number 16 on Richer's list of landscape zodiacs appears as follows:

16) *Système gaulois de Mediolanum (Saint-Benoît-sur-Loire)*

He identifies a zodiac around the town of Saint-Benoît-sur-Loire, considered as the centre of what is now mainland France. Its Roman name, Mediolanum, bears witness to this ancient tradition. It lies on

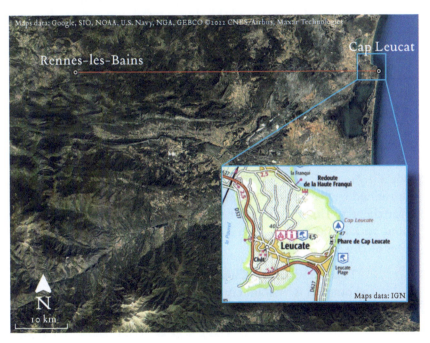

Figure 58: 'Leucate' on the coast in both Greece and France.
Above: Leucate due west of Delphi. Details shown here in Google Earth are taken from a map of Greece in Jean Richer's *Aspects ésotériques de l'œuvre littéraire* (Paris, Dervy-Livres, 1980).
Below: GoogleEarth map of south-west France showing Cap Leucate due east of Rennes-les-Bains, with detail from IGN map. In both instances, the place name based on the root leuké, denoting whiteness, represents the position of the equinox in the local landscape zodiac.

meridian 2°18' E, the longitude of the meridian through Lavaldieu. By 1970, therefore, Richer was aware of another landscape zodiac, in the centre of France, on a meridian that runs through the locale of Rennes-les-Bains.

Unfortunately, nowhere in any of his books does Richer address in detail his work on the sacred geography of France. In the introduction to *Géographie Sacrée dans le Monde Romain*, he states that he has reserved all such discussion for a future dedicated volume which, sadly, never appeared. Nevertheless, he made it clear that he had found the same landscape zodiac formats in France as he had identified throughout the lands of the Mediterranean basin. In *Delphes, Délos and Cumes*, he writes:

> "Periodical trips back to France especially to Brittany, opened my eyes to a deep analogy between what I was observing in Greece and the almost unknown civilization that erected the megaliths."[77]

While we do not have Richer's intended volume, there are a few snippets of information in *Géographie Sacrée dans le Monde Romain* which can offer a glimpse. In several places, he refers to the grand meridian of Gaul, the axis of ancient France.

Richer offers a list of the significant sites through which it passes, as shown in Figure 59. These occupy a narrow band between longitudes 2°18'E and 2°26'E. The modern Paris Zero Meridian, (of longitude 2°20'14.03"E) lies within this range. Whilst these details are not mentioned in his earlier 1970 work, *Delphes, Délos et Cumes,* he does refer there to this meridian of Gaul in the context of a discussion about an ancient site in Algeria, the Tomb of the Christian. Richer notes of this site:

> "The 'Tomb of the Christian' is where the meridian of the omphalos of Gaul cuts the coast of Africa."[78]

The meridian of Gaul and this monument, the Tomb of the Christian, also known by its Latin name *Monumentum Regia Gentis*, are shown in Carte 13 from *Sacrée dans le Monde Romain*, reproduced here in Figure 59, and in Carte 20, shown in Figure 61. These maps show that the meridian is part of a much wider system of extended alignments that encompass the entire Mediterranean world. Richer's reference in *Delphes, Délos et Cumes* shows that he was aware of the meridian of Gaul

---

77   Jean Richer *Delphes, Délos et Cumes*, op. cit. p. 23.
78   Jean Richer *Delphes, Délos et Cumes*, op. cit. p. 217.

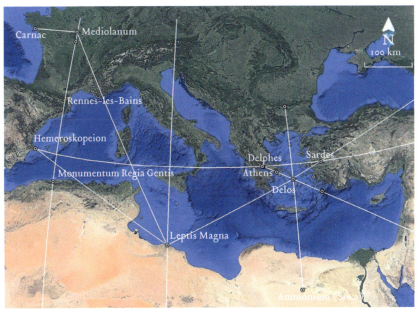

Figure 59: Place-names and alignments depicted in Carte 13 of Jean Richer's *Géographie Sacrée dans le Monde Romain (Paris, Guy Trédaniel, 1985), p.192.*

and that it passed through a landscape zodiac at ancient Mediolanum (modern Saint-Benoît-sur-Loire), for which he gives a longitude of 2° 18'24.12"E. As Rennes-les-Bains falls on longitude 2°19'10"E, it obviously cannot have escaped his notice that it too lies within the bounds of the meridian of Gaul, and therefore forms part of the same system.

Considering all these hints and blatant clues, the conclusion is unavoidable: in 1970, Richer was very well aware of the presence of a zodiac in the landscape at Rennes-les-Bains and that it was yet another example of a format he had found at many sacred centres around the Mediterranean. He was not revealing his hand openly, but he was leaving a trail of undeniable breadcrumbs for someone to eventually come along and follow.

Serpent and Symbol

We've now unpacked several layers of significance relating to the serpent which the narrator encounters in the Sagittarius stanza of the poem *Le Serpent Rouge*.

Firstly, there is a physical layer, in which the serpent manifests in the landscape as a visible, tangible form. Specifically, this is the "serpentine" path of red soil that winds up the south flank of Pech Cardou.

Locations on the meridian of Gaul, from south to north	Latitude	Longitude
1 TOMBEAU DE LA CHRETIENNE	36°34'N	2°31'E
2 L'îlot de Dragonera à l'ouest de Majorque	36°35'	2°19'
En Espagne		
3 l'ermitage de Santa Fé, dans la Sierra de Montseny	41°46'	2°28'
4 Nuestra Senora de la Salud	42°04'	2°29'
En France		
5 Le Pic du Canigou 2,786m	42°31'	2°27'
6 Carcassonne	43°13'	2°21'
7 Conques-sur-Orbiel (Aude)	43°16'	2°24'
8 Pic de Nore	43°26'	2°28'
9 Conques (Aveyron)	44°36'	2°24'
10 Aurillac	44°56'	2°26'
11 Bruère-Allichamps	46°46'	2°26'
12 Meillant (Mediolanum)	46°47'	2°30'
13 Bourges	47°05'	2°24'
14 Saint Benoît-sur-Loire (Mediolanum)	47°49'	2°18'
15 Paris	48°52'	2°20'
16 Saint-Denis	48°56'	2°22'
17 Amiens	49°54'	2°18'
18 Dunkerque	51°03'	2°22'

Figure 60: Locations on the meridian of Gaul, as listed in Jean Richer's Géographie Sacrée dans le Monde Romain (Paris, Guy Trédaniel, 1985), p. 336. These positions are shown by number on the meridian of Gaul on the map in Figure 61. Rennes-les-Bains (latitude 42°55' longitude 2°19'10") falls within the range of longitude values, to the south of Carcassonne.

Confirmation that this initial interpretation is correct is provided by the poem and the landscape themselves. The poem specifies exactly where the serpent appears: the original French phrase in the poem reads *"face à moi"*, which translates as "in front of me". It is precisely as the traveller crosses over into the Sagittarius segment on the zodiac of Rennes-les-Bains that the serpentine form comes into view, directly ahead. It is indeed, from the narrator's perspective, "in front of me".

Secondly, this physical serpent is overlaid with a symbolic interpretation of its identity, according to the poet. It is described in the poem using precise language taken over from the *Homeric Hymn to Apollo* employed to describe the serpent Python, the original guardian of the oracle at Delphi, who according to the legends was killed by Apollo when he usurped this role.

In turn, according to Richer himself, the serpent Python is itself a representation of the earth, and of the path of the sun around the earth. We can recognise that there is an entire complex of symbolism in play here. Apollo is, of course, the sun god. There is embodied in

Figure 61: Selection of place-names and alignments as depicted in Carte 20 of Richer's Géographie Sacrée dans le Monde Romain (Paris, Guy Trédaniel, 1985), p.345, showing the zodiac of Toledo (Toletum) on the Iberian Peninsula, with the meridian of Gaul. The numbered locations on the meridian listed in Figure 60 and the position of Rennes-les-Bains, have also been added to the map.

the ancient drama of Delphi a narrative of the passing of guardianship of the sacred site from the old earth-gods to the new sun god. Yet, in some sense, the two continue to co-exist in an eternal state of flux.

This by no means exhausts the depths of meaning enfolded into this potent cluster of poetic images, but it offers a solid framework for interpretation in the context of the poem. The path of the narrator in the landscape is the visible and physical analogue of the path of the sun around the earth through the signs of the zodiac.

Thus, we can now also say that the narrator is a representation of the sun itself, or that he embodies this solar archetype. In one sense therefore, the narrator also represents the sun god Apollo. The poet walking around the landscape of Rennes-les-Bains is, metaphorically speaking, playing the role of Apollo as the sun god traversing the zodiac.

The various incarnations of the serpent symbolism lead us to the key insight which inspired Richer: Rennes-les-Bains is patterned after or is a type of Delphi. This identification holds on the surface, or physical level: both ancient sites are situated in cleft valleys, near sacred springs. Both have been places to which people have come for healing and spiritual renewal since ancient times.

But then Richer's leap of intuition and imagination took the identity a stage deeper. He perceived that Rennes-les-Bains lay at the focal point of a twelve-fold division of landscape, based around *cardo–decumanus* axes, just as he had discovered at Delphi. Thus, both in its physical characteristics, its sacred nature and its embedded landscape geometry, for Richer, Rennes-les-Bains was a type of Delphi.

We have come a long way. The architecture of the zodiac format has been revealed and confirmed. The identification of Jean Richer as author of *Le Serpent Rouge* has enabled us to make deep inroads into understanding the poem. Our journey is by no means complete and there remain several riddles to be resolved, but we have made promising initial progress. In the next chapter, we will take a step back from the poem itself and ask whether the Team around Pierre Plantard left any evidence which might confirm the involvement of Richer in the Affair of Rennes. In fact, they left such a trail of clues that it is surprising his role has remained in the shadows for so long.

Chapter Eleven
CONFIRMATION FROM THE TEAM

A Trail of Clues

WHILE WE now have strong indications from within *Le Serpent Rouge* itself to show that Richer must have been actively involved in its creation, can we find support from the writings of Plantard, de Chérisey and de Sède themselves which would confirm it? This might seem unlikely, as his name has rarely if ever been noted in direct connection with the Affair. Yet, in fact, all three members of the Team have left on record a series of very revealing comments which, separately and together, leave no doubt at all that Jean Richer's contributions to the project were central and crucial.

Plantard elevates his role above all other authors when it comes to understanding Boudet's book in an extremely revealing statement in an out-of-the-way location. De Chérisey amplifies his comments and expands on them with fascinating details.

But perhaps it is de Sède who leaves the clearest trail. His references to Richer are often curiously oblique, and do not alert the casual reader to any direct link between the two of them. Taken together, however, and read in context, they amount to the most detailed confirmation of his role by any of the team.

Pierre Plantard's Zodiaque de Rennes-les-Bains

As I've noted, in 1978, there were two separate reprintings of Boudet's 1888 book. The first, published by Belfond, included a preface by Plantard and some other additional materials provided by him, including a map and a lengthy and detailed bibliography. The preface is dated June 24, 1978. The second, published by Bélisane, included an introduction by de Sède.

Le Serpent Rouge is listed in the bibliography Plantard provides, but it is mentioned nowhere else in either of the two volumes. Plantard's preface spans more than thirty pages. It has not been published in English translation to the best of my knowledge. It is very strange – even bizarre – and touches on some fascinating esoteric concepts, but with a certain air of wilful illogicality or, one might say, surrealist logic.

There are extensive references to the Tarot, and certain manipulations of numbers and letters which, on the surface, make little sense and are impossible to follow sensibly. Nevertheless, there is also an air of deliberate method in his madness. After reading this introduction, one is left with the impression that Plantard knows much but is revealing little.

From the standpoint of our investigation, perhaps the most startling thing that leaps off the page from Plantard's introduction is a section with the heading: "LE ZODIAQUE DE RENNES". To the best of my knowledge, this is the first time that this phrase had appeared in print in relation to the Rennes affair. In two pages of remarkable passages, Plantard offers an enigmatic description of this *"zodiaque"*, and then explicitly identifies it with Boudet's cromlech.

The topic of landscape zodiacs around Rennes-les-Bains is not something on which Plantard had been writing in preceding years. The report of the 1964 Rennes-le-Château conference at which he spoke extensively on the entire subject has no reference to such things. It is evident that he only learned these ideas from his encounter with Jean Richer later, in the years 1966/1967.

As ever, Plantard's description conceals as much as it reveals. It is not possible to follow the details of his account and arrive at the true solution, as they are a deliberate mix of plausible-sounding classical references with wacky nonsense. However, now that we have separately deduced the existence of the zodiac, as revealed in *Le Serpent Rouge*, it is perfectly possible to understand what Plantard is alluding to.

He describes this zodiac as comprised of twelve *dépôts*, or deposits, arranged in the landscape according to a *"codage astronomique"* or astronomical code, devised by Boudet. Plantard's introduction also includes a sketch map displaying his zodiac of Rennes-les-Bains. It is worth taking some time to examine this carefully as it reveals both his knowledge of the details and his methodology. The map (see Figure 62) displays two sets of alignments converging on two different focal points at bottom right. The first is Plantard's zodiac. The second is the zodiac we have identified around Rennes-les-Bains.

Let's look at Plantard's zodiac first. Its centre, at the lower of the two focal points, is labelled *Source du Cercle et Fauteuil du Diable*, or Spring of Le Cercle and the Armchair of the Devil. The *Fauteuil du Diable*, a well-known local landmark, is a large boulder, close to the hamlet of Le Cercle, that sits on an elevated position with commanding views across the valley. At some ancient time, a seat has been carved into it facing towards the sunrise in the east.

This is the location that Plantard identifies as the centre of his *zodiaque de Rennes*. The map shows three alignments passing through it and specifies the zodiacal positions of each: the line to Rennes-le-Château is allocated to 27° Capricorn, to Aram, 6° Aquarius and to Rok Nègre 0° Aries. In the accompanying Figure 62, I have superimposed blue lines on these three alignments so they can be easily identified.

These indications, though brief, are sufficient for us to reconstruct the details of Plantard's version of the zodiac of Rennes-les-Bains: it is centred on the Fauteuil du Diable near Le Cercle; the signs run clockwise, and it is arranged so that the beginning of the Aries segment corresponds to the alignment to Rok Nègre.

If we compare Plantard's version to the zodiac form we have derived earlier as the basis for *Le Serpent Rouge*, we can see that whilst the overall concept is the same, his description differs fundamentally in all essential aspects: it proceeds in the opposite direction, has a different allocation of the zodiac signs and is not centred on Rennes-les-Bains.

There is still the second focal point of alignments to consider in Plantard's map, however. Slightly north of the first focal point, this one is labelled *croix du cercle*, or cross of the circle. Immediately to its right is shown the bell tower (*clocher*) and cemetery (*cimetière*) of the Rennes-les-Bains church.

Clearly, this second focal point corresponds to the centre of the village of Rennes-les-Bains, adjacent to the church. Three alignments are shown passing through it. Two of these correspond to the ancient axes of the *cardo* through Pech Cardou and the *decumanus* aligned to Rennes-le-Château. The third line leads to Rok Nègre and Château Blanchefort to the north. This is the familiar 345° bearing from Rennes-les-Bains, the Pisces alignment in the *Le Serpent Rouge* zodiac!

Hence, Plantard's diagram here depicts three of the six alignments which make up the correct *Le Serpent Rouge* zodiac. In the accompanying Figure 62, I have superimposed red lines on these alignments. Observe what Plantard has done. His map purports to display the "*zodiaque de Rennes*" converging on the lower focal point, but this is a "fake" version of the genuine zodiac. Meanwhile, immediately above the false version, he has depicted the correct zodiac, but without explicitly identifying it as such or even calling attention to it.

Plantard shows here that he knew perfectly well the details of the true zodiac by swapping out its characteristics for their opposites! The correct version runs anti-clockwise, is centred on the intersection of the *cardo* and *decumanus* axes in the heart of Rennes-les-Bains, and the sign of Pisces is allocated to the alignment through Château Blanchefort and Rok Nègre, just as it says in *Le Serpent Rouge*. (Amusingly, and perhaps deliberately, the true and false zodiacs both allocate Pisces to the same segment of landscape.)

Plantard is aware of all this but masks his knowledge under the guise of a carefully crafted alternative version, concealing the truth beneath a veneer of apparent confusion. All the elements are present, but the details are deliberately altered so that it is not possible to recreate the correct solution from the information supplied. However, for those who have arrived at the solution by some other route, as we have, this sophisticated game provides ample confirmation of the truth.

As all of this occurs in the context of an introduction to Boudet's book about his cromlech, there is no escaping the conclusion that Plantard is establishing a connection between the cromlech and this *"zodiaque de Rennes"*. In turn, this inevitably suggests that he has *Le Serpent Rouge* in mind, as the original source of the zodiac idea in the local landscape. Viewed in this manner, the purpose of Plantard's map becomes apparent. It is a deliberate textual strategy designed to reveal and conceal at the same time. We will encounter this use of deliberate errors being deployed to create a distracting narrative again before we have reached the end of our story.

Elements for a Bibliography

Following the essay and map, Plantard included an extensive bibliography with his introduction to the 1978 edition of *La Vraie Langue Celtique*. If his clues about the relevance of Richer's work and ideas were somewhat veiled in the earlier materials, here Plantard makes a very revealing admission. The bibliography covers some thirteen pages and includes around one hundred titles. It is divided into three sections. The first is an index of books that Boudet mentions in the text. The second is a list of contemporary works that he may have consulted. The final section covers volumes written since Boudet's time that Plantard considers of value in coming to an understanding of the text.

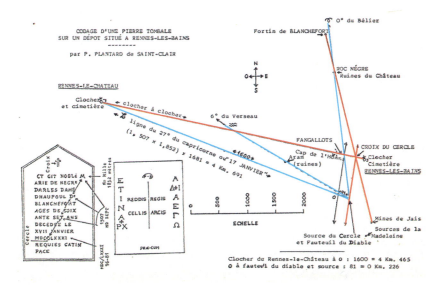

Figure 62: Map of Rennes-les-Bains in Pierre Plantard's Introduction in the 1978 reprint of Henri Boudet's La Vraie Langue Celtique. The blue lines (added) indicate alignments that are part of Plantard's (false) zodiac of Rennes-les-Bains, with its centre at "Source du Circle et Fauteuil de Diable". The red lines (also added) show the alignments of Richer's true zodiac of Rennes-les-Bains, including the cardo and decumanus of the village. Image credit: Pierre Plantard, La Vraie Langue Celtique et Le Cromleck de Rennes-les-Bains (1886), (Paris, Belfond, 1978), p. 39.

The bibliography begins with a title page, headed *"Eléments pour une bibliographie de l'Affaire de Rennes"*. Below this appears a single short paragraph in which Plantard singles out just one book for special mention:

> "Elements for a Bibliography of the Rennes Affair
> The majority of the titles that we report deal directly with the enigma of Rennes-le-Château and its region. We have also quoted some works relating to Gisors and Stenay, believing that their 'mysteries' fully participate in our subject. Even so, we have not hesitated to include in our list a few general texts such as *Sacred Geography of the Ancient Greeks* by Jean Richer, on which the curious reader may profitably meditate."[79]

---

79 Henri Boudet *La Vraie Langue Celtique et Le Cromleck de Rennes-les-Bains*, (Paris, Belfond Press, 1978). Foreword by Pierre Plantard. p. 41.

It was not until 2013 that I encountered this passage for the first time, and it was a truly satisfying moment. This 1978 edition of Boudet is a very rare book. Whilst the Bélisane edition with de Sède's introduction has been re-issued in recent years, and even been translated for an English edition, the 1978 Belfond edition with Plantard's introduction remains a stubbornly difficult text to obtain. It was a thrilling moment for me therefore when I chanced on a copy in Mutus Liber, the bookshop in the village of Rennes-le-Château itself one day. Here at last was the certain proof of the connection I had laboured to establish for all those years between Richer, Boudet and the Team.

Coming on the heels of thirty pages of inspired misdirection, Plantard lets through a moment of plain unadorned truth, delivering to the reader who has persisted through the confusion, finally, the genuine passkey to unlocking the door of the mystery of Boudet's book. It is Jean Richer's *Sacred Geography of the Ancient Greeks,* "on which the curious reader may profitably meditate."

Consider, on the surface of it, the strangeness of Plantard's comment. Richer's book is about landscape zodiacs in Greece. Absent the suggestions put forward in these pages, what could it possibly have to do with Boudet's book, with its absurd linguistic theories and off-beat local histories? There are no external visible points of connection between the books of Richer and Boudet that any uninformed reader could ever discern. Why then would Plantard single out Richer's book as the crucial reference work to consult to truly understand Boudet's agenda? He can surely only be suggesting one thing: Boudet's cromlech is an example of what Richer has discovered and described in *Géographie Sacrée du Monde Grec*: namely, a sacred centre around which lies a zodiac format, a twelvefold regular division of the landscape into segments allocated to the signs of the zodiac. He does not need to spell this out, but the implication is inescapable. Besides, de Chérisey later expanded on this point in an extremely revealing passage that has only relatively recently become available in print.

Phillipe de Chérisey

In 2008, a "new" work of Phillipe de Chérisey's, *Un Veau à Cinq Pattes,* (or A Five-legged Calf) appeared in a privately printed edition in France.[80] The book contains a previously unpublished 80-page manuscript he had completed in 1984, the year before his death, in which he reflects on his involvement with the affair. The edition includes both

---

80 Philippe de Chérisey, *Un Veau à Cinq Pattes* (Paris, France Secret, 2008).

a facsimile reproduction of the author's original handwritten pages, together with a transcription. In this work, de Chérisey discusses a very wide array of topics relating to Boudet's book.

In the midst of this, the reader comes upon two pages headed *Géographie Astrologique*, which are devoted to a discussion of the work of Jean Richer. It begins by requoting Plantard's recommendation of *Géographie Sacrée du Monde Grec* as a key text for understanding Boudet. De Chérisey then embarks on an extended discussion of Jean Richer and his work.

> "Astrological Geography.
> *'We did not hesitate to include Sacred Geography of the Greek World in our list. The curious reader can meditate on it to advantage'*. This advice which appears at the beginning of *Elements for a Bibliography* marks the work of Jean Richer as a special case and serves notice that there is a certain situation here. Long before 1978 one saw tourists in Rennes-les-Bains consulting in the street a hardback book recognizable by its blue cover engraved with golden letters."[81]

As we have noted in an earlier chapter, Professor Richer's first book was published by Hachette under their *Guides Bleus* imprint. The works in this series had a standard binding: they were published in hardback, with a sky-blue cloth cover and embossed gold or silver lettering. This included Richer's *Géographie Sacrée du Monde Grec* as can be seen in the photograph of its front cover shown earlier in these pages.

This is the book which de Chérisey states one could see tourists consulting in the streets of Rennes-les-Bains before 1978! But Richer's book, of course, is about landscape zodiacs of ancient Greece. It is not a guide to the thermal baths and other delights of the small village of Rennes-les-Bains in the Haute Vallée de l'Aude. No one would seriously consider using Richer's book as a literal travel guide for Rennes-les-Bains, because clearly it is not. What could de Chérisey have meant?

He is telling us that Richer's book can be used as a guidebook for Boudet's cromlech. And, of course, it is a guidebook: it was published under the *Guides Bleus* imprint of Hachette! Richer's *Géographie Sacrée du Monde Grec* may be considered, allegorically, as a *Guide Bleu de Cromleck de Rennes-les-Bains !* It is clear that as far as the Team is concerned, Boudet's cromlech is an example in the South of France

---
81   Ibid. p. 117

of what Richer found in Greece and elsewhere, namely a landscape zodiac. De Chérisey continues:

> "The appeal of J. Richer's investigations lies in the simplicity of the subject. It involves applying a zodiac on the Greek world map and using this fact that every zodiac is an oriented circle. We know medieval zodiacs that deemed it necessary to be accompanied by a compass rose, where the twelve signs playing on both space and time correspond to the directions as to the hours of a clock face. (...) More than a book which just makes you think, *'Sacred Geography of the Greek World'* entices the reader to a deeper reflection, and to how it relates to LVC (*La Vraie Langue Celtique*). This is how we should understand the discourse prior to 'Elements for a bibliography'; no more than a motif for meditation, no more."[82]

He even refers explicitly to Plantard's "brilliant reconstruction of the cromlech-zodiac" in these very passages, unambiguously bringing these elements together into a single unified format.

> "The brilliant reconstruction of the cromlech-zodiac by the author of the preface [i.e., Plantard] makes it possible to see that in fact the structure has tilted, as suggested by Jean Richer, and by almost a right angle. Where one would have expected north to coincide with Capricorn, it is close to 0° of Aries, which is marked by the Rokko Negro."

Here in one concise passage, we have everything we have been labouring to connect: Boudet's cromlech, the zodiac, Plantard and Richer. De Chérisey also provides some useful commentary which can help us further understand the allocation of zodiac signs in Plantard's zodiac, as shown in his map sketch (Figure 62). The structure has "tilted", says de Chérisey, according to a suggestion of Richer! What does he mean?

If we consult any of Richer's diagrams of landscape zodiacs, we inevitably find that the north-south axis is allocated to Capricorn–Cancer. These are often referred to in esoteric astrology as the Gates of Heaven. Hence, typically Capricorn occupies the position to the north, as de Chérisey correctly notes. However, as always the system was by no means rigid or inflexible.

---

82 Ibid. p. 119.

We have seen that in the case of Rennes-les-Bains, the main *cardo* axis was offset at an angle of 15° east of north, rather than running due north-south. For some reason, when he proceeded to allocate the zodiac signs to the twelve-fold division of landscape that he had discovered, Richer chose a different approach. Perhaps he based it on an earlier source of which we are unaware. For whatever reason, he assigned the Gates of Heaven to the 45° axis: Cancer in the south-west and Capricorn in the north-east.

The signs then fall into place in an orderly manner, anti-clockwise, as we have documented, with Aquarius occupying the due north position and Pisces beginning on the Château Blanchefort – Rok Nègre – Rennes-les-Bains alignment.

Thus, in Richer's scheme, as used for the template of *Le Serpent Rouge*, the natural orientation of the zodiac wheel has indeed been "tilted", with Capricorn moved clockwise from its traditional position at the top to the north-east position.

We've already mentioned the curious silence from the Priory team surrounding Richer's involvement with *Le Serpent Rouge*. We haven't looked closely at de Sède's remarks yet but when we do later in the next section, we will see the same reluctance there to bring Richer's name into any connection with the poem. Yet here in these quotes from Plantard and de Chérisey, their apparent silence speaks volumes. Some restriction appears to be in place. They are willing to connect the dots from Rennes-les-Bains and *La Vraie Langue Celtique*, to Richer, but not to *Le Serpent Rouge* itself. Why not?

Gérard de Sède

Of all the Team, it was de Sède who was most likely to be on the same wavelength as Jean Richer. He was genuinely interested in the notion of sacred landscape, and displays a sensitivity, knowledge and awareness of the topic in many passages in his writings. This is especially true in *Le Trésor Cathar* which immediately preceded *L'Or de Rennes* as I have briefly touched upon.

While Richer's name is not cited in de Sède's earlier book, the final chapter or epilogue is a detailed discussion of ancient landscape practises that overlaps closely the topics and subject matter in *Géographie Sacrée du Monde Grec*, also published in the same year, 1967. It includes quotations describing the purpose and origin of landscape zodiacs. The following passage, from a section headed *Le Blason Astrologique de Toulouse,* is a good representative sample.

> "The keystone of all celestial architecture, the cross of the cardinal points sharing the rose of the zodiac was also that of the architecture of the temples and the cities, built in the image of the sky: it was the lines of the decumanus marking the four doors of the circular temple whose location, fixed on the ground by the augurs, was then ploughed to delimit the space of the city to be born."[83]

This statement would not have been out of place in *Géographie Sacrée du Monde Grec*, published that same year, 1967, as de Sède's *Le Trésor Cathar*, which shows that they were both thinking and writing on this same narrow topic at the same time. The question is: did they know each other in 1966? Certainly, they must have met sometime before 1970, when Richer's *Delphes, Délos et Cumes* was published *"dirigée par Gérard de Sède"*, but is it possible to show definitively that de Sède knew Richer, and his work, as early as 1966/67, when he was working on *L'Or de Rennes*? If so, this would confirm direct contact between Richer and the Team at the time when *Le Serpent Rouge* was written.

There is no direct mention of Jean Richer in either of Gérard de Sède's two 1967 books but, incredibly, de Sède does cite his name in several later works, including books published in 1977 and 1988. These references appear inconsequential to the casual reader; de Sède has dropped them into the texts as if they are little more than passing comments, yet they are both highly revealing.

In particular, there is a sequence of observations made by de Sède spanning multiple books over twenty years, culminating in a direct reference to Richer, which lifts the lid on their relationship and reveals that they had been in communication from 1967 at least.

It begins in *L'Or de Rennes*, in which de Sède includes several pages on Boudet's *La Vraie Langue Celtique*. He gives some background information on the priest and summarises the book's contents. He describes Boudet's very strange theory that English was the original language from which all other tongues were developed. He also discusses Boudet's cromlech and declares plainly that it simply does not exist in the way it was described.

He also provides a brief explanation for the madness of Boudet's book. He states that what one reads there as apparent nonsense is merely the surface layer, that the text is coded, and that this encoding is based on the principle of puns or word-games. In light of these remarks, let's go back to Boudet's book for a moment. Chapter III of *La Vraie Langue*

---

83   Gérard de Sède, *Le Trésor Cathare*, op. cit. p. 187. (English translation).

Celtique bears the title *Langue Punique*, or in English, Punic Language. The word Punic is the English word, derived from the Latin, which is used to describe anything that refers to the Carthaginians. The Punic language was therefore the language of the Carthaginians and was a dialect of Phoenician.

However, Boudet's discussion of the *Langue Punique* bears no resemblance to the historical Carthaginians, or their tongue. Instead, he presents various spurious words, phrases and derivations which have come entirely from his own fertile imagination.

With this in mind, we now look at de Sède's comments written in 1967 about Boudet's book being in "code" in more detail. Here is what he wrote in *L'Or de Rennes*:

> "In principle, the coding system employed by Boudet is very simple. It was the author himself who exposed it to us in a transparent manner in the guise of a dissertation about a language that is Pseudo-Punic; a pure product of his fancy.
> 'Note', he writes, 'with what ease the Punic language can, by a play on words, manage to create the proper names of people...'. (...) It cannot be said more plainly that the work is coded by means of a process that is dear, more particularly, to the hermeticists; that of the pun and the play upon words."[84]

There is a fascinating subtlety at play here. These passages by de Sède and Boudet were of course originally written in French and appear here in translation.

In English, the word pun is a valid pun on the word Punic, but this is not the case in French. There is no French word *pun*, so the word *Punique* does not offer any possibility of a pun in this way for the French reader of the text.

For an English reader encountering this passage in translation, it might seem that de Sède is merely stating the obvious when he suggests that Boudet's uses of the term *Punique* involves a pun on the word pun, but in fact it is quite the opposite. There is no possibility for any suggestion in French that *Punique* offers such an interpretation. Hence, de Sède's suggestion, in fact, far from being obvious, is incomprehensible in the original French. Given this, we may well ask: from where

---

84 Gérard de Sède, *L'Or du Rennes*, as translated in Bill Kersey *The Accursed Treasure of Rennes-le-Château*, (Worcester Park, DEK Publishing, 2001). p. 106.

did he obtain this idea, that Boudet's "Punic language" is a "process dear to hermeticists", and based on pun? What was his source? De Sède does not offer any references for his assertions, or any discussion on where he might have obtained such perceptive insight into the possible literary antecedents or templates for Boudet's unusual writing style.

He does however revisit the topic in 1978, in the introduction which he wrote for the re-issue of *La Vraie Langue Celtique*. He offers more detail on the curious coding system, but still does not venture any suggestion as to where Boudet obtained the method, or how he himself arrived at this confident understanding of the priest's coding technique.

Then, in 1988, he published *Rennes-le-Château : Le dossier, les impostures, les phantasmes, les hypothèses*. Written some twenty years after his earlier book, *L'Or de Rennes*, had set the wheels in motion of the Rennes affair, in this later work de Sède revisits many aspects of the mystery and offers various updated comments.

In this account, he repeats his assertions about Boudet's *Punique* code from 1967 and 1978, but now he adds a crucial, stunning detail. Again, this nugget of pure gold is dropped into the text nonchalantly with little hint of its explosive implications, but it is the crucial clue to unveiling the relationship between himself and Richer. For the first time, de Sède reveals where Boudet obtained his coding method. The following quotation comes from the English translation of de Sède's 1988 book:

> "Was the curé of Rennes-les-Bains the inventor of this ingenious coding system? By no means. He had quite clearly borrowed it, erudite scholar that he was, from two little-known works by another ecclesiastic who also possessed a sense of humour: Jonathan Swift (1667 – 1745), the celebrated author of "Gulliver's Travels"."

At last, the answer is revealed. According to de Sède, the punning system that Boudet employed comes from the English writer and satirist, Jonathan Swift. He then proceeds to name the two specific texts in which the code is described.

> "So, Henri Boudet's absurd linguistic theory comes straight from the first of these two books, the '*Discourse in Proof of the Antiquity of the English Language, showing by various Examples that Hebrew, Greek and Latin have been derived from English.*'"

This title strikes a familiar chord: this notion of the English language as being the original tongue from which others were derived is one of the main themes of Boudet's book as we have seen. Yet, we are still not quite there with the details of the system. De Sède continues:

> "The *Discourse* is likewise a coded book, the true meaning of which is incomprehensible if one does not have access to the various keys. But Swift has provided us with these keys in a second work, published in 1719 under the significant pseudonym of Tom Pun-Sibi, entitled *Ars Punica* and sub-titled *The Art of Punning; or, the Flower of Languages; in Seventy-nine Rules*. There it is: the "Punic language."[85]

In de Sède's 1988 book, he reproduces the original 1719 title page from Swift's *Ars Punica,* with the caption:

> "The coding system of Abbé Henri Boudet is borrowed from the method that was described in the book shown above written by Jonathan Swift, *The Art of Punning, or Ars Punica*."

Now this is all very tantalising indeed, as we can readily see even just from these titles alone that de Sède seems to be onto something! If we look closer at Swift's system, it is readily apparent that he is correct. The coding system described in the *Ars Punica* is exactly the method that Boudet adopted. Of course, as Swift wrote in English, his pun on the word Punic makes perfect sense in this context, to English-speaking readers. We get his joke–the "Punic language" is the language of puns – but it is lost in the translation into French!

This is truly a marvellous insight: far from being mere nonsense, Boudet's book is an exercise in satire which employs the method devised and described by Jonathan Swift. From where, however, did de Sède obtain this piece of information? Did he stumble upon Swift's work and deduce for himself that it was the origin of Boudet's methodology? He had certainly never previously provided any clues as to how he came to learn it.

Yet this is really such a penetrating observation, one wonders why it took twenty years to share and why de Sède was so reluctant to take credit for it. But then, later in the pages of his 1988 book, de Sède

---

85 Gérard de Sède, *Rennes-le-Château : The dossier, the impostures, the fantasies, the hypothèses*, (Surrey, DEK Publishing 2006). Translation by Roger Kersey of de Sède's *Rennes-le-Château : Le dossier, les Impostures, les phantasmes, les hypothèses* (Paris, Robert Laffont S.A. 1998). pp. 44-46.

reveals the truth at last. The source for the material on Swift's satirical method was: Professor Jean Richer.

> "In his discussion of Swift's two books, Jean Richer, professor at the Faculty of Literature, Nice, has written: 'Anyone who has glanced, however cursorily, at alchemical treatises, knows that the rules of the Ars Punica find therein their constant application.'"[86]

A footnote provides the source for this quotation: it comes from a 1980 book of Professor Richer's entitled *Aspects ésotériques de l'oeuvre littéraire*. In this collection of essays, Richer examined the work of eight authors and poets, including Swift, Victor Hugo, Rudyard Kipling, André Breton and others, to explore the esoteric ideas which underpin their writings.

Turning to his chapter on Swift, we find there a complete, detailed and deeply fascinating treatment of the ideas that de Sède has put forward in his book in relation to Boudet, namely the coding system described in the *Ars Punica*. Richer describes at length Swift's use of the so-called "Punic language" as an esoteric system based on puns and punning.

So, in 1988 de Sède finally revealed that the source for his 1967 claim in *L'Or de Rennes*, that Boudet's book employed a coding system based on puns, was Professor Jean Richer. Now we can see why this particular process was "dear" to the hermeticists: it was found, according to Richer, to be constantly applied in alchemical treatises.

The chapter in Richer's 1980 book first appeared in print in a journal in 1957, so that de Sède could conceivably have become familiar with Swift's punning techniques by reading Richer prior to 1967. If this had been the case, he could have been in the position to deduce for himself that Boudet had based his method on Swift's *Ars Punica*.

However, if de Sède himself had indeed recognised Swift's technique in Boudet, there would have been no need to be so obscure or circumspect in 1967. Quite the contrary: it would have been an extraordinary insight to have recognised Swift's method in Boudet's madness. Yet, he chose to keep the details concealed and did not reveal his source until 1988. Why? Again, we are faced with a curious reticence to directly acknowledge Richer's contribution to this Affair.

The conclusion is all but inescapable: it must have been Richer who recognised that Boudet was writing a satire after the *Ars Punica* method

---

86   Ibid. p. 47.

of Jonathan Swift, and it was he who then discussed it with de Sède. This is how de Sède was able to drop this information into his 1967 book, that Boudet's work was based on Swift's satire and why he could not reveal the source. He was not at liberty to reveal Jean Richer's involvement in the Affair for some unknown reason. Then, twenty years later, de Sède quietly provided the credit to Richer for the original insight.

If this is true, and the trail of texts outlined above do seem to confirm it, then we can conclude that de Sède and Richer were certainly in discussion on these topics, including Boudet's book, prior to the publication of *L'Or de Rennes*.

For completeness, there are some further references to Richer and Nerval scattered through de Sède's writings, which seem innocuous and almost random, but which take on added significance in light of these connections. For example, in his short introduction to the 1978 edition of Boudet's book, de Sède drops the name Gérard de Nerval casually into his text by including him in a list of writers who had spent time in an asylum.

In 1977, de Sède published a book about the Rosicrucians which contains no mention or reference to the Rennes affair at all. It was called *La Rose-Croix*. In a section discussing the myth of Adonis as a solar cult, de Sède refers to *"les anciens zodiaques"*, and supplies a footnote which reads simply: "Jean Richer *Géographie sacrée du monde grec*, 1967".

Here, in the most understated manner possible, in a passing footnote in a book with apparently nothing to do with the Rennes affair, and ten years after the events, de Sède has arranged to leave a direct reference to the crucial Richer text. In *La Rose-Croix,* he also chooses a quotation from Nerval and one from Goethe, as epigrams for this book.

In a minor publication that he authored in 1980 entitled *Saint-Emilion insolite* (or "The Unusual Saint-Emilion"), about the medieval village near Bordeaux famed for its wines and vineyards, he includes an image with the heading *Le Zodiaque de Saint-Emilion*, which depicts a map of the walled town with a superimposed zodiac, as shown in the accompanying Figure 63.[87] In this subtle manner, de Sède quietly signals the influence of Jean Richer on his understanding of historical geography.

Finally, I cannot finish this chapter without mentioning an intriguing confirmation of the particular layout of the zodiac format that I have suggested underpins *Le Serpent Rouge*. It is found in Jean-Luc

---

87   Gérard de Sède, *Saint-Emilion insolite*, (Paris, Pégase, 1980).

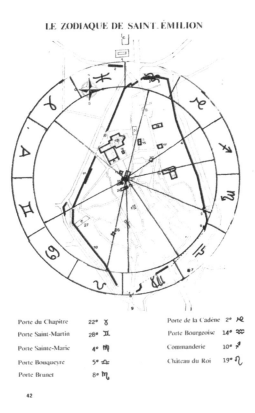

Figure 63: Le Zodiaque de Saint Emilion. From a pamphlet written by Gérard de Sède published in 1981. This image shows a zodiac format superimposed on a map of the ancient French village of Saint Emilion. Image: Gérard de Sède, Saint-Emilion insolite (Villeneuve de la Raho, Pégase, 1980), p.42.

Chaumeil's 2006 book *Le Testament du Prieuré du Sion*, which includes many intriguing items connected with the Affair. It contains a photograph of the cover of what he describes as a "phantom book", the title of which identifies it as the Constitution of the Priory of Sion, dated 1956.[88] It is bordered by a thirteen-sign zodiac, including Ophiucus. The distribution of the signs is identical to the format displayed in the zodiac of Rennes-les-Bains and *Le Serpent Rouge*, except that it has been turned upside-down and then reflected.

The image can be seen in Figure 64, with Aquarius in the centre of the lower side, and Scorpio and Ophiucus divided by the midline

---

88 Jean-Luc Chaumeil, *Le Testament du Prieuré de Sion : Le Crépuscule d'une Ténébreuse Affaire* (Paris, Pégase, 2006), p.112.

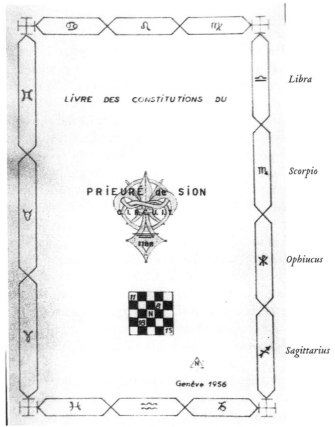

Figure 64: "*Une seule page pour un livre fantôme*", or "*A single page from a "phantom book*". This image appears in Jean-Luc Chaumeil's 2006 book Le Testament du Prieuré de Sion : Le Crépuscule d'une Ténébreuse Affaire (Villeneuve de la Raho, Pégase, 2006), p. 112. Zodiac labels at right have been added.

on the right side, deftly mirroring the manner in which the signs are disposed in the poem.

Summary

In summary, the evidence that the Team all knew of Richer and his work is clear. Plantard privileges Richer's 1967 book above all others in reading and understanding Boudet's *La Vraie Langue Celtique*. De Chérisey makes some equally high compliments about the importance of Richer's insights and perhaps the most extraordinary comment of them all when he says that "long before 1978", people could be seen in the streets of Rennes-les-Bains using *Géographie Sacrée du Monde Grec* as a guidebook. Finally, de Sède lays a trail of clues that confirm Richer's contribution and involvement from at least 1967, yet in such a nonchalant manner as to appear as nothing more than passing references.

The hidden architecture of *Le Serpent Rouge* has revealed itself. The zodiac format is a rich deposit of knowledge, both ancient and modern. It is based on a genuine ancient intervention in landscape, a twelve-fold division inscribed geometrically into the earth. It has its roots in deep antiquity. The Templars and others in the thirteenth century, and possibly much earlier, knew of it and built their châteaux and churches on pre-existing, significant sites in the geometry. Knowledge of these landscape forms, and some of the cartographic properties evidently came into the possession of Henri Boudet, in the late nineteenth century, in Rennes-les-Bains. He wrote about it as the "cromlech", describing what he knew in a veiled fashion, under the guise of an exercise in satire after Jonathan Swift's "punning" language.

Pierre Plantard, Gérard de Sède and others became aware, by some means, of the concealed nature of Boudet's book. They brought it to the attention of Jean Richer, who recognised an example of what he had discovered in Greece and elsewhere: the landscape zodiac. He also recognised the origin of Boudet's literary technique in Swift's works on satire, as he had already written and published on this very topic in 1957.

On a visit to the area, they went for a walk around Rennes-les-Bains, following the route Boudet gives in chapter VII around his cromlech. Later, Richer memorialised the walk in a poem, *Le Serpent Rouge*, in which he drew together various elements of this received ancient knowledge and employed it as background to a prose poem. It retraces the walk around Boudet's cromlech using the zodiac of Rennes-les-Bains to calibrate the progress around the village.

The poem has now begun to open up. The identities within the poem—of *"cet ami"*, of the narrator and of the *"serpent rouge"* itself—and the location of the action, have now been unveiled. Taken together, these revelations demonstrate conclusively that Richer knew, that the Team knew, that Boudet knew and that the Templars knew that there was a very special geometrical artefact inscribed into the very landscape itself in this place.

Yet several of the riddles outlined in Part I remain unanswered, and some of the discussions introduced so far also open up entirely new questions. What are the meaning and significance of the scattered stones, or the 64 squares? What do these have to do with meridians and east-west lines? What is the secret that involves a seal of Solomon?

We have seen that the text of Boudet's book contains layers of hidden meaning, but we have not yet discussed the strange map which accompanied it. If the text describes a landscape artefact, the "cromlech", is it depicted in some manner on the map?

Finally, we still have many questions to resolve relating to the landscape geometry. What are the relationships between the different elements that have been found, including the meridians, the zodiac, and the Sunrise Line?

In Part III I will delve further into these matters.

PART THREE

Solution

Chapter Twelve
GEOGRAPHIC CRYPTOGRAPHY

The Boudet Map and its Curiosities

A BBÉ HENRI Boudet's *La Vraie Langue Celtique* contained a map of the area around Rennes-les-Bains, engraved specifically for the book and included as a separate insert folded inside the back cover. It was printed in three colours and must have represented a major effort and expense. Yet, curiously, there is no mention of it anywhere in the text.

Boudet's engraver has based his work on the 1866 edition of the *Carte d'État Major*, created in the mid-nineteenth century in France as part of a comprehensive national cartographic programme. He has altered it in one significant respect, however. The maps of the *État Major* series were engraved at a scale of 1:40,000. Boudet's version is an enlargement of the original but, surprisingly, its scale is not indicated anywhere.

A large title, reading *Rennes Celtique*, appears at the top in an unusual font that appears to have been devised specifically for the occasion. The design of the lettering and the size and prominence given to the heading make for an odd effect.

The map has further anomalies. There is no grid shown nor co-ordinate system displayed. Lacking any of the elements usually considered essential, this map without scale, grid, or measure does not seem to be a model example of the cartographer's art. Indeed, these shortcomings render it next to useless.

Boudet could not have been unaware of these limitations. Why, then, was it included with the book? Was the cartographer just not up to the task, or was there some deeper agenda at play?

Gérard de Sède addressed these questions when he reproduced Boudet's map in his 1988 book about the Rennes affair. He commented:

> "This book, at first sight ridiculous, is in fact a work of geographic cryptography. That is why it includes a map of the region."[89]

This is an intriguing clue dropped by de Sède. If the reason for the map is that the book is a "work of geographic cryptography", then surely, he can only be suggesting that in some sense: *the map itself is a cryptogram*. This at least suggests a plausible reason for its apparent ineptitude. Perhaps the errors are not unintentional but are part of a game of encryption. If so, what is missing may not necessarily be absent but rather concealed. What else does de Sède tell us about this cryptogram?

> "In other words, the work is a cryptogram. We are thus invited to decipher it, that is to say to transform the apparent nonsense of the text into something meaningful. This operation presupposes the discovery of the key, that is to say the convention used by the author to carry out the reverse operation."[90]

Where might we discover this key? De Sède offers further helpful hints in his 1978 introduction to Boudet's work, even spelling out its precise location.

> "It is a cryptogram containing its own keys, obligingly hidden by the author in the text itself, for the use of those who take the trouble to find them."

The keys are *contained within the cryptogram*. So, now, if the map is a cryptogram, then *its keys must be hidden within the map itself*. Once the concealed key, or "convention" as de Sède describes it, has been found, it can be used to transform the map into "something meaningful". Yet, no such key has ever been found. Boudet's map has resisted all attempts at deciphering its mysteries for over a century.

Consider what kind of information might have been concealed in this enigmatic map, and how it might be retrieved or decrypted. We are looking for something that is missing. What if the encrypted content consisted of just those crucial elements that appear to be absent, namely the usual map details of scale, measure and co-ordinate grid?

---

[89] Gérard de Sède, *Rennes-le-Château : The dossier, the impostures, the fantasies, the hypotheses*, op. cit. pp. 142, Plate C.
[90] Foreword by Gérard de Sède in Henri Boudet *The True Celtic Language and the Stone Circle of Rennes-les-Bains* op. cit. p. xiii. p. xii.

GEOGRAPHIC CRYPTOGRAPHY

Figure 65: The map reproduced in the Belfond Press, Paris, 1978 edition of Henri Boudet's book, La Vraie Langue Celtique et Le Cromleck de Rennes-les-Bains (1886). This edition also includes the Introduction by Pierre Plantard.

To find it we need to discover a key. What if the key was concealed within an element that otherwise seems to serve no apparent purpose? We have already noticed a feature of the map whose prominence seems slightly out of place: the oversized title in a strange custom font. We are now on the very brink of the secret of Boudet's map.

The clue to unlocking the map is hidden in plain sight. It is completely invisible until it is noticed, but it is blindingly obvious once it is seen. It is so simple that a child can grasp it and yet so fiendishly clever that it has remained unnoticed for more than a century. The key is this: the first word of the title, *Rennes*, is *exactly two inches in length and one-half inch in height*. This is the hidden secret of Boudet's map and his book: the measure by which the map is conceived and constructed, and by which it needs to be read, is the English inch.

But how does it help to know that the key is the inch? What do we do with this key? How do we apply it? The answer, again, is embedded in the title. The grid that is obviously missing from the map is a grid of inches, or half-inches. It aligns vertically with the left-hand edge of the title, and horizontally with the top and bottom of the half-inch-high title itself.

There are multiple unambiguous indications that demonstrate this concealed grid is an intentional design layer of the map. With the title providing our cue, we can now rule in the grid on the map. Taking the left and the top edges of the title as reference axes, we extend a grid of inch or half-inch squares across and down the map, as shown in Figure 66. When we do so, the first thing to notice is that the left-hand border shown in the Belfond edition of the Boudet map conforms perfectly to this grid. (This border is missing in the Bélisane edition.)

We also find that the elements of the legend at the bottom right-hand corner of the map are precisely positioned on the grid. The lines of text fall on the sub-division of the inch grid into twelfths, as shown in Figure 67. In effect, the text functions as a kind of invisible "scale".

Exploring the map carefully with the grid made visible reveals a host of correlations. An obvious one is the line of dots marking a ridge which run horizontally across the bottom section of the map, following the grid line as far as it dares without looking too deliberate. Another is the shape and position of Lavaldieu at lower left.

The trail of clues left by de Sède has led us to the solution to Boudet's work of geographic cryptography. The act of inscribing a grid of inches, or half-inches, on the map is the "reverse operation" which restores the map into "something meaningful".

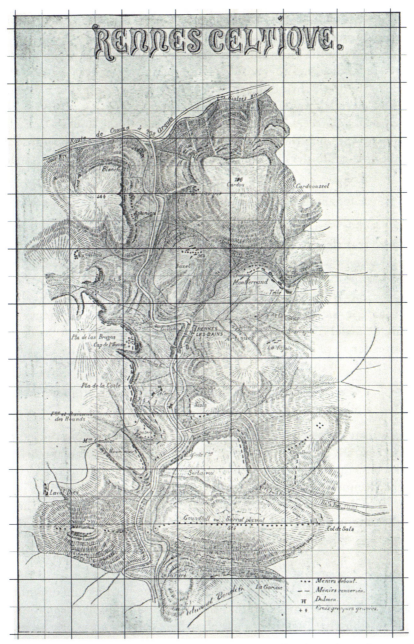

Figure 66: *Boudet's map with superimposed grid of half inches. The grid size and position has been generated using the title as the reference guide. The grid has then been extended over the full map. Notice how it co-incides with the left edge of the frame. There are numerous correlations between grid and map.*

As we will soon see, this insight releases a cascade of concealed information preserved in the map.

Armed with the knowledge that the key is the inch, it is possible to make sense of some otherwise incomprehensible remarks made by de Sède in that 1978 introduction. They occur in the context of a discussion of a well-known local rock formation, which has always been known as *"les Roches Tremblantes"*, or the "Trembling Rocks". De Sède draws attention to the fact that Boudet refers to these by the unusual name of the "Roulers", a word which does not even occur in French.

> "No one except Boudet has called them Roulers. (...) Let us look up what Ruler means in English."[91]

De Sède's remark was of course originally written in French and is here translated into English. He is suggesting to the reader that the reason Boudet chose this word is that he had in mind the English word "ruler", and he urges us to check the meaning. Let's follow his suggestion. What does "ruler" mean in English? A ruler in English is, of course, *a ruler marked in inches*. What a delicious joke and potent clue from de Sède! Immediately following this passage, he then instructs the reader to pay very close attention to the word "Rennes" in the ornate title on the map.

> "Let us next examine the letters of the inscription with a magnifying glass."

His instructions could hardly be clearer without giving the game away entirely. He is directing us to use an English ruler (a ruler marked in inches) to scrutinise very closely (that is, to carefully measure) the word "Rennes". When we do so, the secret is unveiled: the title is laid out in inches. Boudet's map reveals its hidden layer, a grid of half-inches aligned to the title. The missing cartographic elements of this work of geographic cryptography have begun to re-appear.

Doctored From the First Page to the Last

A competent puzzle-maker will always leave a strong and unambiguous confirmation to signal that the correct solution has been found. Is it possible that Boudet has left elsewhere further clues or evidence to support what has been proposed? Here is de Sède again in his 1978 introduction to the Bélisane edition reprint of Boudet's book:

---

91 Henri Boudet, *The True Celtic Language and the Stone Circle of Rennes-les-Bains*, op. cit. p. xvii.

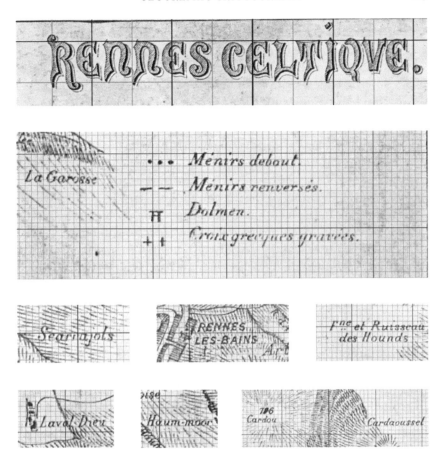

Figure 67: The secret key to unlocking the concealed inch grid of the map is deposited in the position and measure of the title. Much of the text on the map, including the Legend, aligns to this grid when further subdivided into squares of 1/12 inches as shown in the examples above.

> "For this book was in fact composed – and how meticulously – in a way to mislead the majority of people whilst richly rewarding the attentive reader. Like the tombs of the cemetery of Rennes-les-Bains, it is doctored from the first page to the last."

This is very intriguing. The last page of the book is the map and we have seen how that has been "doctored". Is de Sède perhaps suggesting that the first page of the book might also have been treated in a similar fashion? The first page of *La Vraie Langue Celtique* is the title page. De Sède makes some interesting remarks about this page just two paragraphs after the quotation above.

"Let us now look at the title page. The first thing that attracts the eye is the date of publication, 1886. It is placed there – something that is never done – right in the centre of the paper, printed in bold type and surrounded by a loud arabesque pattern. It is clear that in this manner he wanted to attract our attention."[92]

De Sède then follows this passage with the information that this date of publication cannot be correct, as the publisher no longer existed by that time. In doing so, he conveys the clear impression that the reason for the unusual framing of the date on the title page by Boudet was to draw attention to the fact that it was historically problematic. But de Sède is playing an ingenious trick on the reader on this point.

It is certainly true that the date of publication is framed to attract the eye, but it is *the frame itself*, rather than the date, which is the intended target of our attention.

The date on the title page of Boudet's book is framed within an oval. As originally published, the dimensions of this oval are precisely one inch in width and one half-inch in height, as can be seen in Figure 68. Thus, the framing design surrounding the date marks out and defines a square grid based on the inch, subdivided regularly. The arabesque plays the same role as the words *Rennes Celtique* on the map.

Boudet has placed the key to the book, the inch measure, in the most prominent place possible, at the centre and focal point of the title page. If we now extend this inch grid across the title page, we observe several impressive correlations. The small ornamental line beneath the publisher's name measures exactly one-half inch in length and falls exactly on the horizontal grid line two inches below the top tip of the arrow. Hence this innocuous-looking line segment, which seems to be nothing more than a decorative flourish, is in fact solid confirmation of the concealed grid. There is much more to notice on close inspection.

The letters that make up *Langue Celtique* are very carefully positioned in relation to the grid. Each letter in *Langue* is perfectly sized and spaced to fit within the vertical half-inch grid. In *Celtique*, the eight letters have been elegantly disposed across seven grid squares. The name of Carcassonne at the bottom of the page extends over precisely one and a half inches, with the positions of the letters acknowledging the grid lines.

Perhaps most impressively: the author's name, "l'Abbé Boudet" measures exactly two inches in length, just like the key word "Rennes"

---

92   Ibid. pp. xi – xii.

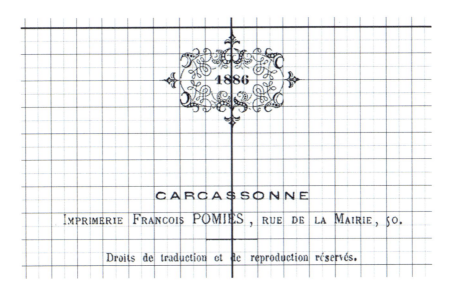

Figure 68: The cover of La Vraie Langue Celtique with superimposed grid of inches, further subdivided into squares of 1/8 inches. The elaborate arabesque that surrounds the date of publication preserves the secret, concealed in its dimensions. The oval measures one inch in width, and half an inch in height. Note also the line below POMIES, one half inch in width and aligned to the grid.

on the map. The first page of Boudet's book has indeed been doctored, and with the same elements as the last page, the map. It too has been covertly arranged on a concealed grid of inches. [93]

Curiously, the facsimiles of the title page in the two 1978 editions differ in the size at which they have been reproduced. In the case of the Belfond/ "Plantard" edition, the width of the ellipse is precisely one and one-eighth inches. In the Bélisane/ "de Sède" edition it measures just under seven-eighths of an inch.

I conclude that the former was enlarged, and the latter reduced from the original, to obscure the key, the inch measure. Whatever the reasons for these alterations in size, the effect has been to thoroughly obscure the most obvious of Boudet's clues to the presence of the inch grid, namely the dimensions of the ellipse around the date on the title page.

It is fortunate that the Belfond map was reproduced at the exact scale as the original, otherwise, the secret would have been lost forever.

---

93   Whilst I have not confirmed the dimensions of the arabesque from an original first edition copy of Boudet's book, based on its published overall dimensions, the claim that the arabesque is based on the inch measure is accurate.

In the case of the map in the Bélisane edition, the border is lacking, as previously noted, and in addition the title has been shifted slightly further away from the map graphic itself, with the effect of obscuring the grid relationship between the title and the map.

The Map-Grid and the Meridians

The hidden grid on the map now revealed to view consists of vertical lines and horizontal lines. I have already identified several meridians physically marked in this landscape. The question naturally arises: is there any relationship between these vertical grid lines on the Boudet map and the meridians marked in peaks?

In the accompanying Figure 70, Boudet's map has been overlaid in Google Earth at correct scale to match the landscape, with the three meridians of Lavaldieu, Bézu and Rennes-les-Bains, and the 45° alignment added for reference. This allows us to directly compare the landscape meridians with the vertical lines of the concealed grid on the Boudet map.

Notice first the Rennes-les-Bains meridian. It falls precisely on the vertical line that passes through the second letter "e" in Rennes in the title. Furthermore, the Rennes-les-Bains church itself lies on the intersection point of this vertical line with the horizontal line six inches below the top edge of the Rennes Celtique title.

The second to note is the Bézu meridian: it is exactly coincident with the vertical line one-and-a-half inches to the left, or west, of the Rennes-les-Bains meridian. This line marks the left-hand edge of the title, Rennes Celtique.

If we now turn attention to the vertical line on Boudet's map one inch to the left of the Bézu meridian, we see that it passes through Lavaldieu, at lower left. This vertical line is the Lavaldieu meridian.

This meridian through Lavaldieu on longitude 2°18′00″E also passes through another very significant Templar site in the landscape just a few miles to the south. The spectacular, impregnable Château Puislaurens was built on the peak of a dramatic mountain at the entrance to the valley. This is a site of major importance—culturally, historically, architecturally and strategically – and its position on this critical meridian in the geometry cannot have been accidental or unnoticed.

We have identified five meridians in close proximity. From west to east, they are the meridians of La Pique, Lavaldieu, Bézu, Rennes-les-Bains and Pech Cardou. Of these five, three are impeccably co-incident with vertical grid lines on the Boudet map, namely those through

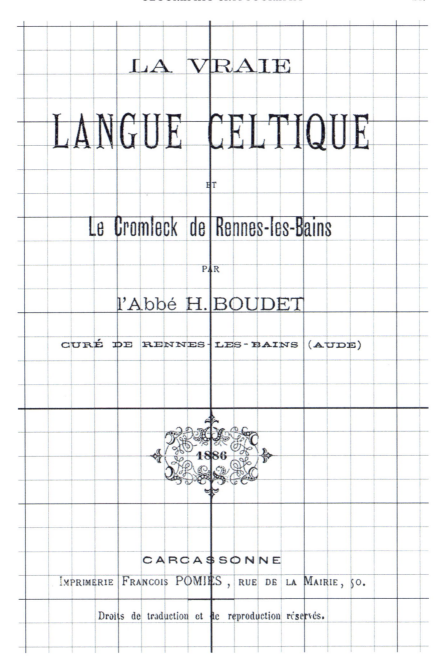

Figure 69: The title page of La Vraie Langue Celtique with a superimposed grid of inches, subdivided further into squares of 1/4 inches.

Lavaldieu, Bézu and Rennes-les-Bains. The horizontal distances between them are respectively one inch and one and a half inches.

Notice that Lavaldieu itself is positioned on a horizontal line of the grid, and therefore on a grid intersection point. Rennes-les-Bains also lies on a grid intersection point. The grid diagonal that passes through both is of course identical to the 45° alignment. Continuing to the north-east, we find Montferrand Château on the same diagonal as well as on an intersection point of the half-inch grid.

From this we can observe that the 45° alignment corresponds with a diagonal through the intersection points on the inch grid on the Boudet map.

But now, if the cartography laid out in inches integrates with the landscape and its embedded features, it prompts again the question of the scale of Boudet's map. What is it?

The Scale of Boudet's Map

As I have noted, Boudet's map is essentially an enlarged copy of the *État Major*, but the scale at which it has been reproduced has not been recorded. It is easy enough to determine, however, by measuring some sample distances between marked locations and comparing these with equivalent lengths on a map of known scale. When we do this, we find that the scale of the Boudet map is 1:25,000, an enlargement by a factor of 1.6 of the 1:40,000 scale of the original source.

First, observe that by a happy coincidence, the scale happens to be the same as that of the modern IGN map. This is very convenient. It means that a measured length on one will be the same length on the other. The same grid of inches on the Boudet map can also be ruled on the IGN map at the same size. Hence there is a regular spacing between these three meridians, both in the landscape and on a 1:25,000 scale map, which is based on the inch measure.

Boudet obviously could not have selected his scale to match that of the IGN map because it did not exist in his day. Why then did he choose to reproduce the *État Major* map at 1:25,000? Without taking into consideration the insights of this chapter, it would seem to be a wholly arbitrary choice.

The table below lists the three meridians marked in the landscape that also happen to coincide with the vertical grid lines on the Boudet map, namely Lavaldieu, Bézu and Rennes-les-Bains. It gives their longitude values, and their relative separations, in arcseconds in the landscape, and in inches in both the landscape and on the map.

GEOGRAPHIC CRYPTOGRAPHY 221

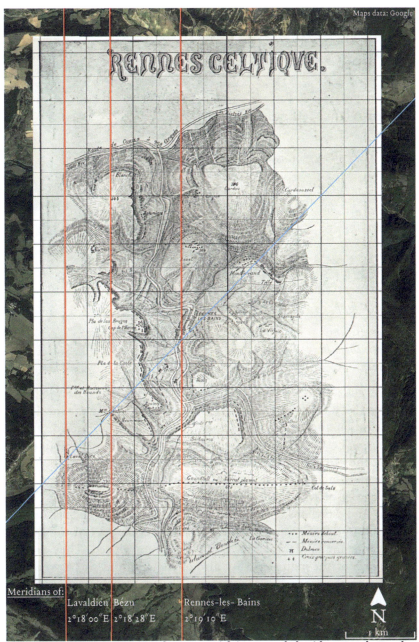

Figure 70: Superimposing Boudet's map with its concealed grid on Google Earth reveals that the three inner meridians of Lavaldieu, Bézu and Rennes-les-Bains coincide with three vertical lines of the half-inch grid. The spacings between them are one inch and one-and-a-half inches respectively on the 1:25,000 scale map.

The Three 'Inch-grid' Meridians and their longitudes:

Meridian name:	Lavaldieu	Bézu	Rennes-les-Bains
Longitude:	2°18'00"E	2°18'28"	2°19'10"

Their relative separations:

	From:	Lavaldieu meridian	Bézu meridian
	To:	Bézu meridian	Rennes-les-Bains meridian
Angular distance between them in landscape		28 arc seconds	42 arc seconds
Linear distance between them on 1:25,000 map		1 inch	1.5 inches
Linear distance between them in landscape		25,000 inches	37,500 inches

From this table, we can see that the distance between the meridians of Lavaldieu and Bézu in the landscape is 28 arc-seconds or 25,000 inches, and between the meridians of Bézu and Rennes-les-Bains is 42 arc-seconds or 37,500 inches. On Boudet's map, these correspond to 1 inch and 1.5 inches respectively, a result which is a direct consequence of his choice of 1:25,000 as the scale for his map.

This suggests that such lengths in whole numbers of inches, or half-inches, and this associated scale, are an inherent part of the knowledge about the landscape that Boudet has reproduced. They are not a property of the map alone but also of the relationship between the features in the landscape themselves. The inescapable conclusion is that these meridians were deliberately laid out to reflect this measure of 25,000 inches and related lengths in the landscape. The use of the inch was intentional.

But if this is the case, then so too must be the scale, because the resolution of lengths to whole numbers of inches, or half-inches, on the map will not occur at any other scale (except obviously for multiples, such as 1:50,000). This would lead us to conclude that the scale of

1:25,000 and the measure of the inch can be considered in this context as, in some sense, a matched pair that are part of the original body of knowledge attached to the landscape geometry. They go together.

Of course, a map of a particular scale does not, in general, require the use of a particular measure. One is usually free to choose any convenient unit of measure to work with, and the map-scale simply provides the conversion factor between distances on the map and in the landscape.

Regarding the landscape geometry of Rennes, and its representation on the Boudet map, it is becoming apparent that we have a special situation. There are certain meridians in the landscape separated by distances that resolve to multiples of 25,000 inches. When these are represented on a 1:25,000 scale map, they resolve to whole numbers of inches.

This pattern is restricted to the meridians that include man-made structures, whilst the meridians of Pech Cardou and La Pique, which are comprised of peaks alone, do not conform to this grid spacing. This may have potential implications for questions surrounding the origin and purpose of these extraordinary meridians.

The knowledge encoded in Boudet's work of geographic cryptography is the blueprint of the geometry woven into the landscape. Now all the disparate elements begin to combine into a single unified scheme, including the map, meridians, key locations and measures.

The concealed grid on Boudet's 1:25,000 scale map is based on the inch. The vertical grid lines on the map are aligned to these same meridians in the landscape. It is apparent that the notion of the inch-grid, and thus by extension also the scale of 1:25,000, is bound up in the original design and construction of the meridians themselves.

This seems to be heading inexorably towards the conclusion that both the scale and the measure are somehow inherent in the landscape geometry interventions, and that this knowledge somehow survived and came into Boudet's possession.

If this is so, then the scale and the measure of the map are not his arbitrary or capricious choices, but vital elements of an ancient knowledge that has been preserved and transmitted down through the ages, and by some means, eventually came into Boudet's possession.

When a cartographer superimposes a grid onto a map, there is, of course, no suggestion that such co-ordinate axes correspond to any objective physical reality, to anything observable. There are no actual lines marked in the landscape. The grid lines are merely convenient notations added to the chart to make orientation and navigation possible. Yet, as we have seen, this is not quite the case in Boudet's map. We've shown that at least some of the vertical grid lines which are found in the concealed layer correspond perfectly exactly with meridians which are physically marked in the local peaks. The grid, in this case, appears to be inherent *in the landscape itself*. The territory *is* the map, and the map *is* the territory.

This is an unsettling observation. How could a co-ordinate grid become embedded in the landscape? There are few if any precedents which could help us to understand such a thing, no maps to this uncharted land. All we can do is marshal the facts and see where they lead us.

Dimensions of the Cromlech

Boudet does not suggest anywhere in his book that his map depicts the cromlech in any manner, yet the curious reader might reasonably wonder whether it is also somehow concealed within his enigmatic work of "cryptographic geography".

Recall that at the outset of Chapter VII of *La Vraie Langue Celtique*, Boudet gives a very exact estimate of the size and shape of the cromlech:

> "Indeed, its mountains crowned with rocks form an immense Cromlech of sixteen or eighteen kilometres circumference."[94]

Evidently, despite the lack of any obvious circular arrangement of the stones, he wanted to emphasise for some reason its geometrical aspect and these exactly specified measures. In a related passage, he also reinforces the same point at some length in a discussion of the deep significance of the symbol of the circle to the ancients.

> "The circles traced by the standing stones had a profoundly religious meaning for the Celts. The Druids, just like the ancient philosophers, regarded the circular figure as the most perfect. It represented to them Divine perfection, immense and infinite, having neither beginning nor end."

I had wondered whether Boudet might have been suggesting that, in some sense, his cromlech could be thought of as a perfect geometrical

---

94   Boudet, *The True Celtic Language and the Stone Circle of Rennes-les-Bains*, op. cit. p. 225 & p. 245.

circle, as if it were a kind of platonic form, rather than, or perhaps in addition to, a tangible collection of stones laid out in landscape.

If so, it occurred to me that such an approach might suggest a possible interpretation of his "sixteen or eighteen kilometres circumference" remark. Instead of representing the physical size of a ring of standing stones, perhaps it was intended to denote the dimensions of a "virtual" circle, considered as a kind of mental construct superimposed on the territory or even on the map. With this in mind, we return to the question of the dimensions of the cromlech.

The obvious question to ask is: what would be the dimensions of these circles if they were represented on Boudet's 1:25,000 scale map using inches as the unit of measure? This is easy enough to calculate. First, let's consider the circle with a circumference of sixteen kilometres in the landscape. This is equivalent to 629,921 inches and corresponds to a radius of 100,255 inches.

On the 1:25,000 scale map, this would be represented by a circle of radius of 4.01 inches. Considering that it would be impossible by eye to distinguish between a radius of 4 inches and 4.01 inches, a difference of just one hundredth of an inch, we can make the following statement:

> *Boudet's cromlech of 16 kilometres circumference in the landscape is accurately represented on a 1: 25,000 scale map by a circle of four inches radius.*

The 16-kilometre circumference cromlech in the landscape becomes a four-inch radius circle on the map!

What of the second value which Boudet offers for his cromlech, of eighteen kilometres in circumference? Remarkably, this translates to a circle on the map of four and a half inches radius, to the same degree of accuracy as above.

The two circles that Boudet cites as the dimensions of his cromlech therefore have radii of four and four and a half inches, respectively. If we now draw these two circles centred on the church of Rennes-les-Bains, they touch the grid lines four and four and a half inches from the centre, above and below, left and right. In addition, these two values supply, in their difference, the measure of the spacing of the grid, namely one half-inch, on the 1:25,000 scale map. Boudet's cromlech, at last, begins to reveal itself. He was referring, I suggest, not to a physical circle of stones, but to a *geometric artefact*.

The circles of these dimensions, the grid and the meridians comprise an integrated geometrical system that has been inscribed, or inserted, or imposed on the physical landscape as a historic reality at some earlier, presumably ancient time. By some means, Boudet had apparently come into possession of some fragments of the knowledge of the ancient geometrical design, and he has chosen to record it, albeit in veiled form.

His map is the record of the ancient landscape geometry that he has called the cromlech. The cromlech circle of 16 kilometres circumference in the landscape is represented by a circle of radius 4 inches on a 1:25,000 scale map. Its centre falls on Rennes-les-Bains, which, on Boudet's map, is the central grid intersection point. Hence on the map, the cromlech is represented as a four-inch radius circle, centred and aligned with the grid. Thus, this circle sits within a square of eight inches by eight inches, as shown in Figure 71.

The Sixty-Four Stones

Now the complexity melts away to be replaced by a single simple image. These are one-inch squares on the map, or 25,000-inch squares in the landscape. This grid of squares correlates with major meridians in the landscape. The cromlech circle measures sixteen kilometres in circumference in the landscape, and when represented on a 1:25,000 scale map has a radius of four inches, and, incidentally, a circumference of sixty-four centimetres! A single unified geometry and measure underpins both the geometry inherent in the landscape and the Boudet map, and this is witnessed by *Le Serpent Rouge*.

The simple grid of inch-squares is a compact memorandum that encapsulates the measures of the complex. This format is the solution to another of the riddles of *Le Serpent Rouge*: we have re-assembled the sixty-four scattered stones. They are the sixty-four inch squares that make up the map grid of the cromlech, the four-inch radius circle on the 1:25,000 scale map. We have fulfilled the instructions given in *Le Serpent Rouge* to the letter. The Leo stanza speaks of the "sixty-four scattered stones", and the Gemini stanza instructs us to:

> "Gather the scattered stones, work with square and compass to put them in regular order, look for the line of the meridian by going from the East to the West, then looking from the South to the North, finally in all directions to get the sought solution ..."

No special pleading is required. Regular order has been restored.

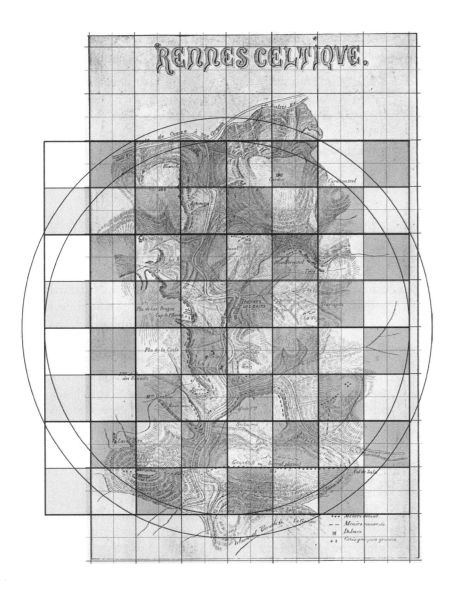

Figure 71: Boudet's cromlech of 16 kms in circumference is represented by a circle of eight inches diameter on a 1:25,000 scale map. The circle of eight inches diameter naturally suggests an eight-by-eight chessboard of inch squares. Here, at last, are the "sixty-four scattered stones".

The Earth as Chequerboard

The concept of the surface of the Earth being divided into squares was one held by many ancient cultures. In India, such forms were called *Ashtapada*. Nigel Pennick, in *Secret Games of the Gods* writes:

> "In Vedic India, the eight-by-eight chequerboard was called *Ashtapada*. The word *Ashtapada* was used to describe the grid employed in land survey. The French writer Bernouf, in *La Lotus de la Bonne Loi*, an obscure work published in Paris in 1854, cites a passage from a northern Indian Buddhist text where the planet herself is described as 'The Earth on which *Ashtapadas* were fashioned with cords of gold.'"[95]

According to Pennick, cities have been laid out on grids from earliest times, all over the world, and he cites multiple examples.

> "From this conceptual layout of sacred space come temples design, holy city and fair layout, and the patterns of gameboards. There is a direct connection and direct correspondences exist. In the Hindu tradition, which has a venerable antiquity, this layout is based upon the *Paramasayika* grid, which is a square composed of 81 smaller squares (nine by nine). At the centre of the grid is the Square of Brahma, the creator's nine squares which are one-ninth of the total area. This nine-square area is seen as the quintessence of existence, the central core from which all space time can be accessed."

Godfrey Higgins' *Anacalypsis* associates the cardo with the ancient practice of dividing the Earth into squares.

> "A Cardo would run through every capitol or principal town, which of course would divide the land into parallelograms of different lengths. The whole world, I do not doubt, was divided into squares or parallelograms, and a cross was fixed at every intersection, as Italy is described, by Niebuhr, to have been. These crosses pointing to the four cardinal points could never be removed or mistaken, for they corrected each other."[96]

---

95 Nigel Pennick, *Secret Games of the Gods*, (Maine, Samuel Weiser, 1989) p. 145.
96 Godfrey Higgins, *Anacalypsis*, Vol. II. op. cit. p. 413.

In another esoteric work from the late nineteenth century, J. Ralston Skinner in *The Source of Measures* wrote:

> "The block of 6 x 6 inches = 36 square inches, four of which is a square foot. This was the favourite subdivision amongst the ancients and the basic square. Therefore, take as the base 6 blocks in length of 1 square each. Kabbalistically, the measure of the Earth is represented as in alternate black and white squares, the white signifying male and the black female."[97]

This reference to an esoteric tradition in which the surface of the Earth is conceived as a chessboard of black and white squares of measure six points to a widespread ancient practice of Earth measurement.

This selection of quotations offers only the briefest of glimpses into the subject but it is sufficient to suggest that related ideas were held universally across all cultures and deeply informed humankind's early relationship to landscape.

The Cromlech as World Map

Boudet described the cromlech as a pair of circles with circumferences of 16 and 18 kilometres; when I transferred these to the 1:25,000 scale map, I found that they correspond to circles of four and four and a half inches radii, respectively.

Normally, one would not mingle metric units with imperial in the same set of values. To have the circumference expressed in kilometres, and the radius in inches is therefore an unusual combination which would usually not be encountered together. Very curiously however, there is a historical precedent for this mismatched coupling, and it relates to the measurement of the size of Earth.

The metre had its origin in modern times as the result of an attempt by the French authorities of the late eighteenth century to derive an Earth-commensurate unit of measure. They ordered a programme of extensive and careful surveying work to determine as accurately as possible the distance between the equator and North Pole along a defined meridian. The metre was then defined as one ten-millionth of this length.

Thus, by this definition, the length of the polar circumference of the Earth was equal to 40,000,000 metres, or 40,000 kilometres. For

---

[97] Ralston J. Skinner, *The Source of Measures,* (San Diego, Wizards Bookshelf, 1982), p. 62.

various technical reasons, some small errors crept into the process. The modern value accepted today for the polar circumference is given as 40,009 kilometres, an excess of just one part in five thousand. Nevertheless, the goal of the French metric project was admirable: to create a system of weights and measures which were commensurate with Earth measure and dimension.

It was thought that such an approach offered advantages to the older systems of measures, which seemed to have arisen haphazardly. For example, the inch was commonly thought to have arisen from the width of the king's thumb, the foot from the length of his feet, the ell from the spread of his arms. Yet such stories are almost certainly apocryphal and do not convey the true origins of these measures. Such considerations are beginning to diverge from the core scope of this book, so we will not dive deeper into the fascinating and yet highly contentious field of ancient metrology, except to note one relevant fact.

The polar diameter of the Earth may be given accurately as 500,518,200 inches. Suppose that we resolved to define a unit of measure by taking this polar diameter and dividing it into 500,000,000 equal segments. The unit that we would arrive at via this process would differ in size from the modern inch by very slightly more than one part in a thousand. So now, if we are willing to put aside any speculations about origins, we are presented with some simple facts about the dimensions of the Earth.

If we round our modern Earth diameter value down by a mere one part in one thousand, and the circumference value by one part in five thousand, the values can be expressed as follows:

> Polar diameter: 500,000,000 inches
> Polar circumference: 40,000 kilometres.

Observe that we have a pair of whole numbers representing diameter and circumference values of a spheroid, with no trace of pi in sight! And look: we also have a circle with the diameter measured in inches and the circumference in kilometres, just as we have found in Boudet's cromlech as depicted on his map.

Comparing the dimensions of the cromlech in the landscape with these results for the size of the Earth reveals that they are related at a scale of 1:2,500!

> Circumference of cromlech = 16 kilometres
> Circumference of Earth = 40,000 kilometres
> Therefore: relative scale = 16:40,000 = 1:2,500

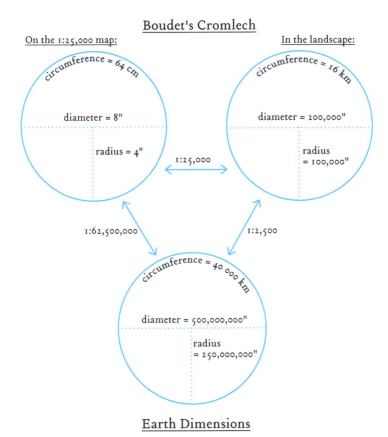

Figure 72: Above: The dimensions of Boudet's cromlech on the 1:25,000 scale map and in the landscape. Below: The dimensions of the Earth.

Boudet's cromlech, whether by design or coincidence, happens to be an accurate 1:2,500 scale representation of the Earth!

A simple relationship of scale therefore links:

(1) the cromlech on the map,
(2) the cromlech in the landscape, and
(3) the circle of the Earth.

From (1) to (2) there is an expansion of scale 1:25,000, and from (2) to (3) there is another expansion of scale 1:2,500. We can therefore combine these two transformations to derive the scale enlargement from the cromlech as depicted on the map to the circle of the Earth. The result is:

[1:25,000] x [1:2,500] = [1:62,500,000]

We can also state the result in this way: the cromlech as drawn as a four-inch radius circle on Boudet's map is a representation of the Earth at a scale of 1:62,500,000, to within an accuracy that is surely impressive for such simple whole numbers. Boudet's cromlech is a map of the world.

Was this outcome intentional, or merely a coincidence? I would not want to press the point too far, but Boudet does make a remark that might suggest he was not unaware of the connection.

In Chapter VII of *La Vraie Langue Celtique*, he informs the reader that there are in fact two cromlechs in the landscape around Rennes-les-Bains. We will return to this point later in the book. The details are admittedly a little vague and ambiguous, but he does state that they are of different sizes, that one lies within the other, and that:

> "The created world is represented here by the smaller circle enclosed within the large one."[98]

At face value, Boudet is saying that *the smaller circle is a representation of the created world*. We have found that the cromlech, as a 16-kilometre circumference circle in the landscape, or a four-inch radius circle on his 1:25,000 map, is a scale map of the Earth. If we assign this format to Boudet's "smaller circle enclosed within the large one", then his description is a perfect match to what we have found. We can even state the scale at which the representation has been made: 1:2,500 from the cromlech in the landscape to the "created world", or 1:62,500,00 from the cromlech on the 1:25,000 scale map.

The idea of the cromleck as world map is admittedly speculative, but what about the broader notion of the cromlech as geometric circles? Are there any hints amongst the Team's writings which would show that they believed Boudet intended his cromlech to be understood as two circles drawn on a map? Remarkably, the answer is yes. Plantard, in his 1978 introduction to *La Vraie Langue Celtique*, draws attention to a seemingly insignificant snippet of local lore in the book.

On page 241, Boudet reports the discovery of an ancient iron millstone by local workers near a stream called La Ferrière, where smelting had been carried out in the past. He describes it as slightly concave and measuring 15 or 16 centimetres in radius. Boudet does not associate this artefact in any way with his cromlech. Nevertheless, Plantard in his usual elliptical style reads Boudet's description of the millstone

---

98   Henri Boudet, *The True Celtic Language and the Stone Circle of Rennes-les-Bains*, op. cit. p. 246.

as coded allusions to the cromlech, or "zodiaque" as he calls it. In the following passage, he specifically interprets the dimensions of the millstone as a reference to a pair of circles drawn on the map:

> "'Twelve palaces were enclosed in a single precinct', with the allusion (referring to Pliny) that these monuments were dedicated to the Sun, this is how the zodiac is described on page 84. Page 246 determines the centre of this zodiac and page 241 gives the dimensions to draw on the map these two circumferences: 15 and 16 centimetres in radius."[99]

Plantard demonstrates here his mastery of conveying the essence of the solution without disclosing the exact details, thereby both revealing and concealing this material at the same time. The "twelve palaces" in a single enclosure, dedicated to the sun, clearly allude to the zodiac format in the landscape. He locates the centre of this zodiac at Le Cercle, as Boudet states on page 246.

But it is the final statement that is the most surprising: Plantard interprets Boudet's reference to the iron millstone as instructions to draw two circles on the map of radii 15 and 16 centimetres. This seems entirely pointless, and hardly what Boudet can have intended by his report of the millstone discovery but observe what Plantard has done. Indeed, circles of these size will not even fit on the Boudet map! Though he has supplied incorrect details regarding the location of the centre and the dimensions of the circle, the act of drawing two circles on the map is indeed the crucial step to unlocking and combining the various hidden layers of the solution to the cromlech.

The correct centre is the church of Rennes-les-Bains. The radii of the two circles are not given by the iron millstone dimensions, but by the cromlech dimensions that are provided in Boudet's text as measures in the landscape, namely, 16 and 18 kilometres of circumference. The "trick" is that these must be scaled at 1:25,000 and converted to inches. These resulting 4 inch and 4½ inch radius circles can then be readily drawn on the map, and we find that they integrate seamlessly with the grid of inches which has been uncovered from the measures concealed in the title. Thus, Plantard has provided the instructions for drawing the cromlech on the map, but in the most ingenious manner, which has ensured that the secrets have remained firmly sealed.

---

99   Foreword by Pierre Plantard in Henri Boudet, *La Vraie Langue Celtique et Le Cromleck de Rennes-les-Bains*. op. cit. p. 28.

De Chérisey repeats and summarises these remarks of Plantard's in the passages about Jean Richer in his book *Un Veau à Cinq Pattes* introduced earlier in Chapter Ten. There he notes:

> "The author of the preface [i.e., Plantard] reveals three examples from *La Vraie Langue Celtique* which, beneath veils, suggest an astrological coding:
>
> Page 84: The Egyptian labyrinth could have been the work of twelve kings (Herodotus) or one (Pliny). A single enclosure contains twelve palaces. Fifteen hundred palaces are arranged around twelve main rooms. It can be—A burial, or a solar temple.
>
> Page 241: We discover in the cromlech of Rennes-les-Bains the fragment of a millstone capable of grinding perfectly and which would have measured from 15 to 16 centimetres in radius.
>
> Page 225: The Rennes-les-Bains cromlech is made of two concentric halos, made of stone, measuring "from 16 to 18 kilometres in circumference". From which it will be deduced that the interlayer of the two circles, lends itself to the insertion of the zodiac signs."[100]

In this passage, de Chérisey brings together in point form the examples making up the *"codage astronomique"* in *La Vraie Langue Celtique* according to Plantard in the Preface. The first is the palace of twelve enclosures as a solar temple. The second is the fragment of millstone of 15 to 16 centimetres in radius. The third is the cromlech as two concentric circles.

When we recall that Plantard took the millstone information as instructions to draw circles on Boudet's map, we can see that de Chérisey is gathering the clues together for us; we just have to combine them. We are to draw two circles on the map, corresponding to the cromlech circles of 16 and 18 kilometres' circumference, and divide it into twelve segments. Here is the zodiac of Rennes-les-Bains, exactly as we have described it. It is also Boudet's cromlech and de Chérisey himself explicitly confirms this for us when he calls it the *"cromleck – zodiaque"*.

De Chérisey's remark is effortlessly informative. If the space between the circles lends itself to the insertion of the zodiac signs, then, clearly he cannot be speaking of circles in the landscape in this case. He is

---

100   Philippe de Chérisey, *Un Veau à Cinq Pattes*, op. cit. p. 119.

surely informing the reader that these circles are to be drawn on Boudet's map to form his cromlech, just as Plantard had stipulated. The zodiac labels can then be conveniently added in the half-inch wide ring between them! However, where Plantard offered (deliberately) misleading details, De Chérisey has provided the correct dimensions by referring specifically to the 16 and 18 kilometre circumference circles in the landscape in this context.

As we have found, if we are to draw these on the 1:25,000 scale map, then they will have radii of 4 inches and 4½ inches respectively, correctly substituting for the decoy 15 and 16 centimetre radii circles which Plantard had proposed. We are beginning to sense the rules of a game in operation.

De Chérisey's Map Grid

If these details seem perhaps somewhat obscure, de Chérisey throws caution to the wind at the end of *Un Veau à Cinq Pattes* and offers a blatantly revealing description of the concealed grid format in Boudet's map. He even provides a sketch of the grid, together with instructions for constructing it by ruling certain lines in reference to various map features, including the title and legend of the map, and the village of Rennes-les-Bains. These are all reprinted in his original handwriting in *Un Veau à Cinq Pattes*, and transcribed into typeset text and professional graphics.

In this instance, de Chérisey has provided instructions that are perfectly coherent and compatible with the scheme proposed in this chapter, except that he has replaced one critical term with another. Now that we have the correct solution to the Boudet map, it is possible to understand what de Chérisey was communicating, although it is yet again a case of the familiar pattern of revealing and concealing at the same time. Nowhere does he disclose the true key, the inch. And yet, for those who have already arrived at these insights by other means, as we have, his descriptions are very revealing and, with a single substitution, they make complete sense.

The crucial clue that signals de Chérisey is in possession of the correct solution is the dimensions he gives for the map of 14 cm x 28 cm. These do not correspond in any meaningful way with the actual size or extent of the full map, but if we convert them to inches, a surprise awaits. They are almost exactly equivalent to 5½ by 11 inches,[101] and these figures assimilate to the hidden map grid in a very satisfying manner.

---

101 14cm / 2.54 = 5.5118" ; 28cm / 2.544 = 11.0236".

The width of 5½ inches is the horizontal distance between the grid-line of the Lavaldieu meridian and the end of the *Rennes Celtique* title as terminated by the full stop. The height of 11 inches is the vertical distance from upper edge of the title to the bottom of the map, at the lower edge of the grid squares containing the legend. The resulting rectangle is a perfect fit to the concealed map grid, as may be seen in Figure 74.

With this initial framework in place, assimilated to the inch grid, the rest of the instructions can be clearly interpreted and make perfect sense if we are willing to make just one simple substitution. All that is required is to replace each occurrence of the word "centimetre" with the word "half-inch".

Below are de Chérisey's instructions in full, reproduced in English translation. I have divided and numbered them into steps. My comments follow, in which I expand on de Chérisey's remarks for each step and offer my interpretation of what he intends us to understand beneath the light surface layer of concealment. Figure 73 reproduces de Chérisey's sketch accompanying his instructions, and its interpretation by his editors, which, in my view, betrays a misunderstanding of his intentions. Figure 74 displays the final result after following de Chérisey's instructions, on the Boudet map, with the hidden grid also shown.

> "1. The Boudet map, 14 cm wide and 28 cm high, is susceptible to three manipulations, all three of which take as their vertical axis the meridian which passes 7/7 from top to bottom, between RENNES/ CELTIQUE (top) to the d of Edmond (bottom).
>
> 2. The first manipulation consists in ruling a line on the map dividing the height in the middle, as 14/14; this gives us the line that borders on the north of the Rennes-les-Bains agglomeration.
>
> 3. The second is to rule a line on the map at the top of the legend, taken as the base. We then obtain a line higher than the previous ruled line.
>
> 4. The third is to extract the title from the height according to the same measures as for the previous operation. Considering that the legend is 2cm high, we will also measure 2 centimetres for the title.
>
> 5. In ruling the map, the same way at the halfway line, we obtain a line that runs close to the south of the city."[102]

---

102 Philippe de Chérisey, *Un Veau à Cinq Pattes*, op. cit. p. 135.

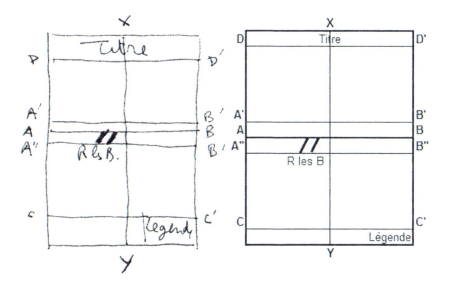

Figure 73: Left: De Chérisey's version of the hidden grid in Boudet's map. It appears in his book *Un Veau à Cinq Pattes*, alongside a facsimile of his hand-written notes. Right: This image shows the interpretation of this sketch provided by the editors to accompany the transcribed version of de Chérisey's notes. Image Credit: Philippe de Chérisey, *Un Veau à Cinq Pattes* (Paris, France Secret, 2008), pp. 135, 337.

Comments:

1. The dimensions of 14 cm x 28 cm are almost exactly equivalent to 5½ by 11 inches. Rule a rectangle of this measure on the Boudet map, with its upper side aligned with the top of the Rennes Celtique title, and right-hand side aligned to the full stop at the end of the title.

The vertical meridian XY is drawn to coincide with the end of the word Rennes. It touches the second "d" in the signature of Edmond at the bottom of the map. In this case, the vertical axis of "7/7" refers not to centimetres, but grid squares. It is not the vertical midline of the 14-centimetre-wide rectangle, but the central vertical line dividing the 14 squares of the complete width border to border of the map, as shown in Figure 74.

2. This is straightforward and correct. This is line AB, which divides the height of 28 cm or 11 inches, into two measures of 5½ inches.

3. This is line CC'. It is one inch above the bottom side of the initial rectangle, the "previous ruled line". The legend occupies this one inch high rectangle. It also further subdivides into half inches, and one-sixth inches.

4. Neither the legend nor the title are in fact two centimetres high, throwing the puzzle-solver off the scent. However, we can easily get back on track by simple substituting every instance of the word "centimetre" with "half-inch". So, for 2cm, we read two half-inches, that is, one inch in total. The Legend is indeed two half-inches high, and so too is the title of the map.

Using "the same measures" then, we rule line DD', one inch below the top line, and another horizontal line to divide it in half. This line defines the bottom of the words Rennes Celtique and confirms that we are on track.

5. De Chérisey here intends that we rule line A'B' one half-inch above the midline AB, and line A"B" one-half inch below the midline. In that way we will have created a one inch high section divided into two half inch high sections, "the same way at the halfway line".

This is exactly as he has depicted it in the sketch he has drawn in his manuscript. It clearly shows the total height of the two middle bands, between A'B' and A"B", is the same as that allocated for the legend and the title: one inch.

Notice also that the lines AB and A"B" neatly frame the built-up area of the village of Rennes-les-Bains north and south. Again, comparison with the sketch shows that this is exactly as he intended.

I suggest that de Chérisey's posthumous editors misunderstood his intent when they produced their version of his grid diagram. They have rendered each of the two central bands as equal to the bands at top and bottom. Careful inspection of de Chérisey's sketch, however, confirms that the width of the two central bands together should be equal to the width of the top and bottom bands. This is entirely consistent with our interpretation, as shown in the accompanying figures.

De Chérisey's instructions make perfect sense as long as one is in possession of the key. He has described the method accurately, yet perfectly securely, by simply omitting the crucial piece of information which the puzzle-solver must supply: the inch.

Summary

The true hidden riddle of Boudet's book is the map, and the secret it contains is the concealed inch grid which also corresponds to the meridians which are present in the landscape. Boudet, it would seem, had knowledge of the landscape geometry, and he recorded it for posterity in encrypted form in this most ingenious of maps, his work of geographic cryptography.

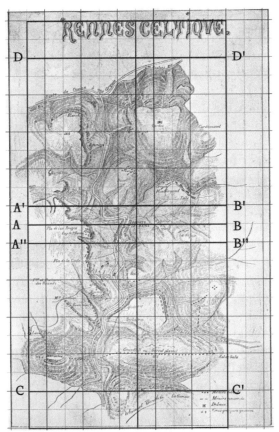

Figure 74: This Figure shows the results of following de Chérisey's instructions to rule a grid on the Boudet map, according to my interpretation. The hidden grid of half inches is also shown for comparison.

The map contains a hidden grid of inches that is by no means an arbitrary artefact of the map-maker's work but stands revealed as a true account of meridians and other alignments in the landscape. This in turn suggests that Boudet was aware of these landscape alignments, and that he arranged for the concealed grid to be encoded in the map to preserve this knowledge, albeit in cryptic form.

Finally, we have arrived at the entrance to the hidden treasure concealed in Boudet's map. It is the confirmation of the persistence into the modern world of knowledge of an ancient landscape form, of great antiquity, and of exquisite construction.

This knowledge was somehow preserved, or revived, as it is apparent that Boudet knew of its existence and was in possession of certain details which described it. Perhaps this information was passed down orally through the generations. Perhaps Boudet had access to private texts or instruction. We shall probably never know. But somehow, he learned of an ancient format in the local landscape.

He called it the cromlech and constructed his map and his book as a vehicle for ensuring the continued transmission of the information. He devised a means to accomplish this which achieved the unlikely twin goals of simultaneously concealing the information from public view and preserving it for private instruction.

Richer refers to the same form as the zodiac. These are different labels for a single underlying phenonemon. It is a complex geometric artefact embedded in the landscape, comprised of alignments, including meridians, marked by prominent peaks and other significant locations.

When Jean Richer visited Rennes-les-Bains, and read Boudet's book, he grasped its full significance. He discovered the landscape division into twelve segments around Rennes-les-Bains, based on the ancient *cardo - decumanus* axis and realised that Boudet must have also known about the ancient geometry on which the town and the surrounding territory was originally laid out.

These insights became the basis of the poem *Le Serpent Rouge*, taking the path of Boudet's walking tour in Chapter VII of *La Vraie Langue Celtique* as the template for a ritual journey around the sacred geography of Rennes-les-Bains, on the zodiac form which echoed that of Delphi.

The zodiac is built on the foundation of a conceptual grid, aligned with landscape features including meridians, which permits very convenient calculation and conversion between inches and arcseconds.

There is a toolkit of mental aids that may be used to conveniently describe the geometry, its elements and how they inter-relate. These led to the creation and communication of visual or verbal maps, in the absence of a culture based on paper or digital mapping. The maps are based on the use of the inch measure and the 1:25,000 scale. The memory of these techniques has been employed in the construction of Boudet's map.

We are by no means finished in our discussion of the map and the meridians, but to proceed further we now need to introduce another piece of the puzzle.

Chapter Thirteen
THE RIDDLE OF THE PARCHMENTS

The Parchments

WE NOW turn our focus to what may be the central riddle of the entire Affair of Rennes: the famous encoded parchments allegedly discovered by Abbé Bérenger Saunière under the altar in the church of Rennes-les-Château around 1896, when he commenced renovation work. It is more than fifty years since these documents were released into the public domain when copies were published for the first time in de Sède's 1967 book *L'Or de Rennes*.

There has been considerable progress since that date in peeling back the layers of obfuscation and mystery that shroud this topic. We can now be certain that the parchments were created in the modern era, and that they do not date from the time of Abbé Bigou in the 1780s, as the *Dossiers Secrets* asserted. This in turn renders it impossible that the parchments as they are known today could have been discovered by Abbé Saunière in the 1890s. Whether he found some other parchments or something else entirely (or nothing at all) when he commenced his renovations of the church of Rennes-le-Château, it could not have been the parchments de Sède reproduced in *L'Or de Rennes* in 1967.

We can be sure of this because the sources from which they were constructed have been found.[103] The text for the large parchment was taken from a Latin version of the Gospel of John, which was first published in 1889.

---

103   Bill Putnam and John Edwin Wood in *The Treasure of Rennes-le-Château: A Mystery Solved* definitively identified the source document for the large parchment as the 1889 Oxford first edition of the Vulgate translation of the Gospel of John, chapter 12: 1-11, edited by Wordsworth/White. Wieland Willker, a German researcher, identified the source of the text for the small parchment as coming from the Codex Bezae, a Greek and Latin manuscript dating from the sixth century AD.

The source for the text of the small parchment has also been identified and confirmed. It was the *Codex Bezae*, a fifth-century collection of New Testament excerpts, written in an ancient script known as uncials. A sample page from this Codex appeared in the *Dictionnaire de la Bible*, by F. Vigoroux, which was published in France in 1895 and was freely available in seminaries and other libraries in the twentieth century. The text of the small parchment happens to be taken from the very same page that was selected as the sample representative page of the *Codex Bezae* for reproduction in the *Dictionnaire de la Bible*!

Each of the two parchments contains a message hidden within its text. The first, in the small parchment, is obviously intended to be relatively easy for an alert reader to notice. It is spelled out in certain letters that have been written slightly higher than the others. When these are strung together, the result is a short sentence, written in French, which begins, "*A Dagobert...*". Henry Lincoln noticed this straightforward coded message when he read *L'Or de Rennes* in 1967, and it became the spark that ignited his interest in the topic.

On the other hand, the larger parchment contains an encrypted message that has been concealed with extraordinary complexity. It consists of 128 letters, which have been inserted into a passage from a Latin version of the Gospel of Matthew as every seventh letter. The first task for a successful decryption, therefore, is to correctly extract these letters from the full text. The resulting block of 128 letters must then be decoded using a convoluted method to yield the final output.

Far from resolving the problem, however, the resulting message only serves to open a new set of questions. It comprises a string of recognisable French words and phrases, but the meaning is ambiguous at best, and incomprehensible without further interpretation.

The instructions for performing the decoding were made public for the first time in 1974 in Henry Lincoln's BBC documentary *The Priest, the Painter and the Devil*.[104] However, oddly, the final decoded message itself had already been published in the *Dossiers Secrets* in 1964, several years before the parchments themselves were revealed to the world in 1967.

The entire complicated process of decoding the encrypted text of the parchments, as outlined by Lincoln in 1974, and by others elsewhere, does not seem to help solve the mystery, but rather only deepens it. This is where the problem of the parchments has remained stuck for over half a century. How can we find the solution to this puzzle, when all

---

104 See Appendix 1 for a detailed analysis of the parchment texts.

attempts to date have failed? And how will we know if we have found the correct answer?

To begin, we need to ensure that we are asking the right question. Only then can we set foot on the path which will lead to the answer. So, what exactly is the mystery of these parchments?

Clearly, the parchments are not what they purport to be, but this does not mean that they are devoid of value and that we should merely label them as a hoax, discarding them as of no interest. Far from it. Rather, it might be more useful to consider whether they might be an example of what Philip K. Dick referred to as "fake fakes". Someone has clearly gone to a huge effort to create these fascinating documents. There must have been a reason to justify it, and probably more than one.

In fact, as we demonstrate in Appendix I later in these pages, the coded messages, the decoding process and the entire "stage business" of the text of the parchments are intended as distractions, designed to intrigue the reader and to deflect attention away from their true concealed content, the better to keep it protected for as long as possible.

The genuine message has not been camouflaged in the text. It is not comprised of letters or words but written in an entirely different language. The secret hidden beneath the fascinating surfaces of these mysterious documents is inscribed in the eternal language of geometry.

The Target for the Correct Solution

The first step in the quest to solve the puzzle of the parchments must be to establish, at the outset, the conditions which a successful solution must satisfy. This is the initial hurdle at which, frankly, all efforts to date have stumbled. Typically, the parchments have been approached as if they were some kind of treasure map, or equivalent, and contain encoded instructions designed to lead the intrepid researcher to a location where some precious or explosive secret lies concealed, waiting to be discovered. Yet all such considerations have led nowhere.

A completely different perspective is obviously required if this puzzle is to yield its solution. We need to have a clear idea of what it is we are trying to achieve when we set out to "solve" the parchments. In order to do so, it is necessary to go back to the source material and consider an intriguing reply from Pierre Plantard when he was questioned about the discovery of the parchments, as reported in Lincoln, Baigent and Leigh's *The Messianic Legacy*,[105] the sequel to *Holy Blood, Holy Grail*.

---

[105] Baigent, Lincoln and Leigh, *The Messianic Legacy*. (London, Jonathan Cape, 1986).

As Plantard recounted the story, Saunière found a total of four parchments under the altar during his renovation of the church. Three of these were genealogies and a "testament" of Blanche de Castille and the Hautpoul family. Then he described the fourth parchment as follows:

> "The fourth parchment, he said, was the original on the basis of which the Marquis de Chérisey had devised a modified version. According to M. Plantard, there was one coded message on each side of the page. In some way, apparently, the two texts interacted with each other. – if, for example, they were held up to the light and viewed, as it were, in superimposition. Indeed, it was suggested that M. de Chérisey's chief "modification" had simply been to reproduce the two sides of the same page as separate pages, and not to the original scale."[106]

This is a fascinating revelation and a deep clue. According to Plantard, the parchments were not separate and distinct documents in their original configuration, but two sides, front and back, of a single sheet of paper. He further suggested that the two designs interacted with each other in some manner when the page was held up to the light.

In reproducing the two sides onto separate pages, and altering the scale of one, the original relationship of both position and size between the two designs had been discarded and lost.

Here is the puzzle of the parchments, and it could not have been more clearly spelled out by Plantard. The problem is to *restore the parchments to their original double-sided layout and dimensions*.

This provides us with a clear benchmark test that any correct solution must satisfy. It must be capable of describing precisely how the two parchments interact as two sides of a single piece of paper, including their absolute and relative sizes, orientation and position.

That is the target which determines whether a solution to the parchments is valid or not. If a proposed solution can describe fully the relative and absolute size and positions of the two designs, it is correct and complete. If it fails to do so, it is not the solution. Nothing more, nothing less. Simple as that. With this as our clear-sighted goal, however modest it might appear before we set out, we proceed.

The parchments were described in Henry Lincoln's 1991 book *The Holy Place*. His account states plainly that they were created by Bigou in the 1780s and discovered by Saunière in the 1890s. To be fair on Lincoln,

---

106   Ibid, p. 245.

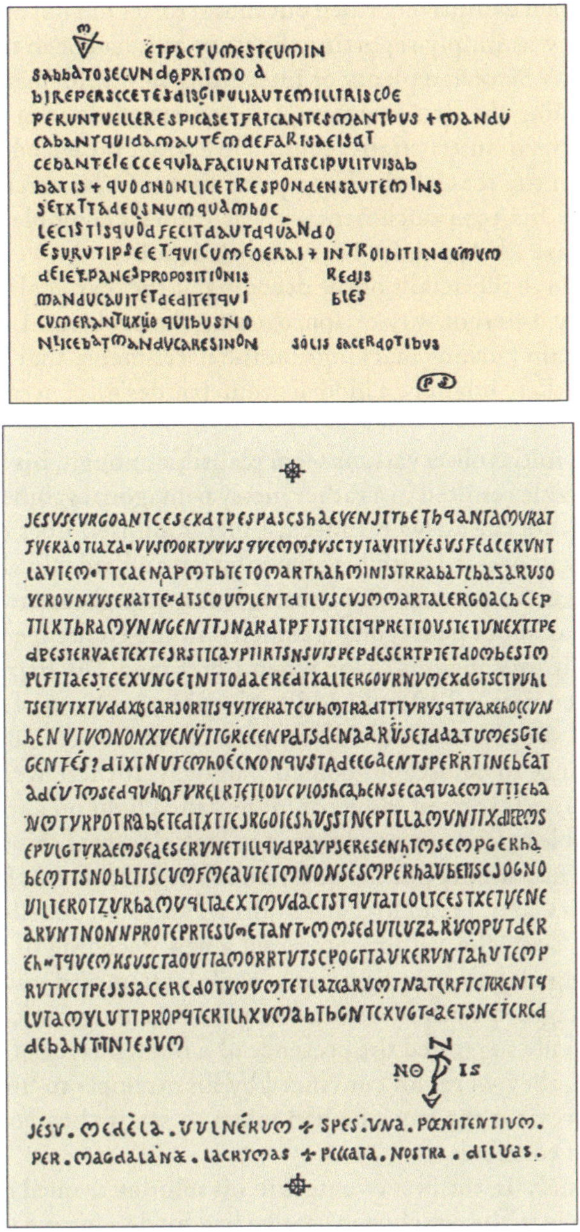
ETFACTVMESTEVMIN
SABBATOSECVNDOPRIMO A
BIREPERSCCETESDISCIPVLIAVTEMILLIRISCOE
PERVNTVELLERESPICASETFRICANTESMANIBVS + MANDV
CABANTQVIDAMAVTEMDEFARISAEISDT
CEBANTEIECCEQVIAFACIVNTDISCIPVLITVISAB
BATIS + QVODNONLICETRESPONDENSAVTEMINS
SETXTTADEQSNVMQVAMHOC
LECISTISQVODFECITDAVIDQVANDO
ESVRVTIPSEETQVICVMEOERAT + INTROIBITINDOMVM
DEIETPANESPROPOSITIONIS REDIS
MANDVCAVITETDEDITETQVI BLES
CVMERANTVXVO QVIBVSNO
NLICEBATMANDVCARESINON SOLIS SACERDOTIBVS

NO IS

JESV. MEDELA. VVLNERVM + SPES. VNA. POENITENTIVM.
PER. MAGDALANAE. LACRYMAS + PECCATA. NOSTRA. DILVAS.

Figure 75: Reproductions of the two parchments alleged to have been found by Abbé Saunière in 1896 in the church of Rennes-le-Château. In fact, they are twentieth century concoctions. Above: the "small" parchment. Below: the "large" parchment .

research which definitively ruled out those stories did not appear until later and he was simply reporting the story as it had been told. In any case, *The Holy Place* had plenty of interesting new information on the topic, including the first ever complete correct account in English of the decipherment instructions required to extract the encoded message concealed in the text of the large parchment. (The version he had presented in his 1974 documentary was (almost) complete, but not correct. Please see Appendix I for further details.)

In addition to the details of the decoding of the text, it also included a completely different way of approaching the problem. Lincoln had noticed certain curious markings on the parchments that led him to suspect that there might be a hidden geometric design concealed behind the text.

While his initial observations seem plausible enough, his attempt to solve the puzzle resulted in a rather messy pentagonal geometry which is obviously not the answer. Indeed, Lincoln himself does not appear particularly impressed with his own solution and quickly abandons it as the book gets underway, simply using it as a springboard to introduce his own ideas about pentagonal geometry in the landscape.

Frankly, it was obvious to me when I first read *The Holy Place* that Lincoln's geometrical efforts hadn't solved the problem of the parchments, so I simply overlooked this part of the book. I couldn't see their relevance to landscape geometry anyway. If these were modern concoctions, as it seemed, how could they shed light on ancient alignments? Unable to find any good reason to look closely at this element of the Affair, I had simply put the question of the parchments to one side.

Some years later, I read Richard Andrews and Paul Schellenberger's *The Tomb of God*. The authors proposed a new decoding of the parchments, taking as their starting point Lincoln's attempted solution from 1991. They agreed with his assessment that the enigmatic marks on the parchments suggested the presence of a hidden layer of geometry. But like me, they were not convinced by his attempts to find it. They decided to review the steps he had taken to see if they could find a mistake and correct it.

In their analysis, the first two steps in his solution seemed reasonable, but they came to the conclusion that he had made a wrong turn in the third step. They then laid out an alternative version of how the solution ought to proceed. When I looked at the proposal that the *Tomb of God* authors had suggested instead, I remained equally unconvinced. The outcome of their solution was equally as unsatisfying to me as Lincoln's.

If there was a genuine geometrical layer hidden within these mysterious parchments, then the solution, when it was eventually found, ought to be fully convincing and leave no room for doubt. Neither *The Holy Place* nor *The Tomb of God* had solved it, in my opinion. I forgot about it all over again.

Then, one day in 2008, flipping idly through the pages of *The Tomb of God*, I came across these passages about the parchments, the error that they had pointed out in Lincoln's attempt at the decoding, and their proposed correction.

As I stared at the page, I had another of those sudden moments of clarity that had so far punctuated my journey. I saw in that instant what that third step was intended to be. It was an electrifying feeling, because I could now sense that the complete decoding of the parchment might well be right in front of me. I quickly grabbed a pencil and ruler. Using the reproduction of the small parchment printed in *The Tomb of God*, I began marking lines directly onto the book itself. Ten minutes later, I was staring in disbelief at the solution to the hidden geometry of the small parchment.

Both Lincoln and *The Tomb of God* authors had indeed missed the correct turn-off, by the smallest of margins. They had both taken wrong turns, from which they did not recover. But the solution was so close at hand! Here, now, are presented the sequence of steps which reveal, for the first time, the hidden geometry of the small parchment.

The Solution to the Small Parchment

As *The Tomb of God* authors pointed out, the first two steps in Henry Lincoln's solution to the parchment geometry presented in *The Holy Place* seem reasonable enough, but the third is an obvious error.

After this, his geometrical construction goes astray and is, clearly, not the correct intended procedure. How can we be sure of this? Quite simply because his suggested resolution of the puzzle does not lead to any result of value.

Let's now review the opening moves he presented, identify and correct the false step, and then find the correct solution.

First, let's take stock of the layout of the small parchment. It consists of 14 lines of Latin text into which several additional elements have been added. There is a triangular geometrical device at upper left. There are three upright crosses intermingled within the text on lines 4, 7 and 10. Finally, there is glyph containing the letters P and S at lower right.

Step 1:

The presence of a geometrical puzzle in the small parchment is signalled by the small triangular device in the upper left quadrant, that does not appear to serve any obvious useful function in relation to the underlying text. Lincoln's first step was to notice that the upper, sloping line of this device, if extended, passes precisely through the small cross at the end of Line 4, as shown in Figure 76.

Observe the process: the puzzle-maker has alerted us to the possibility of a concealed geometry by inserting the small triangular device. This is his "call". He wants us to provide a "response" by searching for a reason for the odd design. When we consider whether it might be to extrapolate one of its lines across the page, he gives us a confirmation signal that we are on the right track, namely the fact that the continuation of the upper side of the triangle passes through the cross on line 4.

A dialogue has now been established between the puzzle-maker and the puzzle-solver. We can trust that there is a definite purpose here, and we are being led towards some destination. Encouraged by the successful opening of the conversation, we continue.

Step 2:

Next, we proceed to the second step described by Lincoln. He noticed that a line ruled from the cross in line 4 through the small cross in line 10 creates an angle of exactly 60° with the first line. This is the signal from the puzzle-maker confirming that it is the correct next step. We rule in the second line, as shown in Figure 77.

The first two steps are logical and sensible, and we can hear the voice of the puzzle-maker confirming this to us. So far, so good. We are all in agreement: Lincoln, the *Tomb of God* authors, and hopefully the reader!

The third step, however, is where both Lincoln and the *Tomb of God* authors tripped up. They both surmised that the two lines ruled in so far naturally suggest the two sides of a triangle. It seems a reasonable intuition to suppose that a triangle needs to be completed by forming a third side, beginning from a point on the second line towards the bottom of the parchment text somewhere, and terminating on or around the triangular device, thereby creating a tilted triangle overlaid on the text.

Lincoln's idea is to extend the left-hand edge of the small triangular device down and across the page until it intersects line 2. The result is an irregular triangle, with three unequal sides and angles.

It's not a bad guess, as far as it goes, but completing a triangle in this manner does not result in any obvious gain in the geometry, nor is there

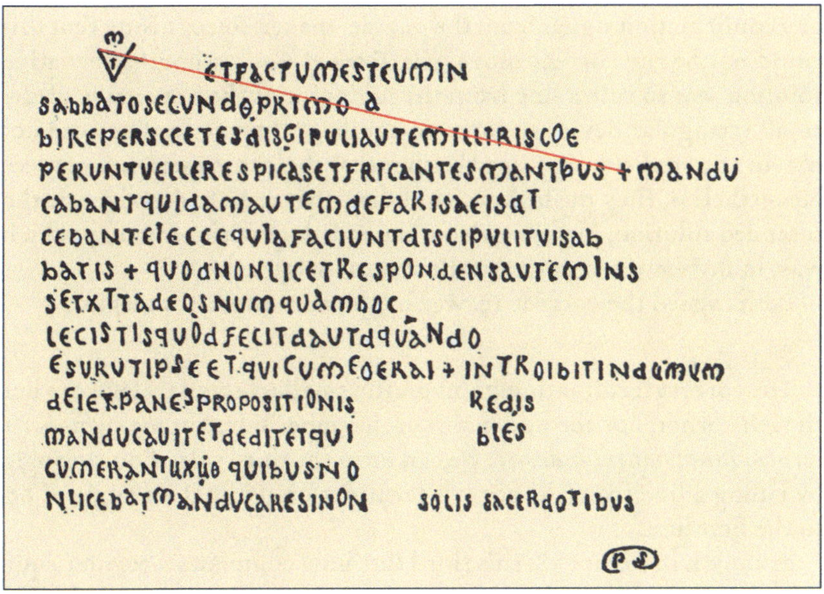

THE RIDDLE OF THE PARCHMENTS 249

Figure 76: Step 1: Rule a line along the extension of the upper side of the small triangular device at upper left. Note that it passes through the cross on line 4.

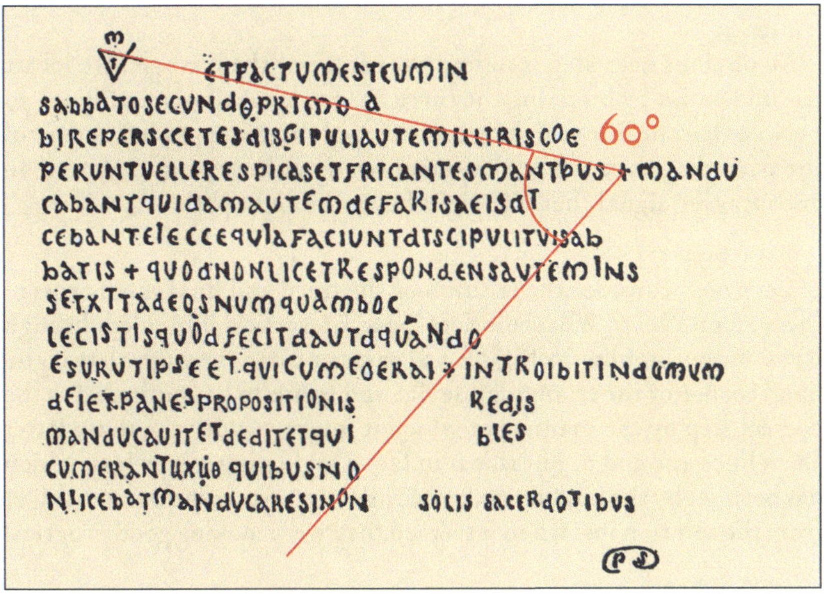

Figure 77: Step 2: A line ruled from the cross on line 4 through the cross on line 10 makes a 60° angle to the first line.

any confirmation signal from the puzzle-maker. Recognising that this could not be the correct move, the *Tomb of God* authors' alternative solution was to rule a line from the midpoint of the upper side of the small triangular device to form an equilateral triangle. They did not obtain any confirmation signal either that their third step was correct, Nevertheless, they pushed on, confident that they had arrived at the intended solution, but they also soon ended up with geometry which was, in my view, a figment of their own imagination.

Both missed the correct answer by the narrowest of margins.

Step 3:

The correct termination point on the small triangular device is not the left corner, nor the mid-point of the upper side, but *the right-hand corner of the triangular device*. We can form the third side of our triangle by ruling a line from the correct termination point at an angle of 60° to the first line.

As shown in Figure 78, this third line now completes a regular equilateral triangle. But how can we know if this is the correct, intended solution? We do not yet have any confirmation signal from the puzzle-maker, so, we proceed with caution.

Step 4:

An obvious next step is to find the centre of the triangle. We locate the mid-point by bisecting the three 60° angles as shown in Figure 79.

Note that the line which bisects the angle at the upper left corner of the triangle grazes the "PS" symbol neatly at lower right. Here is an encouraging signal that we might be on the right path.

False Step 5:

Let's now consider the position of the third and final small cross in the text, on Line 7. What role does it play? A vertical line ruled through it, as shown in blue in Figure 79, passes neatly through that right-hand corner of the triangular device and seems to be confirmed as the correct step by the manner in which it interacts with the three letter Ts on lines 4, 5 and 6. But this is in fact a highly ingenious trap which has been set by the puzzle-maker, designed to divert the puzzle-solver from the solution just when it seemed they were making good progress!

True Step 5:

A neat trick has been used here to conceal the next move. The correct line through the cross on Line 7 in fact passes along the upright stem of the letter T in the final row of text. It also passes along the I in the

THE RIDDLE OF THE PARCHMENTS

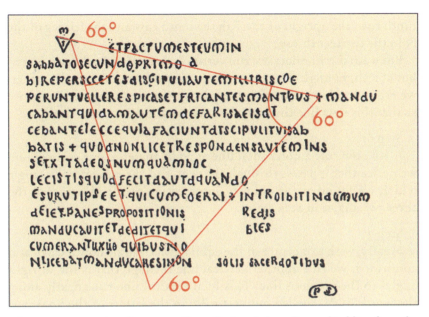

Figure 78: Step 3: Complete an equilateral triangle by ruling a third line from the right-hand corner of the triangular device, at 60° to the first line.

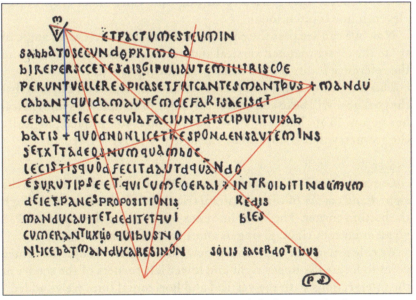

Figure 79: Step 4: Locate the centre of the triangle by bisecting the three 60° angles. False Step 5: It is tempting to rule a vertical line from the cross in line 7 through the corner of the triangular device, as shown. This appears to be confirmed by the alignment of the three letter "T"s over which it passes.

ninth row, the upright of the T in the tenth row and the left leg of the N in the thirteenth row.

But wait: this line does not run vertically in relation to the parchment, but at a slight angle. Here is our next signal from the puzzle-maker: we are to fix this by rotating the parchment a few degrees clockwise so that this new line through the cross on Line 7 runs vertically.

Step 6:
If we now rule a horizontal line at right angles to our vertical line, we notice that it passes through the centre of the equilateral triangle which we found at the point of intersection of the bisectors of the three 60° angles in Step 4.

Step 7:
Finally, with this centre of the equilateral triangle now located and confirmed, we rule another vertical line through that point, at right angles to the previous line. This final vertical line runs neatly along the word SION, spelled out vertically in the letters of the final four rows of text. The result after Steps 5, 6 and 7 is shown in Figure 80.

The alignment to the word SION validates that the grid of vertical and horizontal lines generated through the third small cross on the eleventh line is intentional.

Now we can see that the equilateral triangle is offset at an angle of 15° to the clearly defined vertical and horizontal axes. We have found the reference frame for the small parchment geometry.

This is a clean, sensible, accurate result. It is intuitively obvious that the goal now will be to add another triangle to complete a symmetrical hexagram. From here on, all we need to do is to complete, logically, the geometry that is implied by the figure we have arrived at.

Step 8:
Our aim now is to complete a square. We rule a vertical through the right-hand corner of the equilateral triangle and a horizontal through the bottom corner. The latter line runs neatly along the top of the P-S glyph at bottom right, giving us another confirmation signal.

We rule a diagonal at 45° to our new-found axes through the centre point to locate the upper right and lower left corners of the square at the intersections with the vertical and horizontal lines we've added.

Finally, we add horizontal and vertical lines through these new corners to complete the square, as shown in Figure 81.

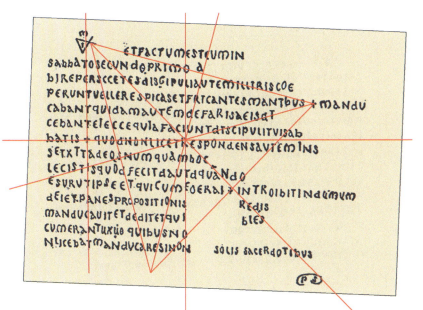

Figure 80: True Step 5: Rule a line through the cross in line 7 and the letter T in the last line. Rotate the parchment a few degrees clockwise so this line is vertical. Step 6: Rule a horizontal line through the cross at right angles to the vertical. Note that it passes through the centre of the triangle. Step 7: Rule a vertical line through the centre of the triangle. Note that it runs neatly along the word SION.

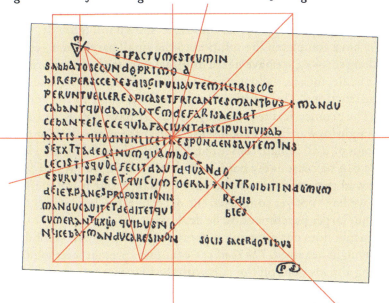

Figure 81: Step 8: Complete the square as shown.

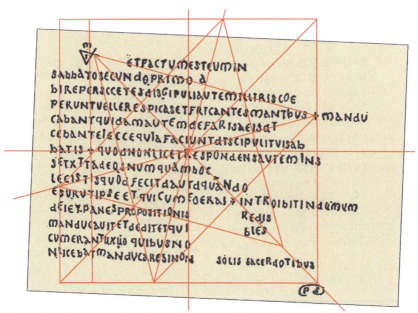

Figure 82: Step 9: Complete the hexagram by joining the points shown.

Step 9:

We have now arrived at the final step: we inscribe a second equilateral triangle, using symmetry, to form a hexagram which sits neatly and symmetrically framed by the square, as shown in Figure 82.

We have completed the solution to the small parchment geometry.

Along the way, we have heard multiple signals of confirmation from the puzzle-maker, validating that we have followed the intended set of instructions. The solution is a hexagram, offset to the vertical by 15°. The geometrical sequence by which we have arrived at this point is straightforward and makes sense even if the motivation or purpose or point of it all remains obscured. It will not remain this way for long.

We can be reasonably confident that we have, so far, successfully followed the intention of the puzzle-maker, but it is equally clear that we have not yet completed the overall task. There is still also the second, larger parchment, to be dealt with.

Lincoln does not offer any suggestions as to how this second parchment might be solved. *The Tomb of God* authors present their solution but having already gone off the rails with the first parchment, they remain lost in the long grass. Their proposed solution is simply not correct: it leads nowhere, offers no insight, and provides no confirmation.

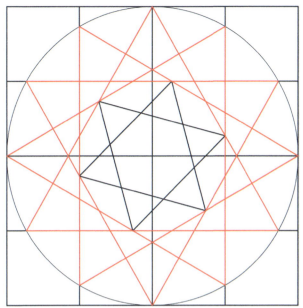

Figure 83: Geometry of the hexagram: *A circle drawn within a 4x4 square generates two hexagrams, shown in red. Another hexagram, in black, sits neatly within the centre 2x2 square, tilted at an angle of 15°.*

Now that we have the solution to the first parchment, we can proceed to investigate how this might interact with the second parchment, to arrive at the full and complete solution. Before we do that however, it will be helpful to review very briefly some simple properties of the geometry of the hexagram.

Geometry of the Hexagram

It is not necessarily obvious that a hexagram is generated whenever a circle is inscribed within a 4x4 square, as shown in Figure 83.

To demonstrate this, we construct a square and divide each side into quarters, to form a grid of 4x4. Now, we draw a circle of diameter equal to the side of the square. The points of a hexagram emerge naturally at the intersections of the circle with the quarter lines. The diagram demonstrates this simple geometric proposition.

Furthermore, this property holds, of course, regardless of orientation: a hexagram with vertices at top and bottom is generated using the intersection points of the circle and the horizontal quarter lines and a second hexagram can also be generated, with points at left and right, by using the vertical quarter lines. In this way a pair of rotationally symmetric hexagrams are generated, from four equilateral triangles. Notice that the horizontal and vertical sides of the four triangles are formed by the quarter lines dividing the large 4x4 square, and furthermore that these make up an 'inner square' of 2x2.

Observe that it is possible to fit a smaller hexagram neatly within this inner square, offset to the vertical by 15°. Where have we seen this familiar form before? It is of course the geometry of the small parchment, with the 15° offset hexagram sitting within the square that contains it!

An obvious thought now presents itself: if the small parchment is based around a 15° tilted hexagram sitting inside a square, and we are searching for a geometric solution which integrates the small parchment with the large parchment in some manner, could it be that Figure 83 provides the blueprint for the scheme we are looking for? Is the design of the larger parchment based around the geometry of the upright hexagrams? Let's explore this possibility.

First, let's inspect the result of superimposing the small parchment on the Figure 83 geometry, as shown in Figure 84. Immediately, we are rewarded with a signal: notice that the right-hand edge of the small triangular device aligns with the side of the large hexagram. This is unexpected, as so far, the larger hexagrams have played no part in the geometry we have extracted from the small parchment. Encouraged we push on to consider now the large parchment and how it might interact with the Figure 83 geometry.

The Large Parchment

Examining the large parchment, we see that it is comprised of several graphic elements. The main body of text forms a large block which is roughly square in shape. There are an additional two lines of text separated from the main body, which appear some distance below. There is a strange insignia or device at lower right incorporating the word SION. Finally, there are two square "target" markers, at the top and bottom of the page.

Considering the question of how the design of the large parchment might relate to the hexagram geometry, an immediate possibility presents itself. Could the "square" of text correspond in some manner to the inner 2x2 square of Figure 83?

To explore this possibility, we first generate a vertical axis for the large parchment by ruling a line through the two target markers. Now we take the Figure 83 geometry and superimpose it on the large

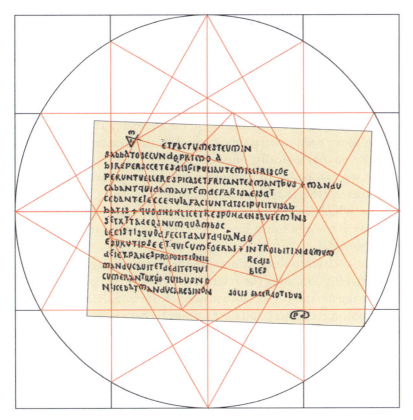

Figure 84: Superimposing the hexagram geometry of Figure 83 on the geometry of the small parchment. Note how the right edge of the triangular device aligns with the side of the larger triangle.

parchment, aligning it to the central vertical axis through the target markers that we've created.

Next, we adjust the relative sizes of the two designs until the right-hand edge of the text aligns with the right-hand side of the inner square, and adjust the relative vertical positions so that the top of the text block aligns with the upper side of the square. We notice now that the left side of the square runs neatly between the first and second letters of each line, an echo of the same geometric gesture that can be observed in the small parchment.

This seems promising. Having achieved a satisfying fit of the hexagram inner square to the left hand and top edges of the square of body text, we now notice something remarkable: the upper "target" marker falls exactly on the intersection of the two hexagram triangles on the central axis above the inner square, as shown in Figure 85. This is more than just a signal: it's a trumpet blast signalling that the geometrical code has been cracked.

The upper "target" marker is the confirmation from the puzzle-maker that our intuition about the presence of the larger, upright, outer pair of hexagrams is valid and perfectly correct. Notice, however, that there is nothing of significance in the geometry at the lower target marker, so we need to keep this in mind. If the first target records the geometry, so too, logically, must the second.

Now we can begin to glimpse how the two parchments might relate to each other, back to back, using the Figure 83 geometry as the template which unites them. This is a remarkable result, but before we can arrive at the finish line, there remain two outstanding issues to be resolved.

The first relates to absolute dimensions. It's one thing to have arrived at a format for obtaining the relative sizes of the two parchments, but what are the overall correct dimensions of the complete arrangement?

The second is the function of the other target marker at the bottom of the large parchment. What does this signify, if anything? So far, it does not fall on any significant geometrical point that we have found or generated.

This bothered me in 2008 when I stumbled upon the parchment solution as I have presented it so far. I tried various possibilities, but nothing worked. I confess I was thrown off for a time because I had fallen into the hidden trap cleverly set by the puzzle-maker which was described earlier in step 5. As a result, the frame of the geometry was slightly skewed, by just enough to obscure the final correct result. But eventually, I found the correct step 5, and the pieces fell into place. Let's now take another look at the pair of upright hexagrams generated in Figure 83.

Geometry of the Seal of Solomon

It is a natural next step to expand the basic hexagram geometry of Figure 83 by joining the outer points of intersection of the triangles, as shown in Figure 86. In this way, a new set of four outer triangles are generated which create a second, outer pair of hexagrams. If we consider them together, the two pairs of inner and outer hexagrams combine to create a pair of *Seals of Solomon*.

Now we have arrived at the full expansion of the basic hexagonal geometry. Notice that all of this has been generated by the

Figure 85: Superimpose the hexagram geometry on the large parchment, by aligning their vertical centre axes and adjusting their relative sizes so that the letter T at upper right defines the corner of the square. Note the upper target marker.

straightforward act of inscribing a circle on a 4x4 square. It is not a matter of arbitrary or personal choices of the geometer, but simply a question of completing the geometry naturally.

The Lower Target Marker

What happens when we add this new pair of outer hexagrams to the large parchment? When we complete the geometry in this manner and superimpose it onto the design we have already determined from Figure 85, we find – oh joy! – that the lower target marker falls exactly on the key new intersection created. This is shown in Figure 87.

The target marker alignment with the hexagram geometry is the next confirmation signal from the puzzle-maker. Two further observations confirm the solution we have arrived at. The second half of

Figure 86: *The geometry of Figure 83 can be extended naturally to create two further 'outer' hexagrams.*

the first of the two lines of text at the bottom of the page runs neatly along the lower edge of the "ribbon" of the Seal of Solomon. Finally, it is apparent that the strange SION device has been carefully positioned in relation to the geometry.

In light of all of this, we can be confident that we have indeed found the hidden framework that underpins the design. Before we can reassemble both of the parchments with the complete geometry however, one final task remains, namely, to determine the correct absolute size of the ensemble.

In order to do so, we need to turn our attention to another element of the enigma, the so-called gravestone slab of Marie de Nègres.

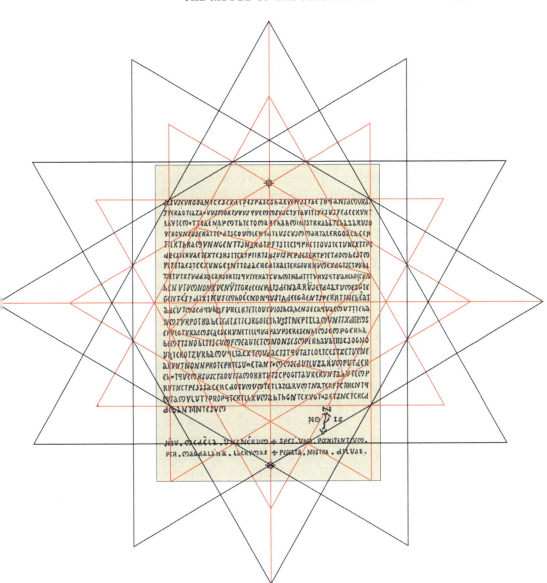

Figure 87: Superimposing the extended geometry accounts for the position of the lower target marker, the first line of the lower text, and the SION device.

The Marie de Nègres Gravestone

Pictured in Figure 88 is the alleged horizontal slab gravestone of Dame Marie de Nègres, as presented to the world in the *Dossiers Secrets*, the publications of the Priory of Sion deposited in the Bibliothèque nationale in the late 1960s. Like the parchments, this too is a modern concoction, and certainly does not date from the 1780s, as claimed when it was published in 1967 in de Sède's book. The extract in which it appears, which is said to have been reproduced from Eugène Stublein's *Engraved Stones of the Languedoc*, was created as part of the same stock of materials as the parchments and at the same time. The proof for these assertions includes the fact that no trace or reference to the book as a genuine volume has ever been located, and that Stublein's signature is incorrect.

If any sliver of doubt remains, it is eliminated by the simple observation that the Aude Historical Society, on a field trip in 1906, noted and sketched the upright gravestone of Marie de Nègres, which certainly *did* exist, but made no mention whatsoever of any horizontal slab. This is because it was devised as part of the parchment mystification in the early 1960s by the team around Pierre Plantard.

The concocted horizontal gravestone slab plays several crucial roles in the assembly of the Priory of Sion version of the Rennes story. Firstly, it contributes to the parchment text decoding process. It provides the key phrase "PS PRAECUM", which rounds out the collection of 119 letters on the (genuine) upright gravestone, to complete the full complement of 128 letters employed in the parchment text encoding.[107] Fabricating the design of the slab provided the opportunity for the puzzle master to choose this combination of nine letters to complete the 128-letter anagram of the gravestone engraving according to his wishes, to arrive at the famous "POMMES BLEUES" message.

But this is not its only function in contributing to the mystery. As we will see, it also duplicates and reinforces the concealed geometrical solution of the parchments and provides a critical piece of information as to dimensions.

Now that we have successfully extracted the Seal of Solomon geometry from the parchments and given that the gravestone slab was concocted alongside them, we might suspect that the same geometry could also be found here. And so it is. The design rapidly reveals itself on inspection. Ruling vertical lines through the centrelines of the two columns of text, and through the vertical midline of the full

107   See Appendix I for further details.

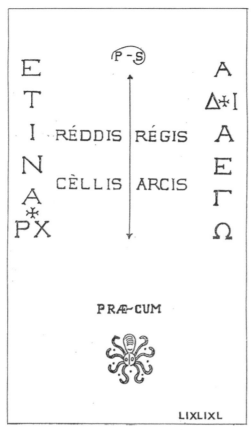

Figure 88: Alleged gravestone of Dame Marie de Nègres (d.1781). The image is said to come from 'Engraved Stones of the Languedoc' by Eugène Stublein, (1923) but no such book exists. In fact, the engraving was created in the 1960s. Image Credit: Pierre Jarnac, Mélanges Sulfureux : Les mystères de Rennes-le-Château, (Couiza, C.E.R.T. Centre d'Etudes et de Recherches Templières, 1995)

design, provides the initial orientation. It is a straightforward task to complete the inner square, and now the rest of the hexagram/Seal of Solomon design falls into place readily, as can be seen in the accompanying Figure 89. There are multiple clear indications that the familiar geometry is present here, too. And again, there is a loud signal from the puzzle-maker that we have indeed located and revealed the concealed geometrical design. The confirmation is the clearly defined top edge of the drawing of the slab.

The slab could have been drawn to be any rectangular proportions, or dimension, but the puzzle-maker has ensured that the top edge coincides with the upper side of the outer triangle in the Seal of Solomon. As always, this is perhaps more easily grasped by reference to the diagram. It is yet again an unmistakable sign from the designer of this enigmatic puzzle that the solution we have arrived at is correct.

Now we can proceed to resolve the one final detail that remains to be determined, namely the absolute dimensions of the two parchments. We know their relative sizes, as together they combine to form the overall hexagram geometry as a matched pair, but this requirement could of course be satisfied at any scale. We need to find an indication which will help us to fix their absolute size. The gravestone slab provides the vital clue that we are looking for.

The document in which the Marie de Nègres gravestone appeared was the fake extracts of Eugène Stublein's *Pierres gravées du Languedoc* published as part of the *Dossiers Secrets* series in June 1966. These was reproduced in the collection *Mélanges Sulfureux: Les mystères de Rennes-le-Château*, with texts re-assembled by Pierre Jarnac, and published by C.E.R.T in 1994.[108] This facsimile reproduction of the gravestone at its original size permits us to retrieve the units of measure of the design. This was not possible in the case of the parchments as they were altered in their relative and absolute sizes when reproduced.

When we do so, we find that the separation between the midlines of the two vertical rows of text is four inches. The Marie de Nègres gravestone slab geometry is laid out in *inches*.

The inner square that contains the tilted hexagram has sides of four inches in length. The complete inch-grid geometry overlaid on the design is shown in Figure 89.

At the bottom right-hand corner of the gravestone is a line of Roman numerals that reads: LIXLIXL. For years I have puzzled over what these enigmatic letters could possibly represent. It wasn't until 2019 when I finally noticed what had eluded me all that time. LIXLIXL is a symbolic *inch ruler,* as shown in Figure 90. The letters are not intended to be read as letters *per se*, but as graphic elements. The forms of the letters mark off segments of one-eighth inches for a total length of 7/8 inches.

LIXLIXL plays the role of a reference measure that establishes the inch, beyond question, as the unit by which the geometry is laid out. This is the thread that runs through the entire collection of documents, linking them together. The purpose of the gravestone drawing is to record and confirm the correct unit of measure, and thereby to

---

108   In *Mélanges Sulfureux* op cit.

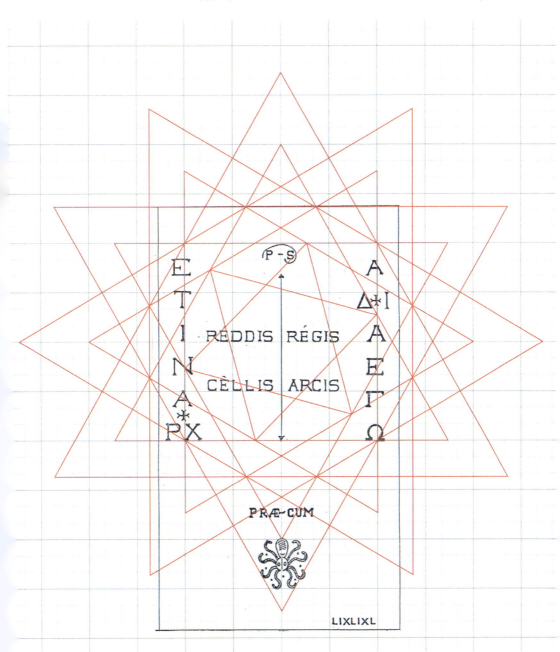

Figure 89: The parchment geometry on the Marie de Nègres gravestone. The dark lines of the construction grid mark one inch squares on document at original size. These are further sub-divided by light lines into squares of one-third inches. Note how the heights of the letters arranged vertically conform to the marked grid.

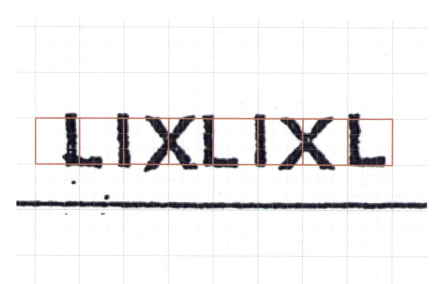

Figure 90: The secret scale on the Marie de Nègres gravestone. The mysterious LIXLIXL inscription is revealed as an inch ruler formed by the shape of the letters. At original scale, the sides of the red squares measure 1/8 inches, and the small grey squares 1/24 inches. As can be seen, the letters, read as graphic elements rather than their alphabetic value, mark out 7 red squares of 1/8 inches.

communicate that this is the measure also employed in its companion documents, the parchment pair.

LIXLIXL echoes Boudet's use of the inch in the map title, and the arabesque on the title page. It also plays the same role as the Legend at the bottom right of Boudet's map, in the same location, which also functioned to record the inch measure and its subdivisions. The parchments and the gravestone drawing and the Boudet map share an identical hidden geometry, and they are based on the inch.

Now we can return to put the finishing touch to the solution of the parchments. The same geometry is found on the gravestone, where as we have seen, it is laid out in inches. As the parchments and the gravestone come from the same cache of materials, they are linked to the same source. Therefore we can confidently conclude that the parchments too must be laid out in inches. The only remaining question is scale. An inner square of four inches is too small to realistically define the size of the parchments. How did the puzzle maker resolve this dilemma? Simply by doubling.

The "inner square" - the framework for the small parchment hexagram, and the large parchment text - is a square of eight inches by eight inches. Figure 91 shows the full and complete solution to the parchment

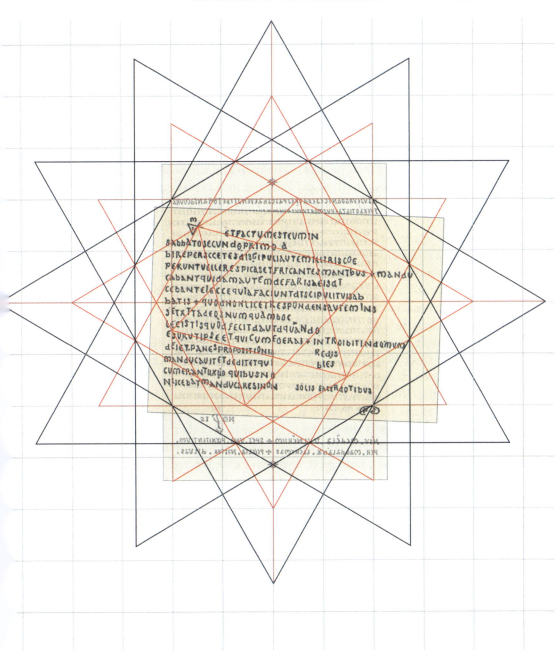

Figure 91: The two parchments, at correct relative sizes and positions, back-to-back, with the hidden geometry shown. This diagram has been reduced from its absolute dimensions to fit the pages of this book. At full size, the reference grid that has been superimposed consists of squares with sides of two inches.

geometry with the correct scale shown by the grid. From geometrical considerations, the distance between the two target markers at this size will be 10.92 inches.

According to the criteria we established at the beginning of this chapter, this answers the riddle of the parchments. We have shown how the two designs interact with each other, on opposite sides of the same piece of paper, with complete details of their relative dimensions and positions. The geometrical puzzle is solved. The two hidden secrets revealed by the parchments are the inch and the Seal of Solomon.

Now, what does it all mean? Why the inch? How could this humble unit of measure play such a significant role in the solution to this mystery? It appears that we might have stumbled upon an invisible thread in the rich tapestry of esoteric history, namely, the use of the inch in concealed geometry.

Why would this be? What is the underlying reason for such an unexpected choice of units, against the demands of the cultural and historical forces that would usually dictate the use of the metric system exclusively in such circumstances?

Keeping in mind that France has mandated the use of the metric system since Napoleonic times, it is evident that some consideration other than common usage must have informed this choice of measure for the parchments, gravestone and indeed for the Boudet map. Indeed, the use of the inch in France is subject to legal penalty, so there can be no reason to expect the design to resolve so cleanly into whole units of this measure, rather than centimetres or millimetres.

There remains a rather extraordinary clue located within the parchments and the Boudet map itself that provides an unexpected and satisfying confirmation that our identification of the role of the inch in the design of these puzzles is correct and fully intended by the puzzle-maker. The clue is concealed in an ingenious manner and provides a final flourish to this most elegant and well-constructed of geometric puzzles. The clue is in the *script*.

The Uncial Script

One of the curiosities, or clues, which exposes the parchments as modern concoctions, rather than eighteenth-century originals, is the hand in which they have been written. Both parchments have been written out in "uncial" script, a form of writing which was commonly employed by Latin or Greek scribes in the fourth to eighth centuries, but which had tapered off by the tenth century. After this date, it fell

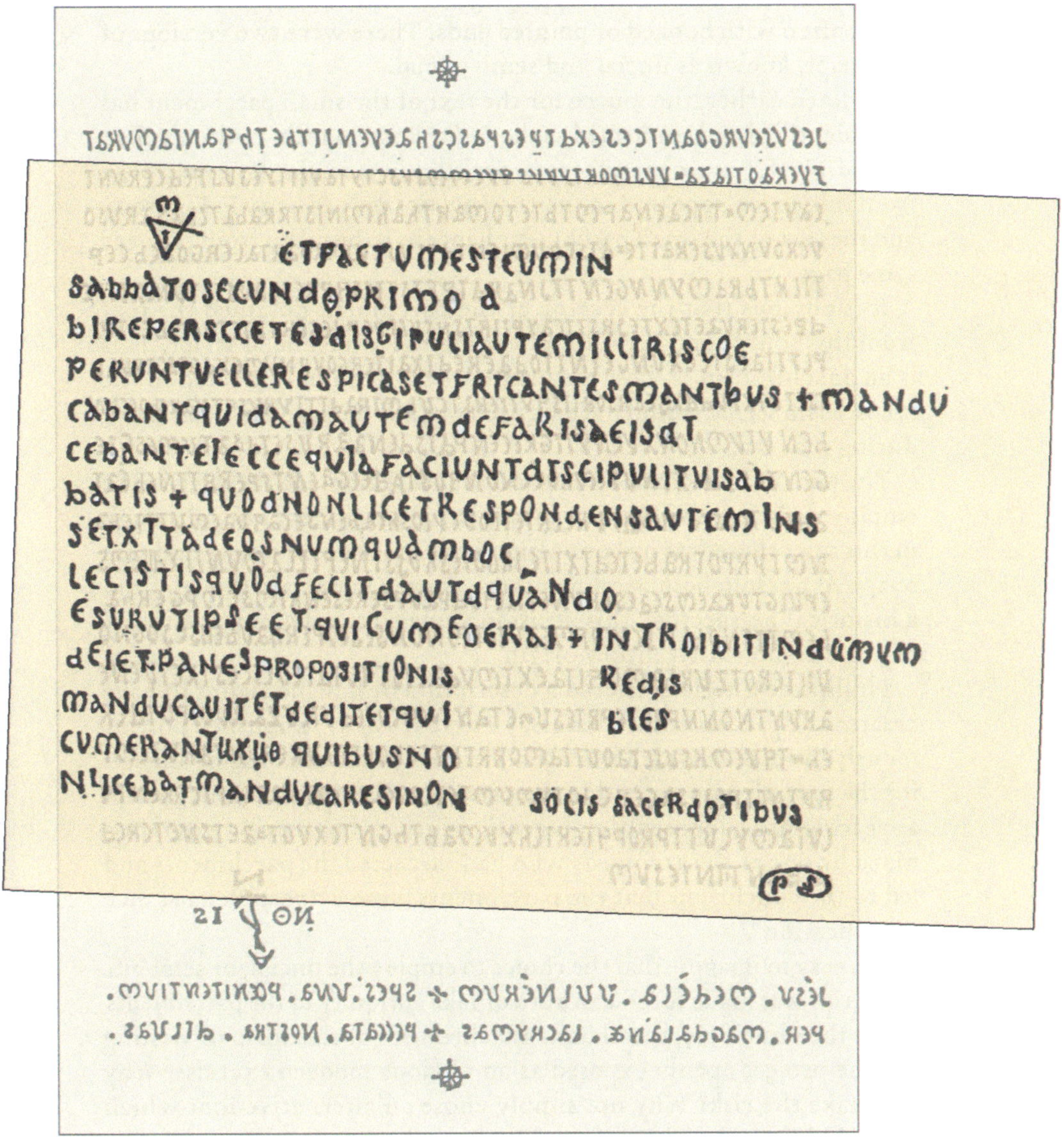

ETFACTUMESTEUMIN
sabbatosecundoprimo a
bIREPERSCCETESAISCIPULIAUTEMILLIRISCOE
PERUNTUELLERESPICASETFRICANTESMANTbUS + MANdU
CabaNTQUIdaMAUTEMdeFARISAEISAT
CEBANTEIECCEQUIAFACIUNTdISCIPULITUISAB
baTIS + QUOdNONLICETRESPONdENSAUTEMINS
SETXTTAdEQSNUMQUÃMHOC
LECISTISQUOdFECITAaUTdQUÃNdO
ESURUTIPSEETQUICUMEOERAI + INTROIBITINDÜMUM
dEIET.PANESPROPOSITIONIS
MANdUCAUITETdEdITETQUI   REdIS
CUMERANTUXŨ QUIBUSNO   bIES
NLICEBATMANdUCARESINON SOLIS SACERdOTIBUS

Figure 92: The two parchments, at correct relative sizes and positions, back-to-back, without geometry. These are reproduced at reduced scale above because of the size limitations of the book. At full correct size, the two target markers will be 10.92 inches apart as noted.

into disuse. Uncial, or semi-uncial script as it was sometime known, consists of majuscule (i.e., upper case) letters that have a rounded form, often with hooked or pointed ends. There were two versions of the script, known as uncial and semi-uncial.

As noted earlier, the source for the text of the small parchment has been identified and confirmed as the *Codex Bezae*, a fifth-century collection of New Testament excerpts, written in uncials. A sample page from this codex appeared in the *Dictionnaire de la Bible* by F. Vigoroux and the text of the small parchment happens to be taken from the very same page.

It is obvious that the creator of the parchment obtained the text from this source and copied it out using the same uncial script. This is an anachronism if these documents were genuinely created in the 1780s, as no scribe would have written out any passage in uncial script during that era.

The person who created the parchments then made the decision to employ the same uncial script for the large parchment also, although in this case, there was no connection between the text source and that script. As a result, both parchments are written in uncials, though from a historical perspective, this decision alone eliminates any possibility of a late eighteenth-century origin.

Phillipe de Chérisey offered a description of how he claims to have created the parchments in his essay *Pierre et Papier*, published posthumously in 1988, in which he makes mention of the use of semi-uncials for the text. He also notes that the presence of an eight-century script appearing in a supposedly eighteenth-century manuscript was recognized by one of the archivists who had inspected the parchments and led to the conclusion that the parchments were a "post-Renaissance mystification".[109]

It is easy to imagine that the choice to employ the uncial, or semi-uncial script was made to lend an aura of fake antiquity to the parchments, but at the same time, the blatant anachronism of this form of writing left the project openly exposed as an obvious modern exercise. Why then take the risk? Why not simply chose an alternative font which offered historical verisimilitude? Perhaps there was another reason.

The word uncial has a secondary definition: it also means "relating to an inch, or an ounce". Indeed, the words are substantially identical in their root forms: uncial/inch/ounce. This secondary definition

---

109 Essay in Jean-Luc Chaumeil, *Le Testament du Prieuré de Sion : Le Crépuscule d'une Ténébreuse Affaire*. (Villeneuve de la Raho, Pégase, 2006), op. cit. p. 97.

has also crossed over into the first definition; the uncial script is often spoken as being made up of letters that are an inch high. Similarly, the semi-uncial script was often considered to consist of letters of half an inch in height. In practical terms, the direct association is loose, and letters written in an uncial script were not necessarily one inch in size, but the association has persisted. For example, William P. Hatch in his 1935 essay on "The Origin and Meaning of the Term 'Uncial'", stated:

> "What did the word *uncialis* originally mean when it was applied to letters? It is obviously derived from the substantive *uncia* and it is generally believed to have denoted letters measuring an inch in length or height."[110]

The association between the word *uncial*, describing the script, and the word *inch*, as denoting inch-high letters, has persisted in the literature, and is a well-attested notion. It is intriguing therefore to observe that the parchments are written in uncials, and that they also conceal a geometry and grid that is laid out in inches. Could the choice of script have been informed or influenced by the secret of the concealed measure?

With this question in mind, let's turn attention back to the Boudet map, which is also laid out in inches. In this case, the letters of the title, *Rennes Celtique*, are also precisely one half-inch in height, so that, in the literal secondary meaning of the term, they are indeed "semi-uncial".

Let us pay closer attention now to the forms of the letters themselves, which are very curious, and unlike any usual form of lettering from the 1880s or any other era. They are all upper-case. They have rounded forms. And they feature hooked or pointed ends. But wait: these are precisely the identifying features of uncial script!

It is apparent that Boudet himself has created his own bespoke version of an uncial, or semi-uncial script! The reason for the strange appearance of the letters is that they have been deliberately designed to mimic, or invoke, the notion of uncial script. In this case, the purpose has nothing to do with creating any false aura of antiquity, and everything to do with the secondary meaning of the term. These are indeed semi-uncial letters, in both senses of the word. They imitate the rounded, hooked form of the upper-case uncial script, and they are written to be one half inch in height. Boudet's specially designed script is his signal to us that our recognition of the importance of the

---

110   William P. Hatch, "The Origin and Meaning of the Term 'Uncial'", in Classical Philology, Vol. 30, N. 03, July 1935, p. 251.

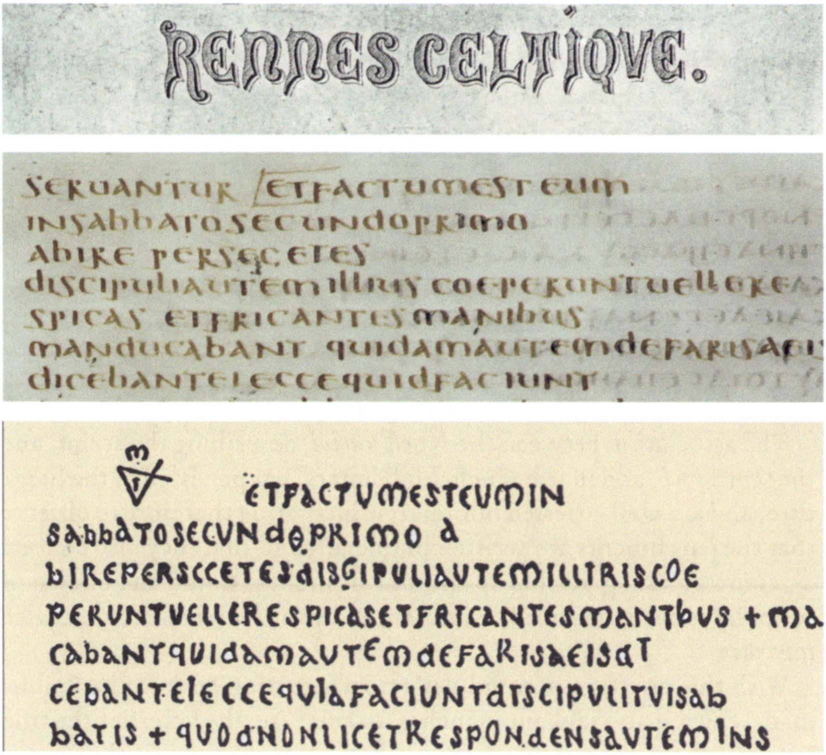

RENNES CELTIQUE.

SERUANTUR ETFACTUMESTEUM
INSAbbATOSECUNDOPRIMO
AbIRE PERSEGETES
DISCIPULIAUTEMILLIUS COEPERUNTUELLERE
SPICAS ETFRICANTES MANIBUS
MANDUCABANT QUIDAMAUTEMDEFARISAEI
DICEbANTELECCEQUIDFACIUNT

ETFACTUMESTEUMIN
SAbbATOSECUNDOPRIMO A
bIREPERSCCETESdISCIPULIAUTEMILLIRISCOE
PERUNTUELLERESPICASETFRICANTESMANTBUS + MA
CAbANTQUIDAMAUTEMDEFARISAEISdT
CEbANTELECCEQUIAFACIUNTdISCIPULITVISAB
bATIS + QUODNONLICETRESPONDENSAUTEMINS

Figure 93: Uncial and semi-uncial script
Above: Title from the Boudet map in custom font, imitating uncial or semi-uncial script. *Centre:* Extract from Codex Bezae in the Grand Larousse Bible Dictionary written in uncial script. *Below:* Extract from small parchment, written in imitation semi-uncial script.

inch, or half-inch, in the layout and design of the map is of prime importance! Now the reason for the choice of uncial script for the parchments becomes clear. It was a reference to Boudet's title, and a major clue to the concealed secret of both the parchments and his map.

Further, it provides a tangible link and connection between the two sets of documents, the large and small parchment on the one hand, and Boudet's book and map on the other. It is becoming apparent that the two sets are intimately related. Both share the same hidden secret, namely the concealed geometry based on the inch. The parchments are simply an alternative rendering of the same underlying idea behind the Boudet map. The parchments, indeed, are a map.

The initial observation that the word *Rennes* in the title of Boudet's map measures two inches in length has released the hidden content concealed beneath the surface of his astonishing book. The key is

Figure 94: More semi-uncials: The letters on the covers of both books measure one half inch in height. *Above:* Extract from cover of Jean Richer's Géographie Sacrée du Monde Grec.(Paris, Hachette, 1967). *Below:* Extract from cover of the Belfond Press reprint of La Vraie Langue Celtique et Le Cromleck de Rennes-les-Bains, (Paris, Belfond, 1978).

contained in the words "semi-uncial", which describe both the script in which the parchments were written and the literal size of the letters.

When Jean Richer's *Géographie Sacrée du Monde Grec* was published in 1967 with its blue cloth cover and gold letters, the book that de Chérisey said was being used by tourists as a guidebook to Rennes-les-Bains "well before 1978", it is intriguing to note that the title was printed on the cover in letters which were precisely half an inch in height!

It almost seems too good to be true that Richer could have left such a strong clue pointing to Boudet's semi-uncial trick. Could it just be a co-incidence? Perhaps, except that the title on the cover of the 1978 republication of *La Vraie Langue Celtique* published by Belfond, with the introduction by Pierre Plantard, is also laid out in letters which are, again, exactly half an inch in height. The two "semi-uncial" titles are shown in Figure 94. The conclusion is irresistible: Richer and Plantard both knew the secret of Boudet's map, the semi-uncial, and neither could resist leaving a tantalising witness hidden in plain sight on the covers of their respective books.

Maps and Parchments

We have seen that the parchments also conceal a hidden layer, and that this too is based on a grid of inches. This opens the possibility of an intentional equivalence between the parchments and the map. We are beginning to see that the parchments have been designed to replicate the concealed layer of the Boudet map. From this we can further conclude that whoever designed the parchment geometry must have known and understood the secrets of the Boudet map geometry.

This observation raises further questions. Can the parchment solution help to shed any further light on the secrets of Boudet's map? Are the geometrical elements that have been revealed in the parchments also found in the Boudet map? We know that they share the inch grid, but what about the Seal of Solomon?

The Seal of Solomon geometry on the Marie de Nègres gravestone has an inner square as we've seen of four by four inches. This corresponds to an outer square of eight by eight inches. The circle drawn on this outer square, with radius of four inches, or diameter of eight inches, then generates the first hexagram.

This is a familiar circle: it is of course the same dimension as the circle of Boudet's cromlech on the 1:25,000 scale map.

A hexagram is always generated by implication as soon as a circle is erected on a 4x4 square. If we divide up the eight-inch square into two-inch squares, to create a 4x4 grid, then the cromlech circle naturally generates the required hexagram. Hence, the circle of the cromlech on the grid of inch squares implies the hexagram geometry, without requiring any additional embellishment.

When we reproduce the Seal of Solomon geometry at the scale of the Marie de Nègres gravestone on the Boudet map, we find it is already present. The same design at the same size and measure is implied by the circle of the cromlech on the grid of inches.

The four-inch radius circle centred on Rennes-les-Bains on the grid of inches generates the same hexagram geometry that has been found on the gravestone, and on the parchments at twice the scale. The accompanying Figure 95 shows these relationships. The parchments have been added, reduced to half scale, conforming to the geometry. The circle that encompasses the Seal of Solomon on the gravestone and

Figure 95: Boudet's map with concealed inch grid and his cromlech depicted as two circles of radii 4 and 4 ½ inches, centred on Rennes-les-Bains Church. The 4 inch radius circle on the inch grid generates a hexagram. The extended hexagram geometry and the parchments (at half scale) have been overlaid.

parchments is the same as the circle of the cromlech.

Meridians on the Parchment

Now that the Seal of Solomon and the inch-grid have been revealed in the Boudet map, demonstrating that it is based on the same geometry as the parchments, we can overlay these different formats at correct consistent relative sizes. That is, we can take the parchments and overlay them on the Boudet map, or the IGN 1:25,000 scale map for that matter, at appropriate relative sizes such that the inch grid template underlying the parchment design corresponds to the inch grid template of the Boudet map. Effectively, we have turned the parchments into a map, which is precisely what they are.

We've already observed that three meridians correspond with gridlines on the map, but there is one more to be discussed. The longitude of the famous meridian, marked in St Sulpice church in Paris, is 2°20′5.7″. If we rule this line on the Boudet map, it is virtually indistinguishable from the grid line falling between the letters "i" and "q" of *Rennes Celtique*.[111]

Notice also that this St Sulpice meridian corresponds to the right-hand side of the square in the hidden geometry of the small parchment, the line marked with the "PS" sign at its lower end. The purpose of this curious glyph stands revealed.

In the Scorpio stanza of *Le Serpent Rouge*, the poet wrote:

> "...the Meridian line, in the very choir of the sanctuary from which this source of love for one another radiates, I turn around looking from the rose of the P to that of the S, then from the S to the P."

The glyph marks the St Sulpice meridian on the parchment considered as an analogue of Boudet's map with its hidden grid. As we have discussed earlier, the path in *Le Serpent Rouge* touches the St Sulpice meridian for the first time precisely in the Scorpio segment, which is the stanza in the poem in which it is mentioned. Jean Richer had evidently plotted the St Sulpice meridian on his map. He noticed that the walking trail circuit touches this line and briefly follows it in the Scorpio segment of his zodiac.

We can only assume that he also noticed that it happened to coincide with one of the grid lines of Boudet's map. By setting this stanza in

111   The grid line 2 inches east of the Rennes-les-Bains meridian (2°19′10″E) is on longitude of 2°20′06″E. The St Sulpice meridian differs from this value by an insignificant 0.3 arc-seconds, and corresponds to just 22 feet in the landscape.

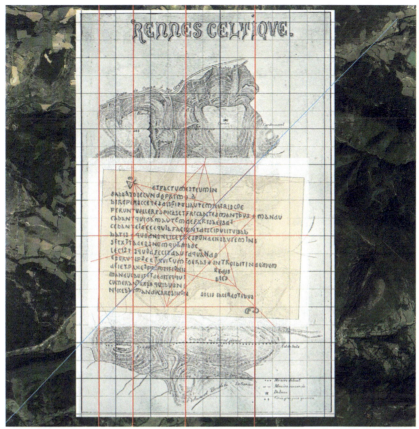

Figure 96: *The layers can now be combined by simple scaling. The figure shows the superimposition of the small parchment onto Boudet's map, with grid, on Google Earth. Note the St Sulpice meridian at right, (on the grid line touching the I of the title) and its interaction with the PS glyph on the parchment.*

the church of St Sulpice in Paris itself, the meridian is brought into the poem naturally, in the correct place, without drawing attention to the concealed reference.

This does not lead to any suggestion that the St Sulpice meridian was known in ancient times. According to the historians of the church, it is a modern creation and, Dan Brown's novel notwithstanding, there is no evidence that I'm aware of to suggest otherwise. The fact that it happens to very nearly align with one of Boudet's gridlines may be considered a lucky strike.

This self-authenticating conjunction of elements ties together the meridians, the geometry, the poem and the parchments into a single, cohesive, unified design. They are all expressions of the same format.

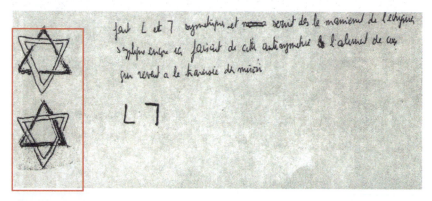

Figure 97: Trails of confirmation.
Above: Extract from the Dossiers Secrets.
Image: Pierre Jarnac, *Mélanges Sulfureux : Les mystères de Rennes-le-Château*, (Couiza, C.E.R.T. Centre d'Etudes et de Recherches Templières, 1995).
Below: Marginalia from the manuscript of de Chérisey's Pierre et Papier.
Image: Jean-Luc Chaumeil, *Le Testament du Prieuré de Sion : Le Crépuscule d'une Ténébreuse Affaire* (Villeneuve de la Raho, Pégase, 2006).

They are all maps of the ancient landscape geometry of Rennes.

Confirmation of the Parchment Solution from the Team

The secret of the geometry of the parchments has been revealed as a Seal of Solomon. Is there any confirmation of this result from the Team on record? There is. In multiple locations. Firstly, de Sède stated it plainly all the way back in *L'Or de Rennes* in 1967, when he said of the parchments that

> "In addition to the coding proper, the author has added rebuses"[112]

A rebus is a picture puzzle. He could not be spelling it out more plainly. And what exactly are these rebuses? Elsewhere, the answer has been provided. As shown in Figure 97, in one of the documents amongst the *Dossiers Secrets*, next to a small sketch of a Seal of Solomon

---
112   Gérard de Sède, *L'Or de Rennes* op. cit. p.103.

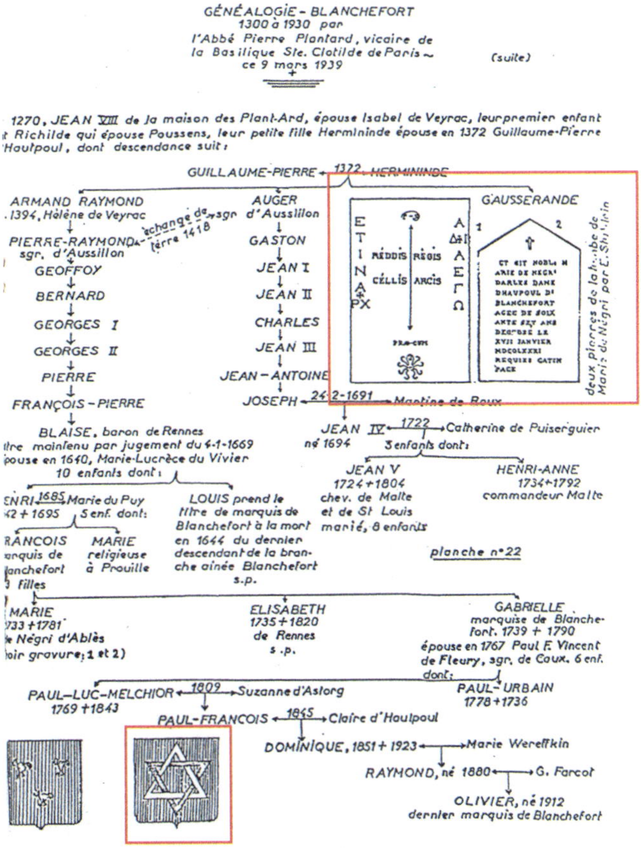

THE RIDDLE OF THE PARCHMENTS 279

Figure 98: Page from the Dossiers Secrets. <u>Upper right:</u> *The fake and genuine Marie de Nègres gravestones.* <u>Bottom left:</u> *Coat-of-arms with Seal of Solomon made from triangles of two colours, echoing the hidden design of the parchments. Image: Pierre Jarnac,* Mélanges Sulfureux : Les mystères de Rennes-le-Château, *(Couiza, C.E.R.T. Centre d'Etudes et de Recherches Templières, 1995).*

appears the following words in French, which state the plain truth:

> "Six doors or the Seal of the Star. Here are the secrets of the parchments of Abbé Saunière."[113]

Some deep hints are also to be found in the hand-written manuscript *Pierre et Papier* by Philippe de Chérisey.[114] The author has doodled several Seals of Solomon in the margins, also shown in Figure 97.

---

113   *Le Serpent Rouge.* p. 5.
114   Reproduced in Jean-Luc Chaumeil *Le Testament du Prieuré de Sion : Le Crépuscule d'une Ténébreuse Affaire.* (Villeneuve de la Raho, Pégase, 2006).

Notice that these Seals of Solomon are depicted with intertwined dark and light triangular bands.

Given that the ostensible purpose of this paper is to disclose the details of the parchment decoding, these are very telling additions. These marginalia, far from being merely decorative, offer telling clues as to the specific intended meaning of these glyphs, as we will be discussed in more detail in Chapter 15.

Elsewhere in the *Dossiers Secrets*, on a page reproducing various genealogies relating to certain noble houses of France, the two gravestones (one fake, one real) of Marie de Nègres are depicted (see Figure 98). At the bottom of the same page is a coat of arms that bears the Seal of Solomon. Yet again, it is shown in the format of two triangular ribbons in contrasting colours. Here is the hidden solution of the parchments, displayed openly, in juxtaposition with the false gravestone, in plain sight. A very similar heraldic device appears in *Holy Blood Holy Grail*, where it is identified as the coat-of-arms of Rennes-le-Château.[115] The same design of the Seal of Solomon can also be seen carved into the woodwork in the Château Hautpoul in the village in a recently published photograph online.[116] It is depicted next to the coat-of-arms of the Hautpoul family and Marie de Nègre d'Ables.

But it is Gérard de Sède yet again who leaves the most blatant confirmation on record of the solution to the parchments. In 1977, he re-issued his original 1967 book, *L'Or de Rennes*, under a new title: *Signé: Rose + Croix*. In English, this translates as: "Signed: Rose + Cross". This is a very curious title to choose as it does not seem to have any explicit relevance to the Affair. What could he have meant by this? Why would he say of the affair of Rennes that it is "signed" by the addition of a rose and a cross?

He offered a clue the following year. In 1978 he published a new book, *La Rose-Croix,* about the history of the Rosicrucian order. The text contains no reference whatsoever to the Rennes affair but as previously pointed out, it does contain a footnote which mentions Jean Richer and *Géographie Sacrée du Monde Grec*. Just a few pages after Richer's name, the illustration shown in Figure 99 appears. Captioned the "Rose of the Winds on the Cross of the Seasons" the accompanying text explains:

> "Around the Cross of the Seasons is found the crown of the twelve signs of the zodiac. Now this figure, which it is necessary to know how to orient correctly, bears the name

---

115 Lincoln, Baigent, Leigh *Holy Blood Holy Grail* op.cit. Figure 3.
116 Online here: http://www.rennes-le-chateau-archive.com/hautpoul.php

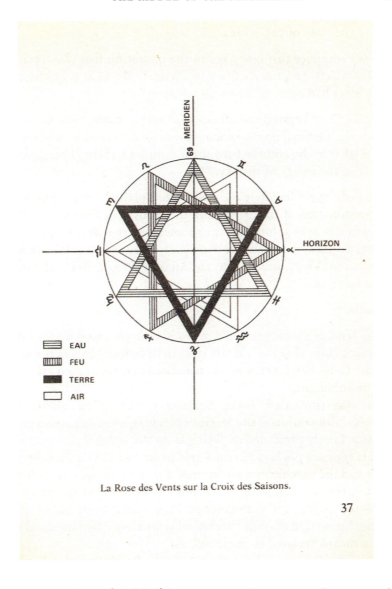

La Rose des Vents sur la Croix des Saisons.

Figure 99: Page from Gérard de Sède's *La Rose-Croix*. "*La Rose des Vents sur la Croix des Saisons*" may be translated as the "*Rose of the Winds on the Cross of the Seasons*". This image of the zodiac and the seasons as the rose and the cross echoes the title of his 1977 work "*Signé: Rose + Croix*", or *Signed: Rose + Cross* According to de Sède's title, the Affair of Rennes-le-Château has been "signé", or signed, with two Seals of Solomon on a correctly oriented zodiac. And indeed, this is the precise description and depiction of the solution to the combined problems of the parchments, the map and the landscape geometry.

the Rose of the Winds."[117]

If we compare this language to the quotation from his 1966 book *Le Trésor Cathare* quoted earlier in Chapter Eleven it is apparent that these ideas had been on his mind all along:

> "The keystone of all celestial architecture, the cross of the cardinal points sharing the rose of the zodiac was also that of the architecture of the temples and the cities, built in the image of the sky."[118]

De Sède could hardly have put it more plainly. In his depiction of the combination of the "rose" and the "cross", or the "Rose + Croix", he has laid out the solution to the geometry of the parchments: a circle, its crossed axes, the twelve-fold division and zodiac, and the two Seals of Solomon. This is indeed how the Affair of Rennes has been "signed", as we have seen.

Summary

The famous parchments alleged to have been found by Saunière in the 1890s (but which are in fact twentieth century concoctions) have revealed a hidden layer, a geometrical construction based around the Seal of Solomon.

The same concealed "Seal of Solomon geometry" can also be identified in the illustration of the Marie de Nègres gravestone, also a modern creation falsely presented as dating from the 1780s.

This geometry is laid out on a grid of inches. Given that the gravestone and the parchments are intended as companion designs by their creator, this is a signal from the puzzle-maker that the parchment geometry itself is laid out in inches. This is reinforced and confirmed by the observation that the parchment texts are written in uncial script, which means "related to the inch".

We have also observed that the title of the Boudet map has been written in custom stylised letterforms obviously intended to invoke uncial script, thus providing a crucial link between the map, the parchments and the gravestone illustration.

As it states in the Ophiucus stanza of *Le Serpent Rouge*,

> "Here is the proof that I knew the secret of the seal of SALOMON."

---

117 Gérard de Sède, *La Rose-Croix* , (Paris , Editions J'ai lu, 1978). p. 37
118 Gérard de Sède, *Le Trésor Cathare*, op. cit. p. 187. (English translation).

Figure 100: "*Here is the proof that I knew the secret of the seal of SALOMON.*"

The parchments are a re-imagining of Boudet's exercise in geographic cryptography. The manuscripts are maps. The only difference between the formats is the relative size. All are based on the inch-grid, but the size of the Seal of Solomon varies. Measuring from the centre of the design to the points of the inner hexagram for comparison, and including Boudet's title page, they are as follows:

>   Boudet title page: two inches
>   Boudet map: four inches
>   Marie de Nègres gravestone : four inches
>   Parchments: eight inches

The format is reproduced in these related sizes: two inches, four inches and eight inches. The geometry is invariant to transformation of scale but requires the basic square of 4 x 4 to generate the hexagram from the circle.

The parchments are maps. They were an inspired exercise in recreating the hidden layer of Boudet's map in an alternative format, a textual manuscript. In both cases, what is seen conceals something unseen. This hidden layer is comprised of geometry. It is based on a grid of inches and a four-inch radius circle, generating a pair of Seals of Solomon. It is ultimately a mental map which relates to the ancient landscape complex that Boudet described as his two cromlechs.

Now we can step back and watch the sequence as it must have unfolded. The knowledge persisted into the historical period of the Templars and Cathars. It persisted into the hearing of Henri Boudet. He wove it into his book, in the words and the map, in the most ingenious, elegant manner imaginable.

The key to unlocking his master puzzle was the inch, and the grid. He shared the secret with certain others, and, by some route, it came into the possession of Pierre Plantard and his circle. In turn, they kept the game going by re-encoding Boudet's blueprints to the Complex in the parchments.

The text of the parchments, the elaborate coded message it contained, and the entire backstory apparatus, were all a cover. The genuine content was the hidden inch grid and Seal of Solomon. In retrospect, we can see now that one of the motivations of de Sède's book, *L'Or de Rennes*, was to place the subliminal geometry before the eyes of the reading public by reproducing the parchments.

If the intent was for this to be spread as wide as possible, then they hit the jackpot and surely succeeded beyond their wildest dreams. Henry

Lincoln, BBC screenwriter, actor and filmmaker, read it in the summer of 1969. He went on to make his three films and write various books. The Affair of Rennes became a global phenomenon.

It was Lincoln who presented the parchments to the English-speaking world in these films. And it was in the second of these, *The Priest, the Painter and the Devil*, made in 1974, that he placed the solution to the code hidden in the text of the parchments into the public domain for the first time. This was the BBC documentary which I had come across by accident in my teenage years, and which was the seed for this entire expedition.

In 2018, Lincoln's films became available on YouTube, and I was able to watch them again for the first time since the 1970s. This also provided an opportunity to finally look into a minor problem about the parchments which had been sitting quietly in the back of my mind ever since reading *The Holy Place* in the 1990s.

When I did so, some surprising results came to light with significant implications for the story of the parchments and indeed the entire Affair. This material has been reserved for Appendix I later in the book.

With the core problems of *Le Serpent Rouge*, *La Vraie Langue Celtique* with its map, and the parchments now decisively resolved, it is time to return to the landscape, which has not yet fully given up all its hidden secrets.

Chapter Fourteen
THE ARQUES SQUARE

IN 2018, the final pieces of the puzzle began to fall into place. Sometimes it takes twenty years of staring at something before you see it. My long years of thinking, investigating and meditating on the 45° alignment had revealed the zodiac hidden in the landscape but had not exhausted the geometric treasures concealed within its depths.

As we've seen, this 45° bearing through the church of Rennes-les-Bains also passes through an additional four significant local ancient sites. This line integrates with the historic *cardo-decumanus* axes of Rennes-les-Bains and gives rise to the zodiac format rediscovered by Jean Richer. This complex of five structures, laid out over a distance of just over seven miles, all dating from at least as early as the eleventh-century AD (and possibly much earlier), bears silent witness to a programme of active human intervention in this landscape in earlier times.

The two sites that mark the ends of the 45° alignment are the church of St-Just-et-le-Bézu in the south-west and Château d'Arques in the north-east. Exploring the area in Google Earth one day, I happened to notice that both of these sites fell close to longitude values with whole numbers of arcminutes: namely 2°16'00"E and 2°22'00"E, respectively. This observation led quickly to a second: it implied that the east-west distance between the two endpoints was exactly six arcminutes, or one-tenth of a degree, of longitude. This struck me as a very neat, curious and convenient length.

When I checked the other positions on the line, I found that Lavaldieu also fell on a longitude value with a whole number of minutes, namely 2°18'00"E. Thus, it marked the one-third division of the full length of the line, being two arc-minutes east of St-Just-et-le-Bézu and four arc-minutes west of Château d'Arques. Whilst I thought this was a noteworthy pattern, I quickly realised that I could not, of course, attribute any direct significance to the fact that all three resolved to zero seconds of arc. This must surely be a coincidence as these values are relative to the modern Greenwich meridian. The international co-ordinate system based on this reference longitude did not come into use until the establishment of the Royal Observatory on Greenwich Hill in 1721, whilst the sites on the 45° alignment were all erected many centuries before this date.

Nevertheless, even after putting to one side the absolute values, the relative separation between St-Just-et-le-Bézu and Château d'Arques of six arcminutes, or one-tenth of a degree, remained of definite interest, particularly when emphasised by the division into thirds by Lavaldieu. The units of degrees, minutes and seconds can be considered independent of any particular co-ordinate system. They arise simply from the division of the circle into 360, a practice that dates from Babylonian times and has been universally adopted. If the end points of the 45° were separated by 1/10°, then this was equal to 1/3,600th of the circuit of the Earth at this latitude.

Contemplating this intriguing result, an idea occurred to me out of the blue one day. If indeed the two end points of the 45° alignment had been carefully positioned to mark out a segment of a specific length, I wondered if perhaps the intent of the builders had been to *define the diagonal of a square*. This struck me as a promising suggestion worthy of further investigation.

I decided to explore it by ruling meridians and latitude lines through the two endpoints at St Just-et-le-Bézu and Château d'Arques to create a square, as shown in the accompanying Figure 101. Strictly speaking, the result was a *geodetic* square, laid out on the curved surface of the spherical Earth. I inspected the result. Slightly disappointingly, it did not result in any obvious further clues which might suggest a conscious strategy.

There appeared to be nothing significant located at the other corners or positioned on any of the four sides or along the other diagonal from the north-west to the south-east corners. Nevertheless, this geodetic square had now been brought into existence, if only, at this stage, on my map and in my own imagination. I decided to call it the Arques Square, after the château that marks the north-east corner.

There was another curious and even uncanny reason why I chose this name. As shown in Figure 4 earlier in this book, the floor plan of the Château d'Arques shows that it was designed and built in the

THE ARQUES SQUARE

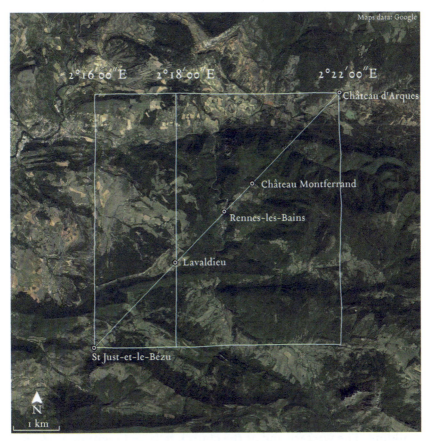

Figure 101: The Arques Square. The endpoints of the 45° alignment, St-Just-et-le-Bézu and Château d'Arques, fall exactly six arcminutes of longitude apart, at 2°16′00″E and 2°22′00″E, respectively. Lavaldieu falls exactly one-third of the way along the length at 2°18′00″E.

form of a square, with a pair of crossed diagonals marking the roofline. Hence, whether by chance or intention, the architectural layout of the château suggested a representation, or even perhaps a map, of my newly conceived Arques Square.

Also, by a charming coincidence to which I attribute absolutely no significance whatsoever, the French name of Arques happens to be an anagram of the English word "square"!

So now I had created a square and given it a snappy, appropriate name. The Arques Square was a geodetic square defined by the 45° alignment as its diagonal. It measured one-tenth of a degree of longitude in width, as I had defined it from observation of the co-ordinates of its endpoints. The question was: did it represent the traces of a genuine intentional intervention in landscape by earlier humans, or was it just an illusion of my own imagination, the result of staring at the map for too long? Was I falling into the same trap I had critiqued in others, creating shapes that did not exist out of dubious alignments, with little or no justification? I began to explore it further, to see if there were any indications that might help to answer these questions.

The Dimensions of the Arques Square

The obvious first step was to determine its precise linear dimensions. For a square drawn on the surface of the globe of a particular angular dimension, one-tenth of a degree of longitude in this case, the absolute value of its width will, of course, depend on the latitude at which it is located. Indeed, the width of such a square will itself vary along its height: if located in the northern hemisphere, for example, the lower, southern side will be longer than the upper, northern side. For this reason, the width of a geodetic square is usually defined to be equal to its width as measured at the latitude of the midline.

So, to find the width of the Arques Square, I had to calculate the latitude of its midline. This in turn was dependent on the precise latitude values of the two endpoints of the 45° diagonal. What are these?

The south-west corner is marked by the church of St-Just-et-le-Bézu, but where exactly should we fix its geographical co-ordinates? We are surely free to choose anywhere within the footprint of the building. The western end of the church falls on longitude 2°16′00″E precisely, so I chose to place the marker for this end of the line segment at that position. It happens to correspond to the door of the church, at longitude 2°16′00″E and latitude 42°52′44.75″N.

What about the north-east corner? At first, I assumed that it must be within the footprint of the Château d'Arques itself, but on closer inspection I realised that this was not quite right.

Château d'Arques is surrounded by a rectangular compound. A square guardhouse occupies the south-west corner, with prominent diagonals also marked on its roof just like the château itself. As noted in Chapter One, the 45° alignment passes precisely through the corner of the compound, runs along the diagonal roofline of the guardhouse, and then continues through the centre of the Château d'Arques itself. Thus, the south-west corner of the compound itself appears to be carefully integrated into the 45° alignment. Using Google Earth, I found that it is actually this south-west corner of the compound that falls

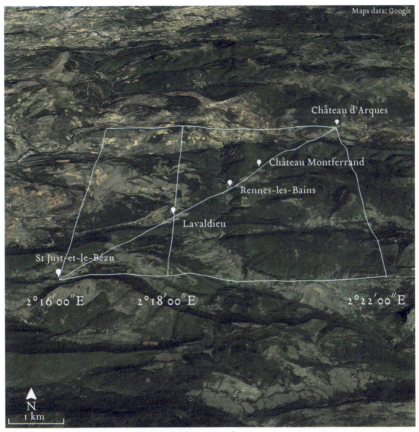

Figure 102: The Arques Square, looking north, in perspective view.

exactly on longitude 2°22′00″E, rather than the tower of the château. Accordingly, I decided to define this point as the north-east corner of the Arques Square. The latitude of this corner is 42°57′10.45″N. Aerial images of the two positions that are now taken to define the NE and SW corners of the Arques Square are shown in Figure 103.

Now that I had fixed the endpoints, I could calculate the midline of the square, equidistant between the two latitude values obtained above. The result was 42°54′57.5″N, or expressed in decimal degrees, 42.916°N.

Using this value we can derive the width of the Arques Square from the known dimensions of the Earth. It is a straightforward task to obtain the length of a degree of longitude at a given latitude by simply consulting a table of Earth measures. Looking up the length of the circuit of the Earth at this latitude of 42°54′57.6″N, using a suitable

42°57′10.45″N

2°22′00″E

2°16′00″E

42°52′44.75″N

Figure 103: Google Earth images showing the co-ordinates of the two endpoints of the 45° alignment. St-Just-et-le-Bézu church and Château d'Arques mark the SW and NE corners of the Arques Square, respectively.

online calculator,[119], we find that one degree of longitude measures 267,886.3 feet. As the Arques Square spans one tenth of a degree, its width at the midline will therefore be one tenth of this figure, or 26,788.63 feet.

For convenience, we will convert this to inches, and the final result is:

Width of Arques Square at the midline = 321,464 inches.

We will adopt this value as the "canonical width" of the Arques Square, equal to one tenth of a degree of longitude at the latitude of the midline, according to modern calculations. Now let's compare this to the physical distance measured in the landscape itself. A square of

---

119 For example, at http://www.csgnetwork.com/degreelenllavcalc.html

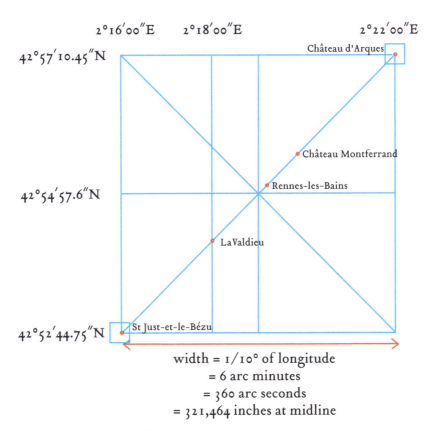

Figure 104: Dimensions of the Arques Square.

sides of length 321,464 inches has a diagonal that can be calculated (by Pythagoras' Theorem) to be 454,618 inches (assuming a flat surface). We can check this against the measured length in the landscape of the diagonal, which is of course the 45° alignment between St-Just-et-le-Bézu church and Château d'Arques.

The distance from the western end of the St-Just-et-le-Bézu church to the south-west corner of the Château d'Arques compound as shown on Google Earth is 455,690 inches. This is within 20 yards of the calculated figure of 454,618 inches. But if we take the measure from the other (eastern) end of the St-Just-et-le-Bézu church it is as close to 454,618 inches as one can measure. This is the distance between the two closest points of each structure marking the endpoints. One might say that the church itself fills the space between longitude 2°16′00″E and the calculated diagonal length of the Arques Square.

Thus, the positions of the church of St-Just-et-le-Bézu and the south-west corner of the Château d'Arques are as close as can be measured to conform to our definition of the Arques Square as one tenth of a degree of longitude in width. Now the position and dimensions of the Arques Square are fully defined and determined.

The Templar Headquarters of Campagne-sur-Aude

Other than the five sites on the 45° alignment itself, I still had not yet identified any other positions in the landscape that appeared to be governed by the Arques Square geometry. I had checked and found nothing at the north-west or south-east corners, or on any of its sides, but what about on the midline? On inspection, I could see there were no marked locations along its length as it crossed the Square, but as I traced its course further to the west, and crossed the Aude River, a surprise awaited.

The village of Campagne-sur-Aude, on the west bank of the Aude, was the *commanderie* (or local headquarters) of the Knights Templar in the Haute Vallée de l'Aude in the twelfth and thirteenth centuries. I found that the village was sited precisely on the midline of the Arques Square.

Campagne-sur-Aude has a rich history, and a fascinating name. The word *campagne* in French means countryside, but it can also denote a military campaign, as the similarity of the words in English and French suggests. In this sense it can also be used to indicate an excursion into a territory.

The design of the village is remarkable, consisting of a series of concentric rings. From the air it has the appearance of a bulls-eye pattern. At its centre lies a tight cluster of buildings, which includes the village church. A circular road surrounds the cluster. The midline of the Square is tangent to the north side of the central ring of buildings and the front door of the church. Of course, without any further indications, this could be a mere coincidence, but it was intriguing, nevertheless, to find the local headquarters of the Templars located on the exact latitude that I had calculated from the endpoints of the 45° alignment.

Now that I had found that the village was marking the midline, the next obvious step was to investigate whether there was any significance in the distance at which Campagne-sur-Aude lay from the western side

Figure 105: Zooming in on Campagne-sur-Aude.
This shows the midline of the Square tangent to the north side of the circle of buildings at the centre of the village, just outside the door of the church.

of the Square. As always, I was looking for any evidence of intentionality as indicated by the presence of whole numbers in suitable measures.

I soon discovered that there is something very remarkable indeed about the geometric relationship between the village and the Square. The bearings from Campagne-sur-Aude to the two western corners of the Arques Square mark significant angles: to the unmarked north-west corner is 50°, and to the church at St-Just-et-le-Bézu in the southwest corner is 130°.

Thus, both lines make accurate angles of precisely 40° to the midline running east-west. And there was more: the distance from Campagne-sur-Aude to each of the two corners measured very close to the whole number figure of 250,000 inches.

Figure 106: The geometry of the Arques Square, showing Campagne-sur-Aude on the midline. Note the two 40° right-angled triangles and how these define the western side of the Square.

To put it another way: the relative position of Campagne-sur-Aude to the Arques Square is defined by a pair of 40° right-angled triangles, each with a hypotenuse of 250,000 inches (as shown in Figure 106). Moreover, the hypotenuse of the upper triangle passes directly through another impressive local site, the Château des Ducs de Joyeuse in Couiza located in a flat stretch of ground on the east bank of the Aude, as shown in the blue close-up square.

This wonderful edifice dates to the fourteenth century and is today fully restored and operating as a fine hotel. I had already noted that there was no permanent structure at the north-west corner of the Arques Square itself, possibly because it falls awkwardly on the flank of a steep ridge. The corner appeared to have been marked instead by the Château des Ducs de Joyeuse.

THE ARQUES SQUARE

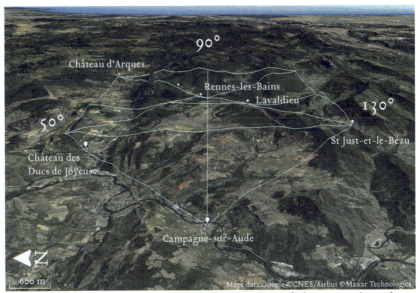

Figure 107: The Arques Square, looking east. Bearings of alignments shown from Campagne-sur-Aude of 50°, 90° and 130°.

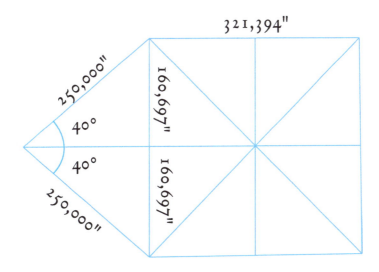

Figure 108: Geometry template of the Arques Square.
The two 250,000 inches, 40° right-angled triangles have small sides of 250,000 × sin 40° = 160,697" inches. Thus, the height of the Arques Square is twice this value, or 321,394 inches by this definition.

The emergent geometry of these very curious right-angled triangles generated by Campagne-sur-Aude offers another method of describing the Arques Square and obtaining its dimensions. If the sides are defined as equal to twice the short side of a 250,000-inch, 40° right-angled triangle, then:

> Height of western side of Arques Square =
> = 2 x (250,000 inches x sin 40°)
> = 321,394 inches

This result is a mere 70 inches short of the "canonical" length of the sides of the Arques Square, or 321,464 inches, as we derived it from tables of values of Earth dimensions when it is defined as one tenth of a degree, or six arcminutes, in width at its midline. The two values obtained by independent methods differ by less than six feet over more than five miles!

Consider that I had proposed a conceptual square based purely on the co-ordinates of the two chosen endpoints on the 45° alignment. With this decision, everything was fixed. There should be no reason to expect any correspondence between this square and other local features if it was nothing but a figment of the imagination. I have found however that Campagne-sur-Aude lies on the midline of the Arques Square, and its western corners lie at 40° angles to the line at a distance of 250,000 inches. Such multiple independent overlapping degrees of order are signals of intentionality.

The Knights Templar had located their headquarters in the Haute Vallée de l'Aude at a precise position in relationship to the Arques Square, to permanently record its presence by marking with great care and accuracy its midline. This ensures that the secret can never be lost. In this most unobtrusive manner, the village of Campagne-sur-Aude was founded here to watch over, protect and preserve the memory of the Arques Square in this landscape. And if this was true, the presence of the geometry and its measures disposed in such orderly fashion also tells us that the design was *laid out using inches as the unit of measure.*

The village of Campagne-sur-Aude marks the midline of the Arques Square as defined by the two endpoints of the 45° alignment, but what about the centre point of the Square itself? There is nothing of particular significance in the landscape at the location itself, but is it indicated or recorded perhaps in any other manner?

The Alet-les-Bains Cathedral Alignment

During my years of living and exploring in the Haute Vallée de l'Aude, I expanded my catalogue of alignments in the landscape far beyond the original limits of the IGN map on which my journey had begun all those years before.

Most of this geometry, I found, was distributed in a relatively wide but fairly narrow region bounded to the south by the high peaks of the Pyrenees, and to the north by the flat gentle plains of southern France. This made sense if indeed such alignments had their origin in intentional interventions by ancient engineers.

From a practical perspective, the foothills of the Pyrenees are an almost perfect landscape in which to execute such work as they provide ready access to a network of elevated sighting positions with valleys between. There was, however, one very remarkable alignment I discovered in February 2012 using Google Earth that was a major exception to this general rule. It ran across the flat plains to the north, passing through an impressive six sites.

The most southerly of these was the only cathedral in the Haute Vallée, at Alet-les-Bains. This impressive ancient structure, now largely in ruins, was until the sixteenth century the seat of the Bishop of Alet. It was also the site of a Temple of Diana in the Roman era. It happens to display Seal-of-Solomon designs in its elaborate windows.

It occupies a commanding position on the banks of the River Aude, at the head of the narrow winding valley by which visitors arrive from the north. The walled village has a special sense about it to this day. Nostradamus lived there for a time and his house, carved with mystical symbols, including, again, a Seal-of-Solomon, can still be found by the curious visitor just off the main square.

The alignment runs at a precise bearing of 330° from the ruins of the Cathedral and passes through an additional five significant sites. As shown on the accompanying Figure 109, the complete list of all six locations is as follows (from south to north):

- Alet-les-Bains Cathedral: now in splendid ruins, on the banks of the River Aude
- Gaja-et-Villedieu Church
- Lauraguel Church
- Cailhau village perimeter
- Cailhavel village perimeter
- Villeneuve-lès-Montréal Church

It passes exactly tangent to the circular outline of the two villages Cailhau and Cailhavel, in a geometric gesture which echoes that of the midline of the Arques Square at Campagne-sur-Aude. This is an extraordinary line, running very precisely through these six sites, including the most important historical church site in the area, the Cathedral of Alet-les-Bains, over near-flat territory on a major compass angle. I was always very curious about how it interacted with the rest of the geometry found in the Haute Vallée itself.

To the south beyond Alet-les-Bains, the alignment passed just a few hundred feet to the south of the church of Rennes-les-Bains itself. This tantalising near miss puzzled me for years until I discovered the Arques Square in 2018. Finally, the true target of the line revealed itself at last: the Alet-les-Bains Cathedral alignment points perfectly accurately, without any adjustment to the line as found, to the precise centre of the Arques Square!

The key to accurate landscape surveying is always triangulation. We can now observe that the centre of the Arques Square is pinpointed by two separate lines meeting like arrows at the target location: the first is the latitude line through Campagne-sur-Aude, and the second is this extraordinary long baseline alignment laid out over the plains from Villeneuve-lès-Montréal through four churches and the cathedral of Alet-les-Bains. Thus the knowledge of the location of the centre of the Arques Square is secured against the ravages of time and memory, permanently marked in the landscape itself at the intersection of the two lines. In this case, X really does mark the spot.

Incidentally, the closest place name to this point on the IGN map is *Le Cercle*, the tiny hamlet which Boudet named as the centre of one of the two cromlechs. He said that one of the two was larger than the other, and contained it. If the two cromlechs correspond to the zodiac and the Arques Square, they would satisfy this description. If not, it is an uncanny twist that the centre of the Square is at *Le Cercle*.

Compass on Campagne-sur-Aude

The wonders have not quite ceased. The five churches on the Alet-les-Bains alignment are not distributed at random locations along that line but mark precise compass angles from Campagne-sur-Aude. Consider first the two churches that mark the ends of this sequence. The most northerly, Villeneuve-lès-Montréal, lies on an exact 345° bearing from the Templar headquarters, while the most southerly,

THE ARQUES SQUARE 301

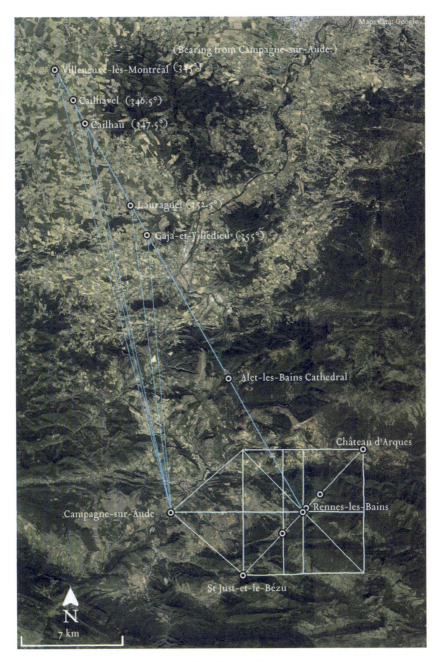

Figure 109: The alignment of six churches or villages on the 330° bearing from the centre of the Arques Square. The bearings from Campagne-sur-Aude to the five villages are also shown.

Gaja-et-Villedieu, is on a precise 355° bearing. This is remarkable: the alignments from Campagne-sur-Aude to these two points marked on the line subtend an angle of exactly 10°, between the bearings of 345° and 355°.

Now let's consider the positions of the three locations on the alignment between the two points, namely Cailhavel, Cailhau and Lauraguel. Incredibly, the three alignments from Campagne-sur-Aude to these sites divide the 10° angle into four segments of 1.5°, 1°, 5° and 2.5°. Together, the five sites mark off calibrated regular subdivisions of a 10° fan of alignments from Campagne-sur-Aude (as shown in Figure 111).

The builders were displaying virtuoso precision in marking out this palette of values. They were showing their ability to divide the circle into regular sub-divisions with great accuracy. There has evidently been a systematic effort to calibrate directions in the landscape around the central observation point of Campagne-sur-Aude. Using Google Earth, it is even possible to gain some practical insight into how this might have been accomplished.

If we look at the fan of lines from Campagne-sur-Aude marking out the churches on the Villeneuve-lès-Montréal to Alet-les-Bains alignment, we find there is a convenient high ridge that offers an excellent vantage-point immediately to the south of Campagne-sur-Aude. It is on the east side of the Aude, at the Gorges de la Pierre-Lys, a spectacular gorge of sheer vertical limestone cliffs carved over millions of years by the river. The ridge runs east-west and provides an ideal platform from which to sight to the north along the valley of the Aude, to Campagne-sur-Aude and beyond.

Meanwhile, there is a line of low hills further to the north, beyond Villeneuve-les-Montréal and the alignment, which could provide suitable elevated positions at the backsight for such sighting alignments.

In this way, each of the lines in the fan from Campagne-sur-Aude to the five positions on the alignment (excluding Alet-les-Bains for the moment) can be traced between a ridge providing elevated sighting positions to the south, and a line of hills also offering high suitable vantage points further to the north, as shown in Figure 111.

We can conceive of teams, working at night, co-ordinating their actions through agreed signals. A central beacon at Campagne-sur-Aude is established. The sighting teams use fires to mark their positions on these sighting ridges and hills and establish particular angles. With care and effort, and assuming there was some powerful reason to do so, it would certainly be possible to lay out such calibrated bearings.

Figure 110: The alignment from the centre of the Arques Square to Ville-neuve-lès-Montréal at 330° runs at a tangent to the boundary of the circular village of Cailhavel, and follows the line of the road as shown.

If this is how these alignments were sighted, then the physical limitations of the geography might explain a curious feature of the geometry. As we've seen, the five alignments range from bearings 345° to 355°. This latter bearing, the alignment from Campagne-sur-Aude through Gaja-et-Villedieu, is just 5° from north. It's a very neat result, but why did the builders stop there? Why not add another point due north of Campagne-sur-Aude on the alignment?

A possible answer to this question presents itself when we look at the terrain in Google Earth. The sighting alignment through Gaja-et-Villedieu is viewed from a position on the western end of the southern ridge. Beyond that point to the west is a sharp sheer drop down the very steep cliffs of the Gorges de la Pierre-Lys. There is nowhere to stand any further to the west of the final alignment, and therefore there is nowhere on the ridge where the meridian through Campagne-sur-Aude could have been sighted.

If we consider these lines in addition to the precision bearings we have already observed to the corners of the Arques Square and the Sunrise Line, it is apparent that an array of sighting alignments passing through Campagne-sur-Aude extends in nearly every direction. It's as though a calibrated compass rose has been engraved into the landscape with the village at its centre.

This realisation finally made sense of the observation I had made years earlier on my sunrise walks to the Fa Tower, the huge monolithic watchtower on a hilltop near the village. The bearing from Campagne-sur-Aude to Fa Tower was 330° and from Campagne-sur-Aude to Esperaza was 30°.

These were fragments of a comprehensive programme of sighting lines converging on the Templar headquarters. They were also related to the ancient signalling network that operated via line of sight to enable communication between the various Templar châteaux, strongholds and settlements that were scattered through these rugged, narrow valleys. Evidently, at least some of these alignments have been laid out according to requirements of angle and measure. How could this have been accomplished?

Geometrical Construction of the Square

Let's consider the challenges that would be involved in marking such a geometric form at this scale in landscape. How would one measure out an accurate length of 321,464 inches, across peaks, valleys, rivers and streams? It cannot be done by unspooling a tape measure. The answer lies in creating geometric figures of definite angle and measure at small scale, and then extrapolating these at large scale across the landscape.

Is there any historical evidence for such practices? The accompanying Figure 112 shows an image from *Pantometria* by Leonard Digges, a book published in the late sixteenth century. [120] It depicts techniques from the Middle Ages, and no doubt earlier, which enabled a surveying team to sight, align and measure long distances in landscape even over mountainous or otherwise inaccessible terrain.

The images clearly show the use of scaling to achieve geometric coverage over large area or long distance. The key to the scheme is a set of relations between simple whole number triangles on the plan

---

120   Leonard Digges *Pantometria* (London, Abel Ieffes, 1591)

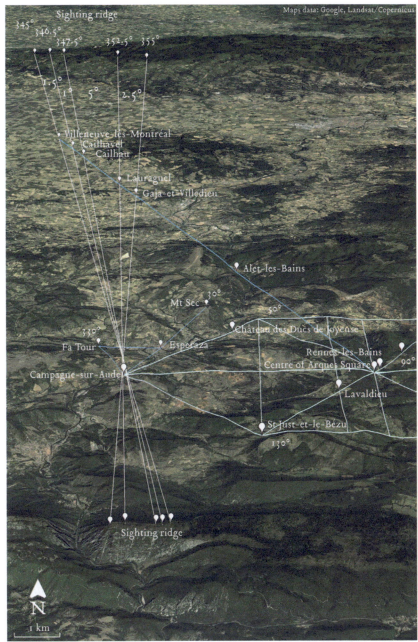

Figure 111: Calibrated Compass Rose of Sightlines through Campagne-sur-Aude. Note suitable vantage points on the high ridge to the south and hills to the north. The five sightlines through Campagne-sur-Aude to the positions on the Alet-les-Bains alignment are separated by angles of 1.5°, 1°, 5° and 2.5° respectively.

and in the landscape, which integrate elements of measure and scale. For example, carefully constructed triangles are laid out on a prepared patch of flat ground, using whole number measures of length and angle, as can be clearly seen in the images. These known figures are then expanded and projected at large scale into the target territory, so that, by geometrical considerations, required distances and bearings can be marked out which otherwise could not be directly measured.

We can take these ideas and apply them to the Arques Square. We have already found evidence for the units of measure involved, namely the inch. What about a scale? Is there any obvious candidate for a scale map of the geometry in the landscape?

Earlier in this book, I wondered whether the ancient landscape engineers, whoever they were and whenever their work was performed, may have employed the equivalent of a "mental map" to guide them in their work. I also considered the possibility, based on the observation of the division of the Bugarach Baseline into three equal segments of 150,000 inches, that such a map might be based on a scale of 1:25,000.

Now we have found a specimen of landscape geometry, as shown in Figure 105, that includes two right-angled triangles with hypotenuse values of 250,000 inches. This is a tantalising number in the context of such a proposal, because it would be represented by a neat length of ten inches on such a scale map. Is it possible that the Arques Square was laid out using a 1:25,000 plan?

To explore this idea, I considered the width of the Arques Square as it would be depicted on such a map. To calculate this, I simply divided the canonical width of 321,464 inches by 25,000 to obtain the width of the Arques Square on a 1:25,000 scale map. The result was:

321,464 inches / 25,000 = 12.858 inches

I did not recognise any significance in this length of 12.858 inches so I searched online to see if it might possibly be related to any historical unit of measure. It did not take long to discover, to my amazement, that this length of 12.858 inches is precisely equal, with no margin of error, to the ancient French unit of measure known as the *pied-du-roi*, or "foot of the king". It is defined as:

1 *pied-du-roi* = 326.6 mm = 12.858 inches

I learned that this was the most widely used "foot" in France from at least the time of Charlemagne in the eighth-century AD, and quite possibly much earlier, until the seventeenth century. It is

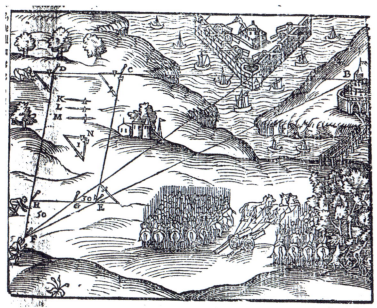

Figure 112: Woodcut engraving from Leonard Digges, Pantometria (London, Abel Ieffes, 1591) illustrating the techniques of landscape engineering practised in the sixteenth century and earlier. Long baseline alignments are laid out using specimen triangles pegged out at small scale on flat ground, which are then enlarged to create known geometry in the wider landscape.

a truly astonishing result that the width of the Arques Square on a 1:25,000 scale map is precisely one *pied-du-roi*. We have now gone way beyond any possibility of co-incidence, or that such relationships could arise by chance. These are unmistakeable signs of intention, and they demonstrate that there is a cohesive system in operation here. It brings together the elements of the inch, the *pied-du-roi*, the arc-second, the scale of 1:25,000 and geometry into a single seamless suite of tools.

A further result quickly follows: it also implies that in the landscape the width of the Arques Square must be precisely 25,000 *pieds-du-roi*![121]

I then considered a 1:25,000 scale plan constructed out of the geometry of the Arques Square. The result is shown in Figure 113. Notice that the two short sides of the 40° right-angled triangles combine to create the left-hand side of the square. We can easily calculate this length using trigonometry as

Side of square = 2 x 10" x sin 40° = 12.856 inches.

---

121   (326.6 mm x 25,000)/25.4 = 321,456.4 inches, less than eight inches from the canonical width of 321,464 inches.

This differs from the value we derived above by an indistinguishable 0.002 inches, or 0.05 mm.

We have generated a perfectly accurate 1:25,000 scale map of the Arques Square, with sides of one *pied-du-roi*, using only a right angle, a 40° angle and a measured length of ten inches. This exercise produces a satisfying result. It is possible to generate the Square, both as 1:25,000 scale map and in the landscape, using whole measures of inches, and angles.

This approach conforms elegantly to the principle of ancient geometry that one should only ever measure out a line segment when it can be expressed as a whole number in some suitable unit. The *pied-du-roi* cannot be laid out, geometrically, as 12.858 inches, or 326.6mm, but it is accurately laid out, at least conceptually using the method above.

But why should this value of 12.858 in inches hold any significance? What does it represent, or equal? The answer is revealed in the ratio of the two values 28 and 360:

360/28= 12.8571+

There it is. This number, taken as a length in inches, is equal to 326.6 millimetres and is therefore identical to the *pied-du-roi*. If it is defined this way then there are 28 *pieds-du-roi* in 360 inches. This is convenient as it permits ready division into sevenths. It is especially interesting in relationship to our geometry, given that the width of the Arques Square is 360 arc-seconds, and the grid separation between the meridians of Lavaldieu and Bézu is 28 arc-seconds. The ratio of these two distances is the same value as the number of inches in the *pied-du-roi*.

This is why the meridian separation distance in inches, namely 25,000, is the same value as the Arques Square width in *pieds-du-roi*. By employing a unit of measure equal to 360/28 inches, the builders availed themselves of an accurate, versatile tool for converting between lengths in different measures with ease which was a perfect bespoke match to the size and location of the Square.

The exact linear distance between any given pair of meridians in the landscape will vary slightly depending on the latitude at which the measurements are taken, because they are parallel lines drawn on the surface of the spherical Earth. Thus, the absolute width of the Arques Square will vary over the range of latitude values which it covers. How much is this variation, in *pieds-du-roi* and inches?

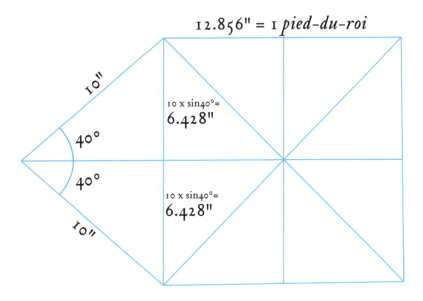

Figure 113: Geometry of the Arques Square on the 1:25,000 map. If the length of the pied-du-roi in millimetres is defined as [2 × 10 × sin40° × 25.4], it is equal to 326.53610 mm, indistinguishable from the modern definition of 326.6 mm.

The Meridians and the Width of Arques Square

The accompanying table lists in the first column the latitudes of the upper side, the midline and the lower side of the Arques Square. In the next column, the width of the Arques Square in *pieds-du-roi* is given at each latitude. Notice that it varies by about thirty *pieds-du-roi* over the full Square, and that the value at the midline is *exactly* 25,000 *pieds-du-roi*, discarding a fraction remaining.

	Latitude (decimal)	Width of Arques Square in pieds-du-roi	Number of inches corresponding to 28 arc seconds
Upper Side	42.953°	24,985.57	24,987.2
Midline	42.916°	25,000.57	25,002.2
Lower Side	42.879°	25,015.49	25,017.2

In the final column is shown the number of inches corresponding to 28 arc seconds, the meridian grid separation difference or angular separation between the Lavaldieu and Bézu meridians. The difference between the corresponding values at the latitudes of the upper and lower sides of the Arques Square varies by only +/-17 inches from the value at the midline, which is 25,002.2 inches.

These values are taken by calculation from modern tables of Earth measure. In practise, these differences are much smaller than even the best surveying could achieve. The midline of the Square could not be any closer to the latitude for which one tenth of a degree equals 25,000 *pieds-du-roi* and 28 arc-seconds measures 25,000 inches.

It is difficult to avoid the conclusion that the Arques Square has been deliberately and intentionally sited in this location to fulfil these conditions of measure, specifically that the meridian grid spacing and the width of the square and its subdivisions lend themselves readily to expression in whole number measures in units of inches, arc-seconds and *pieds-du-roi*.

The Arques Square, the meridians and the cromlech-zodiac of Rennes-les-Bains form a seamless system. The elements are summarised in Figure 114. These are a collection of calibrated distance measurements, laid out in inches, precisely determined as a large-scale piece of geometry using the techniques of ancient surveying described in works like *Pantometria*.

The meridian of Lavaldieu at longitude 2°18'00"E plays a key role in the geometry. It is the element which unifies the different formats of the Arques Square on the one hand, and the field of meridians on the other. It marks the one-third division of the Arques Square. It also acts as the reference meridian for the grid of inches to which the three inner meridians conform.

These results demonstrate that one major purpose of the complex was to establish a series of calibrated meridians for observation of the movements of the heavenly bodies, embedded in a wider geometrical framework relating to the size and shape of the Earth, and its location.

The Sunrise Line and the Meridians

In the next section we will explore how the Sunrise Line, the ancient landscape alignment introduced in Chapter Four and its companion at right-angles, the Bugarach Baseline, relate to the meridians, and we will find again that the Lavaldieu meridian plays a pivotal role. From my first explorations in the map in the 1990s, I was very curious about

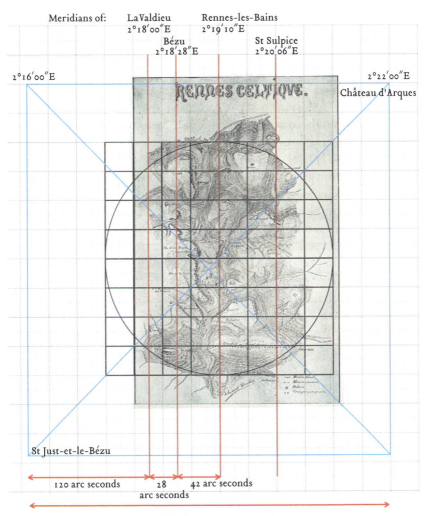

Figure 114: The Measures of the Arques Square: Combining the cromlech, the 8 x 8 inch grid, the four grid meridians, the Arques Square and the Boudet map. The light grey background reference grid is made up of half inch squares, or two squares per inch. The result is a 1:25,000 scale map with one inch on the map equal to 25,000 inches or 28 arc seconds in the landscape.

the role of the Sunrise Line in the geometry. As discussed earlier in these pages, this impressive alignment runs from the summit of the holy mountain, Pic de Saint-Barthélemy, through multiple significant sites, including Campagne-sur-Aude, on a bearing of around 74°.[122]

What was the significance, if any, of this angle, I wondered? It did not fall on a major division of the compass circle. Was the angle perhaps generated by a whole number grid ratio? I explored various possibilities over the years but if there was a geometrical reason, it eluded me.

As described earlier, this angle corresponds to the rising of the sun as viewed from the peak on August 24 each year, the Feast day of St Barthélemy. The local villagers would climb to the summit each year and spend the night at the top to watch the sunrise. This ancient ritual predates the association of the mountain with St Barthélemy, which is relatively modern. This implies that it was named for the saint because of a pre-existing relationship with the date of August 24. In turn, this means that there must have been another reason for this particular bearing to be memorialised in both the landscape and local lore.

The Bugarach Baseline lies at right-angles[123] to the Sunrise Line. It too is an excellent alignment through multiple significant locations, including Château Montferrand (also on the 45° alignment) and Pech Cardou. Together, the Sunrise Line and Bugarach Baseline form a pair of crossed axes tilted at an angle of about 16° to the north-south grid.

Figure 115 shows the relationship between the Sunrise Line/Bugarach Baseline axes and the field of the five meridians.

A Pythagorean Right-Angled Triangle

Extended to the north, the Bugarach Baseline can be traced to the sighting ridge where the viewing positions for the five meridians are found, and in particular, to the point where it intersects the Lavaldieu meridian. The location is labelled Girbes de Bacou on the map. It falls on a prominent local highpoint on the ridge a short distance to the east of L'Homme Mort, the sighting position for the La Pique meridian.

It provides an excellent vantage point, offering sweeping uninterrupted views of the majestic Pyrenees to the south. As it is on the intersection of the Lavaldieu meridian and the Bugarach Baseline, it is perfectly positioned to act as a sighting peak for both. This point, Girbes de Bacou, at the intersection of the Lavaldieu meridian, longitude

---

122   From East to West: Pic de Saint-Barthélemy, Nébias Church, Campagne-sur-Aude Church, Rennes-le-Château Church, Château Blanchefort, Arques Church and Château de Villerouge de Termenès.

123   To an accuracy of about 0.5°.

Figure 115: The field of the meridians, in red and the Sunrise Line and Bugarach Baseline, in purple.

2°18′00″E, and the Bugarach Baseline, is marked G in Figure 116. It plays a very remarkable role in the geometry, beginning with its relationship to Campagne-sur-Aude, the controlling point for the Arques Square geometry, on the Sunrise Line.

Consider first the bearing from Campagne-sur-Aude (C) to Girbes de Bacou (G). It lies at an angle of 36.8° to north. This is a tantalising result, because it is the small angle of the [3:4:5] Pythagorean right-angled triangle. This implies that the line will be the hypotenuse of such a triangle, with the other two sides running north-south and east-west through the endpoints.

These two lines are the east-west line through Campagne-sur-Aude (the midline of the Arques Square), and the Lavaldieu meridian through Girbes de Bacou. If we label the intersection of these lines as P, this implies that CGP is a [3:4:5] triangle. Now if we measure the lengths of the three sides, (using Google Earth as always), we find the hypotenuse of the triangle, CG has a length of precisely 500,000 inches, the shortest side, CP, has a length of 300,000 inches and the remaining side, PG, 400,000 inches, as shown in Figure 116.[124]

It is clearly intended to be a Pythagorean [3:4:5] right-angled triangle with sides laid out in units of 100,000 inches. This is a very neat, astonishing result, but it is just the beginning of wonders.

A second Pythagorean Right-Angled Triangle

Notice that the line CG, Campagne-sur-Aude to Girbes de Bacou, also forms a right-angled triangle with the Sunrise Line and the Bugarach Baseline. This is shown in Figure 117. The intersection of the Sunrise Line and the Bugarach Baseline is labelled Q.

The distance from Girbes de Bacou to point Q along the Bugarach Baseline is very exactly 300,000 inches, and from Campagne-sur-Aude to Q along the Sunrise Line, is 400,000 inches. Hence the sides have lengths of 300,000, 400,000 and 500,000 inches respectively. The right-angled triangle between Girbes de Bacou, Campagne-sur-Aude and Point Q is evidently also a [3:4:5] Pythagorean triangle that has been laid out in units of 100,000 inches.

The Two Angles

The recognition of these two Pythagorean [3:4:5] triangles, aligned to the meridian, and sharing a hypotenuse, allows us to derive the

---

124 A small deviation in practice from theoretically perfect [3:4:5] angles leads to the length CP being slightly less than 300,000 inches, by approximately 4,000 inches in landscape, or 1/6 inch on the 1:25,000 scale map.

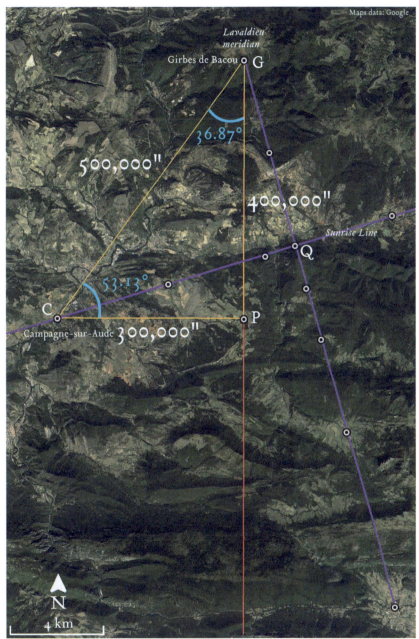

Figure 116: Girbes de Bacou is the intersection of the Bugarach Baseline and the Lavaldieu meridian on longitude 2°18′00″ E. The three points CGP of Girbes de Bacou, Campagne-sur-Aude and Point P form a Pythagorean [3:4:5] right-angled triangle laid out in units of 100,000 inches.

ideal value of the bearing of the Sunrise Line in the geometry. The bearing of the Sunrise Line will be twice the small angle of a [3:4:5] triangle, or 2 x 36.87°= 73.74°. Equivalently, the angle between the Sunrise Lines and the east-west line will be 90°- 73.74° = 16.26°. This is the difference between the two angles of the [3:4:5] triangle, or 53.17°-36.87°= 16.26°. Now the geometrical purpose of the Sunrise Line becomes apparent: its conceptual bearing is 73.74°, and this figure relates to the angles of the [3:4:5] triangle.

If this was the intended design angle of the alignment by the Builders, then it is insightful to look more closely at how the angle of the rising of the sun on 24 August has changed over time.

Over the last thousand years or so, the rising position of the sun on the horizon has been meandering slowly south, a distance of around 2° in total. Today it rises at 74.56°. During the 12th to 14th centuries when the Templars were active in their building programme, it rose directly along the bearing of 73.74°, the conceptual angle of the Sunrise Line obtained from the [3:4:5] triangle.

There is a twist however that must be taken into account in this story. In 1582, the calendar was shifted forward ten days under the reformations of Pope Gregory. The date which was called 24 August after that year corresponded to the date of 14 August in the years before 1582.

If, as I suggest, the ceremony of watching the sunrise on 24 August was correlated to its direction of rising and the Sunrise Line alignment in landscape, then this would imply that the ritual must have been held on the same day of the year according to the sun. This means that it must have been held on 14 August each year before the calendar change of 1582.

Hence it would only have been after 1582, that the date of 24 August would become associated with the mountain. And indeed, it is not until 1629 that the first record exists of the name Saint-Barthélemy applied to a feature on the mountain, some lakes near the summit. It was in 1704 that the mountain was named Pic de Saint-Barthélemy in a book for the first time. Prior to this name change, as noted earlier, it was always known as Mt Tabe or variations on this. Thus the historical record of the name change is consistent with the thesis presented here.

Note this does not necessarily imply that the Sunrise Line could not have been inscribed or laid out in the landscape at some earlier date, but the clear evidence for the observation of the sunrise on 24 August (Gregorian) from the summit of Pic de Saint-Barthélemy, suggests that the line was at least actively in operation during that time.

THE ARQUES SQUARE

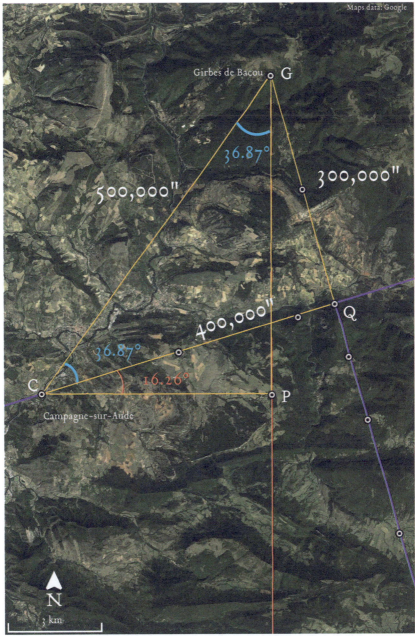

Figure 117: Point Q is the intersection of the Bugarach Baseline with the Sunrise Line. The triangle CQG, between Girbes de Bacou, Campagne-sur-Aude and point Q is also a [3:4:5] triangle laid out in units of 100,000 inches. It shares the same hypotenuse as CPG.

The Geometers preserved this knowledge in an ingenious manner. They determined that it was on 14 August originally each year (later corresponding to 24 August) that the sun rose along this bearing at the latitude of Pic de Saint-Barthélemy. By instituting an annual ritual on this day each year to observe the sunrise from the summit they left a permanent reminder and marker of this special angle. They inscribed this line into the landscape itself, and thereby established it permanently as a frame, or axis, from which to project geometry. Because it was preset to this angle, it automatically generates these extraordinary large-scale Pythagorean [3:4:5] triangles when integrated with a grid aligned north-south.

As the Sunrise Line has passed over the Field of the Meridians, the opportunity has been used to lay out such triangles in units of 100,000". The result is this symmetric pair of [3:4:5] triangles which share the line Campagne-sur-Aude to Girbes de Bacou as their hypotenuse.

More Pythagorean Triangles

In Figure 118, the intersection of the Sunrise Line and the Lavaldieu meridian is labelled Point R. Notice that, by symmetry, the distance from Point R to C, Campagne-sur-Aude, is the same as R to G, Girbes de Bacou. By simple trigonometry, this length CR equals 312,500", and therefore RP equals 87,500". The triangle CPR is also a Pythagorean triangle, with sides of ratio [7:24:25].

Figure 118 shows the result of drawing a circle with centre on R that passes through C and G. Notice that the position of Bugarach church on the Bugarach Baseline falls on the circumference of the circle also. Why should this be?

Recall from all the way back in Chapter One, that the Bugarach Baseline is marked by three lengths of 150,000".[125] The distance from Bugarach church to Q is equal to two of these lengths, or 300,000". It is evident that CQB is the third specimen of the [3:4:5] triangle with the same dimensions. It is the mirror image of the second triangle, CQG, reflected in the Sunrise Line. Its hypotenuse is the Campagne-sur-Aude to Bugarach line, and the other two sides are the Sunrise Line and the Bugarach Baseline. This is shown in Figure 118.

Could such a geometrical design involving [3:4:5] and [7:24:25] triangles, laid out in whole number measures, have been conceived in the minds of humans at such an early time? Are there any precedents for such sophisticated geometry during this epoch?

---

125 See Figure 10.

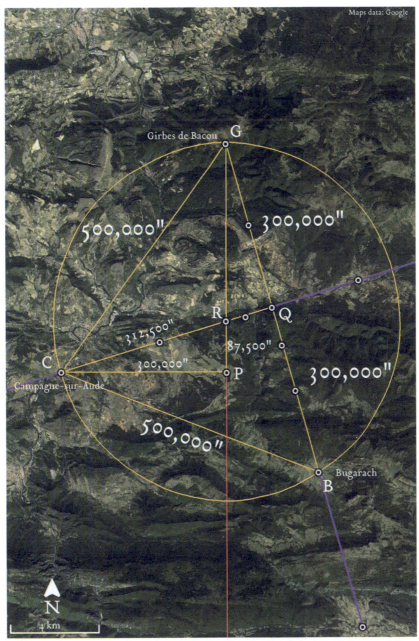

Figure 118: R is the intersection of the Sunrise Line and the Lavaldieu meridian. Notice that CR = RG. Draw a circle of this radius with R as the centre, and it passes through B, Bugarach. Hence RB = CR = RG. Notice also that CPR is a [7:24:25] triangle.

The Susa Tablet

In 1936, a clay tablet was discovered by Mr R. de Mecquenem in the ruins of the Ville Royale in Susa, Iran, ancient capital of the Elam and Achaemenid Empire, during excavations in the Royal Archives. It has been dated to the end of the Old Elamite period, (circa 2,700 BC – 1,500 BC), contemporary with the reign of Hammurabi, sixth king of the first dynasty of the Babylonian Empire (circa 1,810 BC – 1,750 BC). [126] Images of the tablet are shown in Figure 119.

On the tablet, a scribe has recorded a geometrical diagram. It shows a pair of [3:4:5] triangles inscribed in a circle.

Comparison of the geometry in the landscape and on the tablet reveals that they are essentially identical.

Dimensions are given in cuneiform characters. The two main triangles have sides of 30, 40 and 50 units. The scale factor between the tablet and the landscape is therefore 1 unit to 10,000 inches. The hypotenuse of the [3:4:5] triangle, for example, is given as 50 on the tablet and corresponds to 500,000 inches in the landscape.

A Fourth Pythagorean Triangle

On the tablet, the radius of the circle is given as 31¼ units.

Two smaller right-angled triangles are shown. The diagram labels these as having sides of 30, 8¾ and 31¼ units (in hexagesimal or base-60 notation). This is a Pythagorean right-angled triangle with sides in ratio [7:24:25].

This ancient clay tablet from the era of the Babylonian First Dynasty depicts the remarkable geometric fact that the difference between angles of the [3:4:5] Pythagorean triangles is equal to the small angle of the [7:24:25] Pythagorean triangle. This is depicted in Figure 120.

In trigonometric notation:

$$\tan^{-1}(7/24) = 16.26°$$
$$\tan^{-1}(4/3) - \tan^{-1}(3/4) = 53.13° - 36.87° = 16.26°$$

This combination is unique: there is no other pair of Pythagorean triangles which have this relationship[127]. Scribes were recording the knowledge of these triangles at this early phase of history, more than a millennium before Pythagoras himself! Note that the bearing of the Sunrise Line is identical to the hypotenuse of the smaller triangle,

---

126 First reported in *Mem. de Miss. Archéol. en Iran*, Volume XXIX pp 44-62 (1943) and further described in Volume XXXIV *Textes Mathématiques de Suse* (1961).
127 Postscript: This is incorrect. If the height and width of the large triangle are whole numbers, the small triangle will always be Pythagorean.

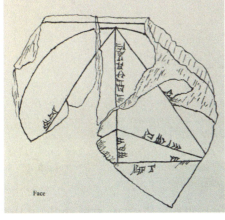

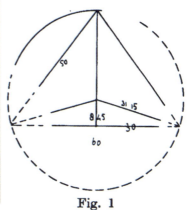

Fig. 1

Figure 119: Clay tablet from Susa, Iran.
Above: Photograph of the fragments recovered.
Below left: Transcription of the geometry and cuneiform characters.
Below right: Recreation of the full geometry depicted on the tablet. The large triangles are shown as [30:40:50] Pythagorean triangles.
The values of the cuneiform numbers are given as they are originally expressed in hexagesimal, or base-60 notation on the tablet.
The expression 31 15 denotes 31 15/60, or 31.25. Similarly 8 45 is equal to 8.75. Hence the values shown in the diagram give the sides of the small triangle as [8.75 : 30 : 31.25]. Proportionally, this can be expressed as [7:24:25].

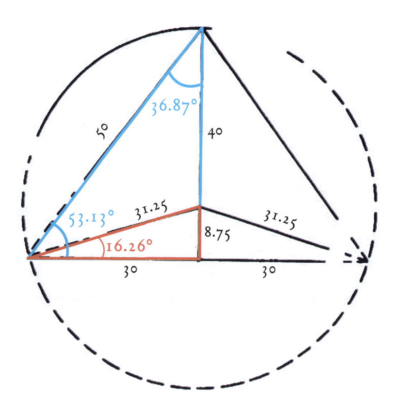

Figure 120: *The geometry of the Susa tablet. On the tablet, the larger blue triangle is labelled with a shortest side of 30, and a hypotenuse of 50. This is clearly intended as a [3:4:5] Pythagorean triangle with the other (unlabelled) side measuring 40. The smaller red triangle has sides labelled 8¾, 30 and 31¼. This can be expressed as a Pythagorean triple in the ratio [7:24:25]. Corresponding angles have also been added in the diagram above.*

the radius of the circle. This at last reveals the true grid angle of the Sunrise Line: it is intended to be the small angle of a [7:24:25] triangle; in grid notation [7:24], or 16.26° to the east-west line.

The existence of this tablet and the geometry inscribed on its surface is startling confirmation that the geometrical design that has been identified in the landscape around Campagne-sur-Aude and the Haute Vallée was known in the ancient world.

Note however that this does not necessarily imply or even suggest that there must be a direct connection between the tablet and the landscape geometry at Rennes, or that the geometers responsible for the Complex must have come from Susa. It is enough to show that this

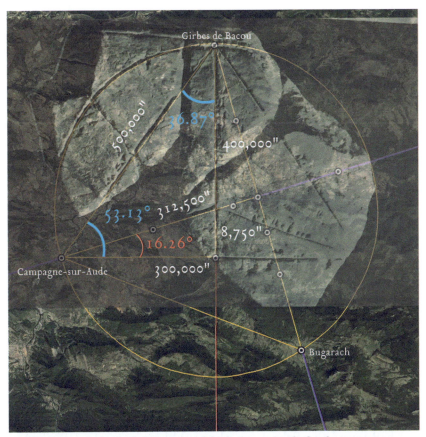

Figure 121: Superimposing the Susa tablet geometry on the landscape geometry reveals that they are identical, at a scale of 1:10,000 inches. The diameter of the circle is 625,000 inches, in the landscape, or 25 inches on a 1:25,000 scale map. Converted to angular measure: the horizontal width of the circle is 700 arc-seconds (= 625,000 × 28/25).

sophisticated specimen of geometric wonder was known in human history as early as the second millennium BC, and was part of the pool of knowledge available to the builders. Nevertheless, this discovery validates in the most remarkable manner the geometry that has emerged from the simple consideration of the interaction of the Sunrise Line with the Field of the Meridians.

If the measures and angles and Pythagorean triangles that have been found were a figment of the imagination, it would be inconceivable that the same identical details would emerge on a tablet from 1,700BC. For completeness, I should clarify that I did not come across the Susa tablet until 2020, long after discovering the details of the landscape geometry.

In light of these spectacular results, I submit that the Susa tablet geometry stands as independent proof and extraordinary confirmation that the geometry I have rediscovered in the countryside around Rennes-les-Bains in the Haute Vallée was indeed the result of a conscious and deliberate programme of active intervention in landscape by intelligent agents at some earlier period.

The Segments of 150,000 Inches

As we've seen, Girbes de Bacou evidently plays multiple significant roles within this remarkable complex of landscape geometry, including the northern terminal peak of the Bugarach Baseline, the sighting peak for the Lavaldieu meridian, and the corner of two Pythagorean [3:4:5] right-angled triangles.

Recall from above that the distance from Girbes de Bacou along the Bugarach Baseline to Point Q is 300,000 inches. As shown in Figure 122, this length is divided precisely in two by Combe Loubière. This implies that the distance from Girbes de Bacou to Combe Loubière is 150,000", and equally, Combe Loubière to Point Q is 150,000".

The reader will remember, all the way back in Chapter One, that we have already encountered this segment of 150,000 inches between Combe Loubière and point Q. Henry Lincoln had pointed out in *The Holy Place* that the distance from Combe Loubière to Bugarach Church along the Bugarach Baseline is divided equally in thirds by point Q and La Soulane.

I had found these three lengths could be accurately described as three lengths of 150,000" in the landscape. This led to the further breakthrough realisation that these segments were each represented by a length of six inches on a 1:25,000 scale map.

Now, more than twenty years later, I realised that there was a fourth segment of 150,000" to be added to the original three equal segments along the Bugarach Baseline, namely the length from Combe Loubière to the sighting point of Girbes de Bacou.

The total distance from Girbes de Bacou to Bugarach Church is four segments of 150,000 inches, for an overall length of 600,000 inches. The origin and purpose of this regular division of the alignment is coming into view. They are the result of generating a fan of alignments from Campagne-sur-Aude, at angles symmetrically arranged around the Sunrise Line, on to the sighting backdrop of the Bugarach Baseline. Furthermore, the angles have been laid out according to simple whole number grid ratios.

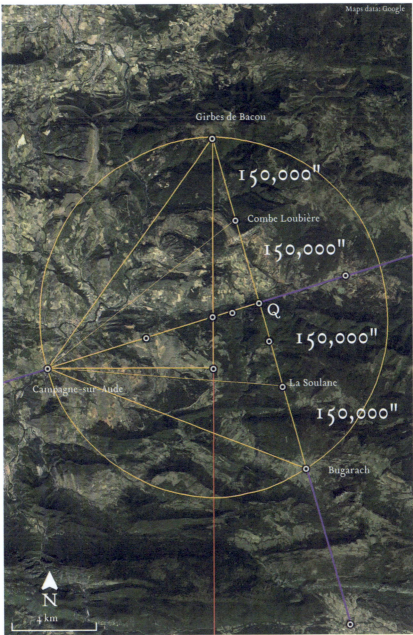

Figure 122: The distance between Girbes de Bacou and Bugarach is divided into four lengths, each of 150,000 inches. These lengths of 150,000 inches are represented as six inches on a 1:25,000 scale map, solving at last the origin of this relationship which I had first noticed back in 1996. Compare to Figure 10.

Sighting Grid on the Sunrise Line Alignment

These equal lengths were created as part of a system of calibrated measures on accurate sightings taken from Campagne-sur-Aude. They are laid out using triangles marked on grids at small scale to generate the required angles, then projected across the landscape at large. I devise a notation to represent the angles as whole number ratios in square brackets. For example, I will refer to the angle of the Sunrise Line to the horizontal, or east-west line as either [7:24] or 16.26° (because $\tan^{-1}(7/24)=16.26°$).

As shown in Figure 123, the positions of the six locations on the Bugarach Baseline can all be marked on such a grid conveniently. The bearings from Campagne-sur-Aude to the locations listed below generate the following grid-angles relative to the Sunrise Line[128]:

	Grid notation	*Angle to Sunrise Line*
to Girbes de Bacou	[3:4]	36.86°
to Combe Loubière	[1½ :4] or [3:8]	20.55°
to Point Q	[0:4]	0°
to La Soulane	[-1½ :4] or [-3:8]	-20.55°
to Bugarach Church	[-3:4]	-36.86°
to Caudiès-de-Fenouillèdes	[-3:2]	-56.31°

By simply constructing such a grid at convenient small scale and aligning it to the Sunrise Line, the correct angles could be read off to give the bearings from Campagne-sur-Aude to the various locations along the Bugarach Baseline. This simple whole-number grid system enabled the landscape engineers to establish a baseline of known angles which could be reliably and easily replicated and checked, by which to ensure the overall accuracy of the full geometric system.

---

128 Where [X:Y] denotes the number of squares in each direction to generate the angle, with Campagne-sur-Aude defined as [0:0]. X denotes number of squares parallel to Bugarach Baseline (positive towards north), Y the number of squares parallel to the Sunrise Line.

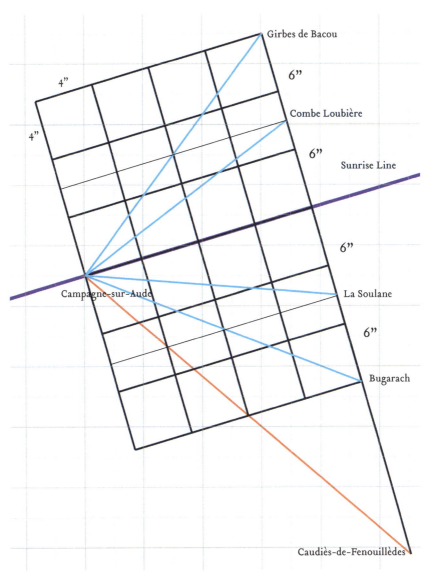

Figure 123: Grid geometry of the Bugarach Baseline. If this diagram is drawn so that the black squares of the tilted grid and the bold grey squares of the upright grid have sides of 4 inches, then it forms a 1:25,000 scale map of the landscape. The diagram acts as a master blueprint for the geometry and can be used as a simple aid to accurately generate the design. The geometer rules the grid in inches, aligns it to the Sunrise Line, and then joins appropriate grid points to create the required angles. In practise, small deviations from the ideal geometry will inevitably occur as it is marked out in landscape, but these do not alter the conceptual accuracy of the underlying design as pure geometry.

The Campagne-sur-Aude to Caudiès-de-Fenouillèdes Alignment

Figure 124 shows the alignment from Campagne-sur-Aude to Caudiès-de-Fenouillèdes. Notice that it passes exactly through the church of St Just-et-le-Bézu. This is of course the familiar 130° bearing to the south-west corner of the Arques Square. Caudiès-de-Fenouillèdes marks the intersection of this line with the Bugarach Baseline.

Now in light of the grid geometry shown in Figure 123, we can see how this angle is generated. The Campagne-sur-Aude to Caudiès-de-Fenouillèdes alignment runs at an angle of 56.31° to the Sunrise Line, which is itself on a bearing of 73.74°. The total of these is 130.05°. The combination of the two grid angles yields the target angle of 130° to an accuracy of 1/20th of a degree.

The position of Caudiès-de-Fenouillèdes on the Bugarach Baseline acts as an anchor point for the entire system. It establishes and records the 130° alignment from Campagne-sur-Aude which defines the Arques Square, so that it can always be resighted and recalibrated.

In effect, there are two grids established with their origins at Campagne-sur-Aude, and a [3:4:5] triangle has been constructed on both. One of these is aligned to the north-south and east-west. The other is aligned to the Sunrise Line and the Bugarach Baseline. This very remarkable line Campagne-sur-Aude to St Just-et-le-Bézu to Caudiès-de-Fenouillèdes integrates the two grids.

We can now begin to glimpse how these different elements interact. The Sunrise Line plays a primary role as the major axis for the geometry. The Bugarach Baseline, at right-angles, creates a framework on which the perfectly accurate work required for the Arques Square has been erected.

In the accompanying Figures, we are gazing upon a dazzling display of virtuoso geometry, based on a sequence of interlocking Pythagorean triangles, marked securely in the landscape of the Languedoc by supreme masters of this ancient science.

It is quite fascinating to consider that Henry Lincoln and David Wood had stumbled upon a fragment of the true solution when they wrote about the Sunrise Line. Some of the details were garbled in transmission, and they had not found the entire scheme by any means, but they had certainly identified some genuine fragments.

It is also truly uncanny to consider that my entire journey in this mystery began with the realisation that the three lengths of the Bugarach Baseline were each six inches on a 1:25,000 scale map. It is

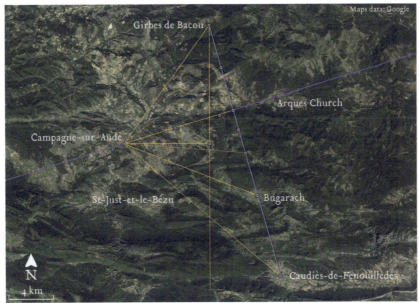

Figure 124: The bearing of 130° from Campagne-sur-Aude. It passes through St-Just-et-le-Bézu and meets the Bugarach Baseline at Caudiès-de-Fenouillèdes.

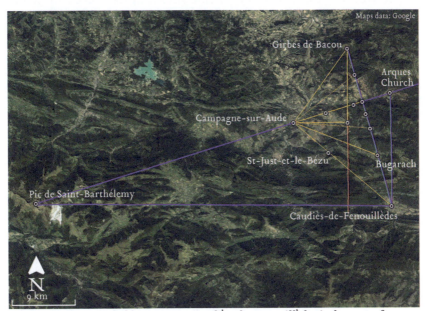

Figure 125: To acceptable accuracy, Caudiès-de-Fenouillèdes is due east of Pic de Saint-Barthélemy, and due south of Arques Church. The three locations form a large [7:24:25] triangle.

impossible to describe the sensation that this insight brought at the time. It was an overwhelming sense of having stumbled upon something of vital importance to this mystery. And yet, it was to be two decades until I was able to understand at last the reason for this repeated measure. It was to establish a series of calibrated sightline bearings from Campagne-sur-Aude as part of laying out the Arques Square.

An entire integrated system of wide-scale landscape surveying begins to come into view, based on these right-angled triangles and marked alignments. The Arques Square dimensions are now shown to be embedded within a wider scheme and to display knowledge of the dimensions of the Earth, its curvature, and the position on the Earth's surface at which this intervention has been laid out. The geometry was based on a system which integrated the inch, the *pied-du-roi* and the map scale of 1:25,000.

Pythagorean Triangles in Ancient Landscape

It might seem unlikely that such sophisticated geometry incorporating Pythagorean triangles could have been employed by landscape engineers centuries or even millennia before the ancient Greeks. There are however a number of megalithic sites that have been shown to have been laid out using such formats, including the [3:4:5] triangle.[129] In Brittany, near a village named Crucuno for example, there is a rectangular enclosure of standing stones that makes for an interesting comparison with the geometry described in the last few pages. It was surveyed by Alexander Thom, who measured it as 30 by 40 megalithic yards.[130] This implied a diagonal of 50 megalithic yards. Hence the rectangle contained a [3:4:5] Pythagorean right-angled triangle.

Thom also found that the diagonal of the rectangle, (or hypotenuse of the triangle) was aligned to the direction of the summer solstice sunrise at this latitude. The Crucuno enclosure is shown in a Google Earth image in Figure 126.

The units of measure are different than in the Campagne-sur-Aude geometry but otherwise, the Crucuno rectangle is constructed from the same elements as described in this chapter: the [3:4:5] Pythagorean triangle, laid out in whole number measures, aligned with an important local sunrise. Crucuno is a precedent for the Campagne-sur-Aude geometry and shows conclusively that such techniques were employed.

129   See for example Anne Macauley, *Megalithic Rhythms and Measures: Sacred Knowledge of the Ancient Britons*, Edinburgh, Floris Books, 2006, and Robin Heath, *Alexander Thom: Cracking the Stone Age Code*, Pembrokeshire, Bluestone Press 2007.
130   One megalithic yard = 2.72 feet

Figure 126: Crucuno, Brittany. 47°37'30" N; 3°07'18" E. *This megalithic stone rectangle is laid out as a [3:4:5] triangle in units of 10 megalithic yards and oriented to the compass angles. (1 megalithic yard (MY) = 2.72feet) The bearing for the diagonal of the rectangle is therefore 53.1°N, corresponding to the the summer solstice sunrise at this latitude.*

After many years of contemplating the landscape geometry of the Haute Vallée, an underlying order had slowly emerged. The critical insight was the realisation that the inch was the primary unit of measure of the Complex. The various alignments weave together to form a unified geometric system in the most exquisite and exact manner, revealing the handiwork of the Master Geometers who created it.

Purpose of the Arques Square

Why would it be of any benefit to construct a geodetic square on the surface of the Earth measuring one tenth of a degree of longitude in width? The answer is to establish one's location in space and time. The transit of heavenly bodies across the two meridians can be observed and the time between them carefully measured. From this data, over time, a great deal of very accurate and useful information about the motion of the heavenly bodies and the size of the Earth can be assembled.

As a thought experiment, imagine that somewhere on the equator, two meridians are marked, separated by one degree of longitude. The distance between them will be 365,224 feet. This value is almost exactly equal to the number of days in a year multiplied by 1,000 – a remarkable, even astonishing, synchronicity.

As one degree is equivalent to 60 arc minutes or 3,600 arc seconds, this implies that the separation between the two meridians of 3,600 arc seconds will be 365,224 feet. This is a very convenient pair of values, as it permits ready conversion back and forth between units of measure in space and time.

In practice, the distance between two such meridians of nearly 70 miles would be too far for reliable and repeatable line-of-sight communication. But if we were to reduce the separation by a factor of ten, to one tenth of a degree or six arc minutes, or 360 arc seconds, it would correspond to a width of 36,522 feet, or around seven miles, which would be manageable.

So, we imagine a very simple observatory comprising a geodetic square of one tenth of a degree in width laid out on the equator. Its two vertical sides act as a pair of calibrated meridians for observations of transits. They are separated by 360 arc seconds, or 36,522 feet.

Similar observatory squares could be laid out on other latitudes. Maintaining the angular width of one tenth of a degree, such squares would be shorter in absolute measure as we moved further north or south away from the equator. This presents an opportunity: we could locate squares at particular latitudes that offered other convenient conversion values.

For example, at latitude 9.73°N, a square of 360 arc seconds width measures 36,000 feet.[131] At this latitude therefore, one arc second has a length of 100 feet.

At latitude 34.86°N, a square of 360 arc seconds will measure 360,000 inches. At this latitude therefore, one arc second has a length of 1,000 inches.

Both of these are obvious choices. Here is another possibility:

At latitude 42.91°N, a square of 360 arc seconds will measure 321,428 inches = 360,000 inches x 25/28.

This is of course the latitude of the Arques Square, and a mere 40 inches shorter, which, over five miles, is essentially identical. The division of 360 by the factor of 28 is of great interest because:

$$360 / 28 = 12.85714...$$

This is identical to the value of the *pied-du-roi* in inches. The division of 360 into 28 was familiar to ancient astronomers and astrologers. In

---

131 36,000 feet = 360 arc seconds, therefore 100 feet = 1 arc second

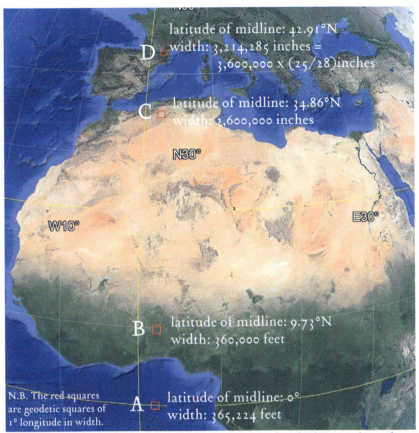

Figure 127: Hypothetical geodetic squares, *A, B, C* and *D*, laid out on the surface of the Earth. The width of each square in angular measure is one degree of longitude. Their midline latitudes and widths in feet or inches are shown above. At the equator, the width of Square *A* is the number of days in the solar year x 1,000. The Arques Square corresponds to Square D, at 1/10 scale, or 1/10° degree wide.

Indian and Chinese astrology, the circuit of the heavens was divided into 28 "lunar mansions" corresponding to the full range of phases of the moon. Each of these mansions or houses therefore had a width of 360°/28 = 12.857°.

In the context of our thought experiment, it would be highly convenient to employ a unit of measure obtained by dividing 360 inches by 28. This permits measures corresponding to lunar house dimensions to be readily incorporated into the system of interchangeable units which characterise our virtual observatory. Is it possible that the *pied-du-roi* may have had its origin as a geodetic measure arising out of and relating to the Arques Square? Was the *pied-du-roi* originally defined as the width, on a 1:25,000 scale map, of a geodetic square one tenth of a degree of longitude, at the latitude where it was equal to 360,000 x 25/28 inches, namely 42.916° N, six degrees south of Paris? If the true origin of the *pied-du-roi* emerged from just these considerations then the Arques Square might be considered as a kind of ancient implementation of the modern Bureau of Weights and Measures, in which reference standards of lengths and other measures are kept.

All of these speculations might strike the reader as rather fanciful and arbitrary, but the geometric reality is that the Arques Square fulfils precisely the requirements of both location and dimension to the letter. Unless we are dealing with a co-incidence of staggering rarity, we can only conclude that the Arques Square was laid out by engineers who knew the dimensions of the Earth.

They employed the very unit of measure which was the standard of length in France from antiquity until the dawn of the modern era, thus exhibiting a mastery of integrating Earth dimensions with landscape surveying with units of measure, though it has been thoroughly hidden from the eyes of the world for millennia.

Finding the position on the Earth's surface which corresponds to a calculated latitude would not have presented insuperable difficulties. Whilst the problem of longitude was not, according to the conventional history of geodesy, solved in the ancient world, we can be confident that determination of latitude was achievable. The discoveries by Professor Richer of lines of latitude on which multiple temples and other sacred sites may be found in ancient Greece is sufficient proof of this.

Summary

My long relationship with the 45° alignment had culminated in it opening like a time capsule to display an entire geometric structure embedded in the surrounding landscape, defining it, recording it, and ensuring that it persists over long time cycles. It's as though the blueprint of the Arques Square is encoded or enfolded in the alignment itself. It took twenty years of mapping it, thinking about it, exploring it, and even dreaming about it, to eventually reveal its hidden contents to understanding.

What is it that has been brought to light? It is a very carefully laid out geodetic square, over five miles in width, accurately surveyed and marked out in landscape. Both its size and the locations of the sites from which it has been formed are highly resolved. The key elements of measure by which it has been laid out are the inch, the *pied-du-roi* and the 1:25,000 scale. It is defined by a segment of the 45° alignment, confirmed by the position of Campagne-sur-Aude and sealed by the Villeneuve-lès-Montréal alignment. It embodies knowledge which reflects an understanding of the size of the Earth and the position of the landscape geometry on its surface.

The landscape, the map, the 45° alignment, the twelve-fold division, or zodiac, or cromlech, and the meridians all harmonise smoothly with the Arques Square. The entire set of elements co-ordinate around the 1:25,000 scale, the inch, the arcsecond and the *pied-du-roi*. All of this can only take place at the specific latitude that corresponds to the midline of the Arques Square, marked by Campagne-sur-Aude.

The Arques Square complex includes a calibrated field of meridians that act as a reference frame for precision observations in the sky and in the landscape. The formats that have been embedded into the territory itself include an entire conceptual apparatus by which their spatial relationships may be conveniently handled.

It is a precision astronomical instrument for the ages, by which the Earth and heavens were observed, measured and more. The Arques Square is an invisible remnant of an ancient architecture of landscape. Its re-emergence into modern times is due as much to dreams, intuition and synchronicity as to the immutability of geometry and stone. It stands as silent witness to the ability of the ancients to inscribe geometry into territory at large scale, at significant locations which demonstrate knowledge of Earth dimensions, and with precision handling of measure and angle.

In the first three parts of this book, we have plotted a route through the tangled woods of the Affair of Rennes by following the trail set by the riddles of *Le Serpent Rouge*. We've come a long way. In Part One, we uncovered the hidden identity of the author and made inroads into the sources for some of the language in the poem.

In Part Two we turned our attention to the geography and setting of the poem and revealed the Cromlech-Zodiac of Rennes-les-Bains.

In Part Three, we've explored certain implications of these discoveries, and brought to light the carefully concealed secrets of Boudet's cromlech, his enigmatic map and the parchments. These have led in turn to the revelation of the Arques Square, an extraordinary intervention in ancient landscape. The poem has unwrapped itself to reveal deep layers of knowledge, and this spectacular sequence of inter-related geometric forms.

The reader can now breathe a quiet sigh of relief, as this brings to a close the more technical elements of this presentation involving geometry, numbers and maps. But we are not quite finished, and there remain several riddles to be addressed. We now know what the poem is about; but what does it all mean? In Part Four, we will turn our attention to the ideas behind the poem.

We will trace the roots of *Le Serpent Rouge* back in time to locate the texts which inspired Richer to write it, and the sources where he found the blueprint on which it has been constructed. We will also follow the trail forward, to examine certain more recent writings which relate to these themes. Our aim will be to investigate how these ideas have been preserved, propagated and eventually recovered, to try to understand how the poem has functioned as a vehicle for the transmission across time of a body of forgotten knowledge.

PART FOUR

Transmission

Chapter Fifteen
GRAND VOYAGER OF THE UNKNOWN

WHERE DID Jean Richer find the literary inspiration for his prose-poem *Le Serpent Rouge*? In this chapter we will try to answer this question by following certain clues provided in the poem itself. Consider, for example, the very first words of the opening stanza of *Le Serpent Rouge*:

"How strange are the manuscripts of this Friend..."

What exactly are these manuscripts and why should they be described as strange? We know who "this Friend" is of course: it is he whose "number is that of a famous seal", Gérard de Nerval. The "manuscripts of this Friend" must surely therefore be *the writings of Nerval*. But why would they be described as "strange"? In a 1957 book about Nerval's life and work, Richer wrote:

"While madness gave his later work a fascinating strangeness, this was not what made him a great writer."[132]

There it is. According to Richer's own account, there was a "strangeness" around Nerval's work, a result of his sometimes precarious and deteriorating mental state. Beyond doubt then, the manuscripts are the writings of Nerval. A little further on, in the third stanza of the poem, we read:

"In despair of finding the way, the parchments of this Friend were for me the thread of Ariadne."

The word *"parchemins"* (parchments) has been used interchangeably here with *"manuscrits"* (manuscripts). For the author of *Le Serpent*

---
132 Jean Richer, *Gérard de Nerval : Poètes d'aujourd'hui*, (Paris, Editions Pierre Seghers, 1957). p. 18.

Rouge, the writings of Nerval have enabled him to find his way through the maze of the poem, just as Ariadne's thread guided Theseus through the labyrinth of the Minotaur. If we compare the line above from the poem with a comment written by Richer in that same year 1967 in which *Le Serpent Rouge* was published, we learn the reason why the manuscripts of Nerval gave him direction:

> "The attentive and sustained study of the work of the poet Gérard de Nerval was a discipline that accustomed me to taking a comprehensive view of complex systems."[133]

Richer has left clear indications, therefore, that enable us to interpret his comments in *Le Serpent Rouge* about the "strange" "manuscripts" of his "friend" playing the role of the "thread of Ariadne": it was in the poetry and prose of Nerval that Richer found the intellectual training that enabled him to navigate this maze and made possible his discoveries in the ancient world.

But what relevance could the literary output of a nineteenth-century French poet have to understanding early Greek conception of landscape? In order to answer this question, we need to turn to Richer's writings on Nerval.

The arc of Richer's evolving understanding of Nerval's thought may be traced in his three major publications on the poet and his work, which appeared at different stages of his career in 1947, 1963 and 1987. None of these have been translated into English. These are relatively obscure works, even in France, and are almost unknown in the English-speaking world.

Richer's first book was *Gérard de Nerval et les Doctrines Ésotériques*, based on his PhD dissertation and published in 1947. The title already signals the focus of his interest: the esoteric. Relatively short, it has chapters with titles including (my translation, with originals in footnotes): "Symbolism and the Kabbalah", "Masonry and Arithmosophie", "The Queen of Sheba" and "Descent to the Underworld and Initiation".[134] *Arithmosophie*, as we have seen in Chapter Seven, is the French term for the serious literary game of turning words into numbers using a simple alphabet cipher. "*La reine de Saba*" is the Queen of Sheba, the "*Reine d'un royaume disparu*", the queen of a lost kingdom mentioned in the Virgo stanza of *Le Serpent Rouge*. The text of Richer's monograph covers topics including Martinism, initiation, Goethe, tarot, the

---
133   Jean Richer, *Sacred Geography of the Ancient Greeks*, op. cit. p. 261.
134   "*Symbolique et Cabbale*", "*Maçonnisme et Arithmosophie*", "*La reine de Saba*", and "*Descente aux enfers et initiation*"

ancient Egyptian goddess Isis, serpent symbolism, the spiral nature of time and the grand myths and allegories of the ancients. Here we can glimpse in embryonic form the elements of *Le Serpent Rouge* already beginning to coalesce in Richer's mind.

In 1963, Richer published a greatly expanded and revised work on Nerval. This huge 700 page tome was entitled *Nerval: Expérience et Création*. It is a deep exploration and meditation on texts from the full range of Nerval's writings including his poetry, fiction, reportage, correspondence, writing for the theatre and much more.

In 1987 Richer offered his final work on the poet, entitled this time: *Gérard de Nerval: Experience vécue et création ésotérique*. Here at the end of his career Richer produced a summation of his lifework on Nerval, revisiting and reworking the themes that had occupied him for over four decades.

These three books offer the opportunity of a series of snapshots of the development of his thought as it evolves significantly over the years. It is not until the final work that we see his fully matured views on display. And it is here, some twenty years after *Le Serpent Rouge* appeared, that Richer placed on record the crucial clues for understanding the architecture of his 1967 poem.

In his 1987 book, Richer introduces a completely new analysis of Nerval's short story *Sylvie*, with elements that are entirely absent from his earlier 1963 book. In his final work, he proposes that the narrative is built around the idea of an initiatory journey around a zodiac. He demonstrates at length that the characters, events and action are based on this hidden architecture, and even supplies the source for this theoretical framework:

> "*Sylvie* describes very exactly a process of psychological regression in the sense understood by Jung and his disciples, and one which takes the form of an initiation on a round of the hours, coupled with a reverse-circuit of the zodiac."[135]

In this passage, Richer's description of *Sylvie* can now be recognised as an exact template of *Le Serpent Rouge*. He even goes so far as to create a diagram of the zodiac structure, as shown in Figure 128. He uses it as a map to navigate the narrative of *Sylvie*, and to reveal the hidden currents which flow beneath the surface of Nerval's story. One glance at this extraordinary schematic of the plot is sufficient to confirm that

---

135   Jean Richer *Gérard de Nerval: Experience vécue et création ésotérique Gérard de Nerval : Expérience vécue et création ésotérique*, (Paris, Guy Trédaniel, 1987).

we have found the prototype for the poem. *Sylvie* is the 'manuscript' or 'parchment' of his 'friend' that helped Richer to find his way. Taking hold of Ariadne's thread, we will now proceed to a close examination of Nerval's charming story.

Sylvie

In Nerval's short novel, first published in 1853, the narrator embarks on a tour of the villages and landscape of his childhood in the Valois region of north-east France. It can be read as a delightful reminiscence of a Parisian writer returning to the countryside of his childhood, but it is also a deeply considered meditation on memory, dream, love, loss and the nature of time. Some critics have considered it to be a minor masterpiece of nineteenth-century French literature. Umberto Eco went so far as to call it "one of the greatest books ever written".[136] Richard Sieburth writes:

> "The finest of these promenades is the love-story of his childhood and adolescence, *Sylvie*, subtitled *Souvenirs du Valois*. It recounts a series of return journeys to Mortefontaine, layering one memory on top of another in a complex time-scheme greatly admired by Proust. ... Yet its psychology, and its symbolic structure, are so subtle and modern as to appear almost contemporary, reminding one of nothing so much as a film such as Ingmar Bergman's *Wild Strawberries*."[137]

In *Sylvie*, via a series of flashbacks, the action of the present and the narrator's memories of his past become intermingled into a single, seamless flow. This creates a rich context to explore his complicated emotional relationships with three women in his life: Sylvie, his childhood sweetheart, still living in the village of their youth; Aurélie, the actress on the Parisian theatrical stage to whom the narrator/author[138] has devoted his affection; and the figure of Adrienne, the blonde girl he met fleetingly as a boy, who sang at a fête he once attended and who he had never forgotten.

The action of *Sylvie* in the first twelve of its fourteen chapters takes place over the course of a single day, 23/24 August, but the narrative

---

136 Umberto Eco, *Six Walks in the Fictional Woods*, (New York, Harvard University Press, 1994), p. 11.
137 Sieburth, *Gérard de Nerval Selected Writings*, op. cit. p. 64.
138 In real life, Nerval had fallen head over heels in love with an actress named Jenny Colon. In Sylvie, she is portrayed as Aurélie.

is punctuated by a sequence of memories and reminiscences so that the past and the present comingle. The effect is to blur the passage of time so that it becomes a challenge to keep track of the distinction between the narrator's account of his journey and his recollections of his earlier life. This is a deliberate strategy of the author, who remains in perfect command of the temporal flow of his story. Sieburth writes about this curious effect of Nerval's treatment of time, quoting Proust:

> "Proust observed that to read *Sylvie* for the first time was to experience a disorientation verging on mild panic. Forced at every moment to leaf back to the preceding pages just to get their textual bearings (particularly during the first seven nocturnal chapters of the novella) the reader – like the narrator himself – feels perilously lost in the woods."[139]

The date on which the story unfolds 24 August, is the Feast Day of St Barthélemy. On this day each year, the inhabitants of the local villages enact a celebration, the *Fête du Bouquet provincial*, featuring archery, dancing, singing and wagons drawn by oxen. The narrator recalls the joyful memory of taking part as a young boy and is drawn back to the village from Paris to experience it again as a young man when he sees some short lines in a newspaper announcing it is to take place the next day. The *Fête* has its roots in the far distant past and represents an unbroken link to the "ancient land of the Valois where, for more than a thousand years, the heart of France has beaten". As Nerval writes at the end of Chapter I, "we were merely repeating from age to age a Druidic festival that had survived all subsequent monarchies and forms of religion."

This date, 24 August, is of course the same day on which the sunrise from the summit of the Pic de Saint-Barthélemy was observed in an age-old ritual, as discussed earlier. It is certainly an uncanny synchronicity that Nerval's story and the Languedoc geometry should share this same important fête. Richer could not have been aware of this invisible thread connecting the two regions at opposite ends of France, and yet, he almost hints at a connection when he refers to the setting of *Sylvie* as "this mystical geography of a Valois of dreams".[140]

---

139 Sieburth, *Gérard de Nerval Selected Writings*, op. cit. pp. 64, 148, 149.
140 Jean Richer, *Gérard de Nerval : Expérience vécue et création ésotérique*, (Paris, Guy Trédaniel, 1987), p. 319.

The Zodiac Structure of *Sylvie*

In this new discussion of Nerval's story in 1987, Richer set out to demonstrate that *Sylvie* is based around the concept of a clock face, inspiring him to construct a model based on a twelve-fold division of a circle. This also functions as a zodiac, setting up a rich field of symbolic associations that fuse *Sylvie*, the life of the poet and cycles of time into a single, cohesive framework. Richer describes the discovery of the specific key that enabled him to reconstruct the circular form of the narrative and chapter structure of *Sylvie*.

He had noticed a short passage near the end of Chapter III, in which the character of Adrienne is identified with a statue of Diana/Artemis accompanied by her deer, mounted on an antique Renaissance clock. Diana, goddess of the hunt, is the female deity who rules over Sagittarius, sign of the bow, and who is also associated with deer. Richer understood this to imply that Nerval was linking Adrienne with Sagittarius, via Diana. Immediately following this passage, a cuckoo clock chimes the time: it is one o'clock in the morning.

From these apparently slender clues, Richer recreates the zodiac format of *Sylvie*, which acts as the underlying framework for the structure of the story. The unfolding action, the characters, and even the mood of each chapter, subtly reflect the symbolism of the zodiacal sign in which it occurs.

The validity of this mapping is supported by clues highlighted by Richer from throughout the narrative. It is also supported by some revealing comments from contemporary correspondence of Nerval, in which he discusses the role of symbolic planetary associations in the invisible architecture of *Sylvie*. These confirm that the zodiac form was not just a figment of Richer's imagination but a conscious narrative strategy of the author himself.

In the scheme that Richer recreates, each chapter is assigned to an hour. The correct alignment of chapters to hours is calibrated by the striking of the clock in Chapter III. Thus, Richer arranges his diagram so that one o'clock corresponds to Chapter III, with the remaining chapters then distributed in turn around the clock face.

Accordingly, Chapter I corresponds to eleven o'clock, Chapter II to twelve o'clock, Chapter III to one o'clock, etc. There is a total of fourteen chapters, so that after one complete round of the clock face, the cycle continues and overlaps to account for the final two: Chapter XIII corresponds to the same hour as Chapter I, namely eleven o'clock, and Chapter XIV to the same as Chapter II, namely twelve o'clock.

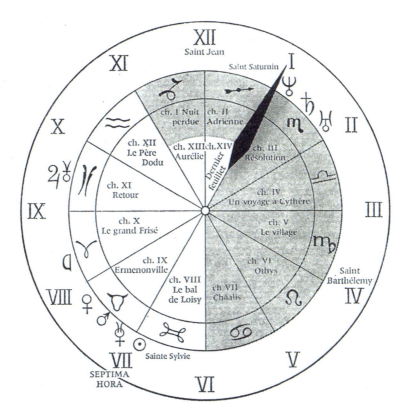

Figure 128: The zodiac structure of Nerval's Sylvie (1853), according to Jean Richer, as it appears in a 1966 paper Saggi e ricerche di letteratura francese, (Pise, t. VII, 1966, pp. 201-37), and then later in 1971 in Nerval au royaume des archétypes Octavie Sylvie Aurélia (Archives des lettres modernes, no. 130, 1971, (11) V, archives nervaliennes no 10), then finally in 1987 in his Gérard de Nerval: Experience vécue et création ésotérique (Paris, Guy Trédaniel, 1987), p.253.

Next, Richer suggests that a zodiac format is also to be overlaid on the clock face. The clue that helps to determine the correct orientation is again provided by the incident in Chapter III, in which the character of Adrienne is associated with the goddess Diana/Artemis, representing Sagittarius.

Chapter II is entitled Adrienne; therefore, says Richer, we allocate this chapter to the sign of Sagittarius. The remaining signs of the zodiac can then be inserted on the clock face, this time proceeding in an anti-clockwise direction. The result is shown in Figure 128.

There is one further subtle detail to be added before the complete map schematic is complete. Richer decides that the hour hand of the clock should be set to the position of one o'clock, to mark the key moment of the striking of the cuckoo clock. With this configuration of clock-face, zodiac and chapters now defined, Richer observes that the hour hand pointing to one o'clock happens to correspond to 0° Sagittarius on the zodiac, which is the exact position of the Ascendant in Nerval's zodiacal birth-chart!

Taking a cue from this serendipitous match, he then proceeds to add the positions of the planets at the poet's birth to the clock-face diagram. Finally, he also marks the feast days of the saints that occur in the story. The result is a richly overlaid tapestry of associations, weaving together the narrative arc of *Sylvie* with the hours of the day, the signs of the zodiac, the birth-chart of Nerval and the feast days of the Saints.

Richer proceeds methodically through the chapters of *Sylvie*, highlighting characters, keywords and incidents that align with the hour and zodiac sign of the allocation. These connections are not so deterministic as to become formulaic. Some of his suggestions are entirely plausible whilst others seem to be perhaps stretching the point. They are unlikely to be noticed by the casual reader, yet Richer makes a compelling case. This zodiac is configured as a symbolic space, rather than as a literal geographical construct, but the resonance with *Le Serpent Rouge* is clear.

It is emphasised by a comparison he draws between the action of *Sylvie* and the layout of a popular board game of the eighteenth and nineteenth centuries, known as the Game of the Goose, or *jeu de l'oie*. For Richer, there is little doubt that the association was intended by the author, and he mentions the point twice:

> "In this Game of the Goose that Gérard takes us through..."
> "The comparison of *Sylvie's* structure with the Game of
> the Goose was imposed on us."[141]

The game board of *jeu de l'oie* depicts a spiral journey of sixty squares, representing the number of months in five years, around which players advance with a throw of the dice. Nigel Pennick discusses this famous game at length in his wonderful compendium of ancient and modern lore, *Secret Games of the Gods*. He places it in a lineage of similar games which date back to the Game of the Serpent in ancient Egypt, and

---

141 Richer *Gérard de Nerval : Expérience vécue et création ésotérique*, op. cit. p. 258 & p. 273.

forward to Snakes and Ladders in more modern times. So, we have yet again the idea of the symbolic course of the sun represented as a spiral or circular path, on which a journey is enacted, which is also linked with the serpent!

Richer's chapter on *Sylvie* includes many other fascinating and relevant items. He chooses to use the word *"pèlerinage"*, or pilgrimage, to describe Nerval's journey in the story, a term with obvious resonance to the *"pèlerinage éprouvant"* of *Le Serpent Rouge*. In summary, it is clear that in Richer's 1987 book he establishes Sylvie as the archetype of the format he employed in 1966 to construct the poem, the notion of a journey around a space constructed on the model of the zodiac.

When Did Richer Discover the Zodiac in Sylvie?

There is no mention of the zodiac specifically in relation to *Sylvie* in Richer's 1963 book. There is considerable discussion about its complex temporal structure based on flashbacks and the tricks of memory, so we can be confident that questions of its relationship with time were very much on his mind. But nowhere in the work is there any suggestion of a cyclical format, much less a zodiac template underpinning the events that are narrated. The inspiration for this idea must therefore have come some time after 1963, but before 1987 when it appeared in his final book. Evidently, Richer's views on Nerval's story went through a major transformation, even an enlightenment, sometime between his two latter works.

This prompts the obvious question: when exactly did Richer's notion of an underlying zodiac format of *Sylvie* first occur to him? Was it around the same time as, or even before, *Le Serpent Rouge*?

Richer published this material for the first time in 1966 in an Italian literary journal in a paper entitled *Sylvie de Nerval, ronde des heures, un récit en forme de voûte étoilée* (or "Nerval's *Sylvie*, a round of hours, a story in the form of the starry vault"). A "story in the form of the starry vault" would be an equally appropriate title for the zodiac-shaped narrative of *Le Serpent Rouge*.

He republished the paper in 1971 as part of a short collection of essays on three of Nerval's works, called "*Nerval au royaume des archétypes: Octavie Sylvie Aurélia*", or "Nerval in the kingdom of the archetypes: Octavie Sylvie Aurélia". [142] Here it appears under the chapter heading "*Sylvie: Ronde des Heures. Le Zodiaque*". This is the origin of the material,

---

142 Jean Richer, *Nerval au royaume des archétypes : «Octavie» «Sylvie» «Aurélia»*, Archives Nervaliennes, no. 10 in Archives des lettres modernes no. 130 (1971).

including the zodiac diagram of the structure of *Sylvie,* which subsequently appears in his 1987 book.

The time-frame of the origin of Richer's novel understanding of the zodiac format of *Sylvie* can therefore be traced back to 1966, the very year of composition of *Le Serpent Rouge*! The notion of a narrative built around a tour of a zodiac was occupying Richer's thought even as he was visiting Rennes-les-Bains and reading Boudet's book after meeting de Sède and Plantard during that time.

Sylvie and Le Serpent Rouge

With Richer's discovery of the zodiac of Rennes-les-Bains, assimilated to Boudet's cromlech, an irresistible literary opportunity presented itself. He would write a prose-poem that took as its action a description of a walk around the village and the landscape zodiac he had found. He would structure the work after the model of *Sylvie,* by allocating each stanza in turn to a sign of the zodiac, thus setting up a symbolic link between form and content.

In this manner, he could write openly about the zodiac of Rennes-les-Bains, without explicitly revealing the details. The description is concealed in the stanza structure, rather than in their content! To unlock this layer, it is necessary to have recognised and identified the twelve-fold division in the landscape around Rennes-les-Bains. With the correct allocation of zodiac signs, the unfolding action of the narrative of the poem can then be mapped to the landscape.

In constructing *Le Serpent Rouge,* Richer has taken the textual architecture which Nerval employed for *Sylvie,* based on a progression around a zodiac, and used it as framework for a retelling of the narrative of Boudet's tour of the cromlech in his chapter VII. This is why the "manuscripts of his friend" were like the "thread of Ariadne". It was by analysing and coming to an understanding of the inner dynamics of *Sylvie,* and similar related works by Nerval, that Richer was able to acquire the means to orient himself within such difficult and challenging material. In particular, *Sylvie* was the guide that helped him to find his way through the labyrinth of the sacred geography of Rennes.

Our trail has led from Jean Richer's *Le Serpent Rouge* to Gérard de Nerval's *Sylvie.* Both can now be understood as examples of stories that feature a conventional narrative structure on the surface but are patterned after the template of a concealed esoteric format, namely the zodiac in this case. But where did Nerval derive the idea for such an approach to storytelling?

Conte Initiatique

For Richer, such works as *Sylvie*, and *Aurélia*, another short novel from this late stage of Nerval's career, fall within a genre which he calls the *"conte initiatique ou du récit de voyage allégorique"*[143], which we might translate rather clumsily into English as the "initiatory tale, or account of an allegorical journey". There are various examples of such works in the history of European esoteric literature.

Nerval draws on this rich tradition of the novel constructed as an allegory of an initiatory quest or voyage. In European literature, the text considered to have established the genre was the celebrated *Hypnerotomachia Poliphili*, (Poliphilo's Strife of Love in a Dream) by Francesco Colonna, first published in 1599. Commenting on Chapter IV of *Sylvie*, Richer remarks that Gérard was inspired in some of these passages by the chapters of the *Hypnerotomachia* which are replete with astrological allusions. One of the most influential of such works however was created by Goethe, the forerunner of the genre known in German as the *Kunstmärchen* or artistic fairy tale.

Goethe's Märchen

Originally published in 1795, Goethe's story *Märchen*, known in English as "*The Green Snake and the Beautiful Lily*", grew out of some correspondence with Schiller, in which he spoke of his desire to represent a certain intuition in narrative form. He wanted to write a work that explored in a profound manner, under the guise of a simple and beautiful fairy-tale, what he described as the freedom of the human soul in coming to a relationship with super-sensible reality.

Nerval was certainly familiar with Goethe's work. Indeed, he owed his career as a writer to his early success in translating *Faust* into French at the age of just twenty. Goethe himself was highly impressed with Nerval's effort. It is said that when he read it, he remarked that it had never been understood so well. This is a remarkable literary compliment paid by the "Great Man of German letters" to a twenty-year-old writer for a debut work and demonstrates the depth of Nerval's insight both into Goethe's masterpiece and his mindset.

In *Nerval Expérience et Création* Richer wrote:

> "He may have read Goethe's *Märchen*. It was there, perhaps, that he came up with the idea of poetically using certain images borrowed from alchemy and tarot."

---

143 Jean Richer, *Nerval Expérience et Création*, (Paris, Hachette, 1963). pp 508,140

Whether Nerval consciously adopted *Märchen* as a prototype for *Sylvie* and other works, in Richer's view there was no doubt that he was working in the same vein as Goethe, and even that the results were comparable. Such stories as these examples may be read as enchanting tales, woven of elements of the everyday life of people of the times, and yet beneath their subtle surfaces lie deep structures that have been carefully crafted out of archetypal material, drawn from elements of tarot, zodiac, alchemy and related topics.

The plot of *Märchen* is too convoluted to summarise here, but a measure of its charm can be had from a list of its characters: they include two will-o'-the-wisps, a ferryman and his wife, their small dog, three vegetables, a magical payment of gold, a benevolent green snake, a prince, the beautiful Lily and a friendly giant. There is an underground temple and a river and a mountain.

In the climax of the story, the temple comes to the surface, the giant is transformed into a huge gnomon at the centre of the city square, with the hours of the day marked as a sundial into the pavement in mysterious symbols, whilst the snake is transformed into a bridge made of precious stones.

We can speculate on the direct extent to which Nerval found his inspiration for *Sylvie* from *Märchen,* but there is certainly a connection that can be drawn between Goethe's masterpiece and *Le Serpent Rouge*.

Le Serpent Vert

In 1935, a translation of *Märchen* into French by the Swiss occultist Oswald Wirth was published, entitled *Le Serpent Vert* (The Green Snake). There are clear resonances between *Le Serpent Rouge* and *Le Serpent Vert,* beginning with the pairing of names and, as we will see, there is no doubt that Richer was heavily influenced by this work.

Le Serpent Vert was accompanied by an extended essay, also by Wirth, which offered a profound analysis of Goethe's mystery tale, illustrating the deep currents of esoteric lore at work in the text. This essay played an influential role in certain strands of twentieth-century French literature. In his analysis, he placed considerable emphasis on an interpretation of the archetypes of the fairy-tale in terms of the tarot.

Oswald Wirth was a leading esoteric scholar, artist and writer, born in 1860. Deeply interested in Freemasonry and astrology, he was a student of the poet and esotericist Stanislas de Guaita, with whom he created a celebrated version of the Tarot deck, derived from the Marseille tarot. We note in passing that the tenth Arcana of this deck, the Wheel of Fortune, depicts a caduceus formed of two entwined serpents. One is green, the other red.

We have, of course, already encountered Oswald Wirth in these pages. It is his line drawing that appears on the second page of the *Le Serpent Rouge* pamphlet, with the caption, *"découvrir une à une, les soixante-quatre pierres"*. We can be certain that Richer was familiar with Wirth's translation, as he tells us in several places, including in *Nerval: Expérience et Création*:

> "The too little-known translation of *Märchen* by O. Wirth is entitled *The Green Serpent*. It is accompanied by an important commentary, which highlights Goethe's borrowings from alchemical symbolism and tarot."[144]

In one passage, he even compares Nerval's work directly to Goethe's classic short story. Here for example, in *Nerval au royaume des archetypes*, his essay on the Nerval works *Octavie*, *Sylvie* and *Aurélia*, published in 1971, he wrote:

> "We wanted to establish more precisely than we had previously done that Nerval, the author of *Aurélia*, endowed French literature with an initiatory narrative, quite comparable to Goethe's *Märchen* whose symbolism, as Oswald Wirth showed, is borrowed from the Tarot."[145]

The antecedents of the poem of *Le Serpent Rouge* can therefore be traced back to Goethe's *Märchen* via two textual paths. In the first place, *Märchen* arguably influenced Nerval to write *Sylvie* in the vein of the *conte intiatique*.

Secondly, and independently, *Märchen* inspired Wirth's translation, *Le Serpent Vert*, and its accompanying essay, which showed how tarot, alchemy and astrological themes reveal the depths of Goethe's masterful fairy tale.

By taking inspiration from these two textual lineages, *Märchen* to *Sylvie*, and *Märchen* to *Le Serpent Vert*, Richer found the base materials to compose his own specimen of a *conte initiatique*, the poem *Le Serpent Rouge*. But there is a further twist here because it is not only in the works of Gérard de Nerval that Jean Richer found the pattern of the *conte initiatique:* it was also in his biography itself. It is impossible to disentangle Nerval's public life from his inner world; there was no

---

144 Ibid, p. 523.
145 Jean Richer, *Nerval au royaume des archétypes*, op. cit. p. 66.

dividing line between the two. His work was the outward expression of a restless, searching imagination in which the streets of Paris merged with the pathways of antiquity, and ancient gods mingled with the patrons of bars and cafés. Within this seamless blend, the boundaries between past and present, myth and reality, waking life and the dream-state, were erased. Hence, if *Le Serpent Rouge* echoes the structure of Nerval's writings, it can also be said to reflect the outline of his life itself.

Richer is by no means the only author to have been gripped by the literary urge to construct an archetypal narrative out of Nerval's lived experience. As *Le Serpent Rouge* was published anonymously, (or at least under false names), with Nerval's identity carefully veiled, there was no question of anyone consciously following his lead. Nevertheless, Richer's scholarly approach to interpreting Nerval's writings inspired at least one other author to attempt something similar.

Nerval's Life as Archetypal Story

In his introductory essay to Nerval's *Collected Works* in the Penguin English translation, Richard Sieburth observes:

> "an entire tradition of scholarship stemming out of surrealism has sought to interpret his (i.e., Nerval's) work in terms of the symbolism of arithmosophy, astrology, alchemy or the tarot." [146]

Here, Sieburth is referring unmistakably to the work of Jean Richer. One author who followed in his footsteps was Richard Holmes, who we have met earlier in this book. Holmes describes in his book *Footprints* how his immersion in the Nervalien universe, illuminated by Richer's work, led him to construct a seven-volume series of notebooks laying out an elaborate symbolic interpretation of his life.

> "I next moved to the opposite extreme and began to interpret Nerval's life almost entirely in terms of the magic world by which he himself was so fascinated. Much of this was influenced by the great Nervalian critic, Jean Richer, and his study *Gérard de Nerval et les doctrines ésotériques* (1949). But I went much further. Everything in Nerval's life came to have symbolic meaning, full of archetypes, alchemical processes, astrological signs, mystic correspondences and invisible harmonies."

---

146   Richard Sieburth, *Gérard de Nerval Selected Writings*, op. cit. p. xxvi.

The work became ever more complex:

> "I came to believe that my own biography would have magic properties, and I started to organise it in the form of a commentary on seven Tarot cards, each one covering a phase in Nerval's life, presented not chronologically but in a series of cycles. It started with La Lune, Les Amoureux and L'Etoile... and ended with La Tour and, of course, Le Pendu—the Hanged Man card that appears also in Eliot's *The Waste Land*." [147]

Until a friend released him from his own spell:

> "One late summer evening, at a cafe in the place Royal where we had our rendezvous, I showed my notebooks— seven notebooks in seven different colours—to my friend Françoise. Her face took on a curious expression. *"Ce n'est pas la peine de te rendre fou, chéri,"* she remarked quietly. *"Ce n'est-pas ta vie à toi, après tout !"* She swept me off to a late-night showing of Les Enfants du Paradis at La Pagode. And slowly I began to realise what was happening." [148]

It is apparent that Holmes was gripped by the same enthusiasm that inspired Richer to write *Le Serpent Rouge*, and to construct his own symbolic biography of the poet which envisions his life as "the Great Work". We shall encounter another author who was also moved to create such a work before this book is complete.

The Itinerary of the Noble Voyager

In this discussion of the allusions to Gérard de Nerval and *Sylvie* in *Le Serpent Rouge*, another significant phrase remains to explore. We have seen how the unnamed poet in the first two stanzas, *"cet ami"*, has been identified as Nerval himself via the *"un sceau célèbre"* riddle in Chapter Seven, but he is also referred to as the *"grand voyageur de l'inconnu"* (the grand voyager of the unknown). Can this phrase be sensibly applied to the poet also? Certainly, it can.

Whilst the phrase is broad and has been used at various times to describe certain writers and other explorers of inner space in French literary and even Masonic tradition, it does also have a specific

---

[147] Richard Holmes, *Footprints: Adventures of a Romantic Biographer* op. cit. location 4907 4935

[148] Richard Holmes, *Footprints: Adventures of a Romantic Biographer* op. cit., "It's not worth the trouble, my dear. It's not your life after all.", location 4935

resonance to Nerval in the writings of Richer. To see this, we first need to observe that in both French and English, the words "grand" and "noble" are essentially identical and interchangeable. Thus, "grand voyager" and "noble voyager" may be taken as equivalent phrases.

In 1951, Jean Richer wrote an essay exploring the literary relationship between Nerval and the eighteenth-century French writer, Jacques Cazotte. It was entitled *Deux nobles voyageurs, Cazotte et Nerval /notes et étude par Jean Richer* (Two Noble Voyagers, Cazotte and Nerval / notes and study by Jean Richer).

Elsewhere, he wrote of the two:

> "Their lives unfolded like allegorical poems. One can recognise in these distant journeys the allegory of a complex spiritual adventure."[149]

Then later in 1987, in his final book, *Gérard de Nerval: Expérience vécue et création ésotérique*, he is bold enough to title the final chapter: *L'Itinéraire du Noble Voyageur*, or The Itinerary of the Noble Voyager. That concluding chapter in the last book he published touches on deep mysteries of Nerval's esoteric and spiritual life. He could not be making it much plainer, having laid down a trail of references from 1951 to 1987, that Nerval is the Noble Voyager. In 1966 he only slightly modified this terminology in referring to the poet in *Le Serpent Rouge* as *"grand voyageur de l'inconnu"*.

This title is more than appropriate for Nerval. His output included much of what today would be called travel writing, with his main work in this genre being *Voyage en Orient*, an account of his travels in the Middle East in the 1830s. But even outside these works, his writings were always "voyages" of one kind or another, whether internal or external, whether of realms known or unknown. So, the phrase "grand voyager of the unknown" is certainly an apt description of Nerval's life and works.

We've now followed a trail of texts from *Le Serpent Rouge* to Nerval's *Sylvie*, and other examples of the *conte initiatique* in European literature. Here we can find the source material that provided the basic template and inspiration for Richer to write his poem in 1966.

Yet, as Richer himself informed us, there was also another writer who had a profound influence on his understanding of Nerval's work, and in particular the notion of the initiatory journey around the zodiac. This was Professor Carl Jung, the Swiss psychologist and prolific author.

---

149   Jean Richer, *Experience vécue et création ésotérique*, op. cit. p. 129.

It was in Jung's works on the subject of alchemy that he found a rich theoretical foundation from which to arrive at an even deeper understanding of Nerval's thought. We will turn to this topic in the next chapter.

Chapter Sixteen
DREAMS, ALCHEMY AND THE OMPHALOS

As I write these words, it is more than twenty years since I had the first curious inklings that *Le Serpent Rouge* must have been written by someone who was familiar with Jung's writings and thought on alchemy.

When I came to the conclusion, based on the evidence of his own works, that the author of *Le Serpent Rouge* was Jean Richer, it was a welcome bonus to find references to Jung scattered throughout his books. These included several quotations from *Psychology and Alchemy*, and in particular from the very pages which had originally piqued my interest. It slowly became apparent that the influence of Jung's alchemical writings on Richer's thought had been profound.

In these works he discovered critical foundation material for his work on Nerval, on sacred geography, and on the enigmatic poem at the crossroads of all these ideas – *Le Serpent Rouge*.

In this chapter, I will explore how Richer's perception was inspired and transformed by his deep engagement with the writings of Jung on alchemy and examine how these works influenced his thinking and research.

In 1963, Richer wrote:

> "Let us remember, moreover, that all the legends, allegories and myths of humanity, as well as the great literary works, have received an alchemical interpretation." [150]

What can he have meant by this? In what sense could the foundation stories of humanity be understood as alchemical, and where might such interpretations be found? A clue to resolving these questions can be found elsewhere in the same work, where Richer also observed that:

---
150 Jean Richer, *Nerval, Expérience et Création,* op. cit. footnote 38, p. 257.

> "The masterpieces of universal literature have zodiac structures."[151]

Several years later, in 1967, he re-expressed the same idea, clothed in slightly different language:

> "All the grand cyclic poems of antiquity are based on a foundation of astral beliefs."[152]

For Richer, the deep stories of humanity are fundamentally *alchemical* and *zodiacal*. He found the inspiration for this view in the works of Jung, and specifically in his explorations of alchemical symbolism. In writing *Le Serpent Rouge*, Richer set out consciously to create a work of literary art based on these foundational narrative principles.

To understand the deeper layers of architecture of *Le Serpent Rouge*, therefore, we need to turn to Jung's alchemical works, and their effect on Richer's unfolding understanding of both Nerval and landscape geometry.

We presented initial evidence for this influence regarding Richer's work on sacred geography in Chapter Four. The idea of the journey of the heroic figure across landscape understood as reflection of the heavens, which Richer presents in *Sacred Geography of the Ancient Greeks*, is directly inspired by his reading of Jung's commentary of certain alchemical texts, as he explicitly acknowledges in the body of the text and the footnotes.

These lead to passages in which Jung explores the theme of the world-journey of the hero as an allegory of the sun's passage around the zodiac. This, as we have seen, is the template for the action of *Le Serpent Rouge*.

In this way, by following the trail of references that Richer left, it is possible to recreate the chain of connections that linked *Le Serpent Rouge* and Jean Richer to Jung's alchemical master works, *Mysterium Coniunctionis* and *Psychology and Alchemy*.

Twenty years later, I am finally able to confirm my original intuition and identify the texts that demonstrate exactly where and how the author of *Le Serpent Rouge*, Jean Richer, found inspiration in the alchemical works of Jung.

---

151 Jean Richer *Nerval : Expérience et Création.* op. cit. p. 257.
152 Jean Richer, *Sacred Geography of the Ancient Greeks.* op. cit. p. 256.

Alchemy and the Quest for Wholeness

> "The point is that alchemy is rather like an undercurrent to the Christianity that ruled on the surface. It is to this surface as the dream is to the unconscious, and just as the dream compensates the conflicts of the unconscious mind, so alchemy endeavours to fill in the gaps left open by the Christian tension of opposites."[153]

As briefly introduced in Chapter Three, Jung's aim in these works is to explore what he understands to be the central goal of the alchemical opus, namely: the *reconciliation of oppositions through the work*, a process also known as the "chymical marriage" or the *coniunctio*. Jung describes it in this way:

> "The alchemist's endeavours to unite the opposites culminate in the 'chymical marriage', the supreme act of union in which the work reaches its consummation."[154]

Thus, for Jung, alchemy's apparent preoccupation with the production of gold masked its true purpose. It was not the outward concern with matter and the material world shown by the alchemists that was the goal of the *opus*, but rather the transmutation that takes place within the alchemist himself, or herself.

In Jung's understanding, this inner work of the soul, projected onto the various physical substances and their manipulations and transformations in the crucible, acted to reconcile certain oppositions or polarities arising within the psyche. This was the heart of the alchemist's daily struggle. Thus, the tasks carried out in the workshop were merely externalisations of these internal soul processes.

Taking as his raw materials for these studies the writings of various alchemists, Jung shows how many of these texts present the nature of the alchemical work in allegorical form as a quest around a sacred centre in a mythical landscape.

The Stages in the Work and their Colours

If *Le Serpent Rouge* is a poetic account of a mythic journey of a hero around a zodiac in a landscape, then it too might be considered as a modern example of exactly the kind of alchemical text which Jung is discussing in these passages. But can it support such a reading? Could

---

153 C.G. Jung, *Psychology & Alchemy*. op. cit. p. 23.
154 C.G. Jung, *Mysterium Coniunctionis*. op. cit. p. 89.

Le Serpent Rouge be considered an alchemical text?

There is at least one overt reference to alchemy in the poem, in the Ophiucus stanza, which alerts us to pay close attention to this aspect.

"the base lead of my writing may contain the purest gold"

This is obviously a reference to the transmutation of lead into gold, or the outer sense of the *opus* in the Jungian reading. Are there any other indications within the poem that connect it with alchemical work?

Jung discusses at length how the different stages of the alchemical process came to be symbolised by colours, a theme he returns to frequently. In the earlier historical development of alchemy, the number of stages and colours was considered to be four. Later, one of the stages was usually absorbed and the process reduced to three stages and their corresponding colours.

He explains that the first stage, *nigredo*, is characterised by the colour black, the second stage, *albedo*, by white, and the third and final stage, *rubedo*, by red, although this is a somewhat simplified description of the entire process. Here he amplifies on the way this symbolism plays out:

> "The *nigredo* or blackness is the initial state either present from the beginning as a quality of the prima materia ... or else produced by the separation of the elements. (...) From this the washing (*ablutio, baptisma*) either leads directly to the whitening (*albedo*) ... or again the "many colours" (*omnes colores*) or "peacock's tail" (*cauda pavonis*) lead to the white colour that contains all colours. (...) The *albedo* is, so to speak, the daybreak, but not till the *rubedo* is it sunrise. (...) The *rubedo* then follows direct from the *albedo* as the result of raising the heat of the fire to its highest intensity. The red and white are King and Queen, who may also celebrate their "chymical wedding" at this stage."[155]

The very first stanza of *Le Serpent Rouge*, Aquarius, offers the vital clue considering this outline of the stages of the process, which confirms that the poem is intended as an alchemical allegory. Here is the passage:

> "for him who knows that the colours of the rainbow form a white unity, or for the Artist who, under his brush, makes from the six shades of his magic palette, gush forth the black."

---

155   C.G. Jung, *Psychology and Alchemy* (London, Routledge, 1968), p. 231.

Notice that the subject of colours has been brought in immediately, but who is the Artist? One might imagine that a visual artist is intended here, and that the colours are a reference to the shades of paint applied, with a brush, to a canvas. Elsewhere however Jean Richer has deposited a clue that permits us to identify the true nature of this Artist without ambiguity.

It occurs in his book of essays, *Aspects ésotériques de l'œuvre littéraire*, which I discussed earlier in Chapter Eleven in relation to Jonathan Swift and the Punic language in Boudet's book. Specifically, it is found in a chapter on the French poet and essayist Oscar Milosz. Richer is quoting from an article by Dom Pernéty on the *"blancheur"*, or whiteness in alchemy, where the alchemist is referred to as the Artist.

> "The whiteness after the putrefaction is a sign that the Artist has performed well."[156]

Richer himself has supplied the key. The title the Artist (*l'Artiste*), in the Aquarius stanza denotes the alchemist himself.

These references to colours in *Le Serpent Rouge* are not to be understood as concerning optics or light, or the pigments applied to a canvas by a painter, but rather the stages to be accomplished in the Great Work. This is the reason that the stanza concludes strongly with the word "black". It is a clear reference to the *nigredo*, the blackness, the initial state of the work. The Artist, or Alchemist, knows that the beginning of the work is characterised by the bringing forth of the black, expressed as *jaillir le noir* in the poem.

Now we can break down the complete Aquarius quotation: the "colours of the rainbow that form a white unity" refers to the *omnes colores* or "peacock's tail" (*cauda pavonis*) that lead to the white that contains all colours. The Artist is the alchemist, and he commences the entire process by bringing forth from his "magic palette", his alchemical laboratory with its furnaces and crucibles and substances, the black, the *nigredo*, the initial stage.

Is this an unintended coincidence or does the author of the poem intend that we should take this reference as explicitly evoking the *nigredo* stage? If we now search the poem for all subsequent references to colours, we find a very curious outcome. Firstly, there are numerous references in the next few stanzas to black and white, always occurring together. The next mention of a colour occurs in the Virgo stanza, where we read of the *"Dame Blanche de Légendes"*, or the "White Lady

---

156   Jean Richer, *Aspects Esotériques de l'œuvre littéraire,* op. cit. p. 210.

of Legend." This can be taken as representing the *albedo*, the whiteness, or second stage.

There are no further colours mentioned until we arrive at Sagittarius, where we encounter the *serpent rouge*, and the first reference to the colour red in the poem. It is mentioned twice in this stanza, with another instance of white in between. Here, then, is the *rubedo*, the redness, the final stage of the process. I will return to pick up this thread later to look at the *rubedo* stage again more closely.

In summary, first the Artist brings forth the black, which is then followed by the white and finally the red. Thus, the progress of colours in *Le Serpent Rouge* matches impeccably with the stages of the alchemical process. Notice that the episode of the encounter with the "*serpent rouge*" in the final quarter of the zodiac is perfectly synchronised with the final stage of the process.

If the narrative of colours follows the pattern of the stages of the *opus*, then we can be confident that this poem about a walk through the forests around Rennes-les-Bains, this initiatory journey or trial, was intended to be understood as an allegory of the work of the alchemists.

Now we can observe how Richer finds the source material in Jung from which he develops these ideas. We will also find out why the zodiac format plays a crucial role in alchemical symbolism. These insights will allow us to observe how Jung's writings on alchemy and the zodiac came to influence Richer's thought so profoundly both regarding his interpretation of Nerval's work, and in his discoveries in sacred geography.

Dream Analysis and Mandala Symbolism

In *Psychology and Alchemy*, Jung takes as his starting point the lengthy review of a large collection of modern dreams described by unnamed patients in his clinical practice, as well as his comments and analyses on these. Using these reports as his raw data, he proposes that there are certain images and symbols that arise universally in dreams, regardless of the age, sex, culture, race or background of the subject. It is as though in our dream world we share a pool of imagery from which our mind can draw, which Jung calls the collective unconscious.

> "The symbols of the process of individuation that appear in dreams are images of an archetypal nature which depict the centralizing process or the production of a new centre of personality."

One of the most prominent of these symbols is the mandala, a word which comes from the Sanskrit for "circle", or "discoid object". It refers to the sacred symbolic diagram which acts both as a map of the universe, and an instrument of meditation in certain spiritual traditions in the East. For the most part, the dreams which he gathers for discussion share this theme.

> "I have put together, out of a continuous series of some four hundred dreams and visions, all those that I regard as mandala dreams."
>
> "The term 'mandala' was chosen because this word denotes the ritual or magic circle used in Lamaism and in Tantric yoga as a yantra or aid to contemplation."[157]

Jung makes a detailed study of this mandala symbolism and its very frequent appearance in significant dreams, reported by subjects of all different backgrounds and ages, under a variety of transformed but always recognizable symbolic guises.

Why does the mandala appear universally in dream symbolism? For Jung, the goal to which the psychotherapeutic process leads is human wholeness, and this process is marked by the prediction of symbols of unity, of which the mandala is a perfect expression. The mandala therefore can be understood as an image of the self. As Jung remarks:

> "The self is not only the centre, but also the whole circumference which embraces both conscious and unconscious; it is the centre of this totality, just as the ego is the centre of consciousness."[158]

Thus, when the mandala occurs in dreams, it signals that the dreamer is occupied with questions relating to the self, or integrating the self, or achieving wholeness. If they should dream of a journey towards a mandala, or around a mandala, Jung would interpret this as a personal expression of the universal task, the struggle of the self, or soul, to overcome division, to strive toward unity.

Under these broad indications, there are obviously many objects and items that might be understood as a variation of the mandala. These include the various circular formats associated with the passage of time, including the clock, the year-circle, or the horoscope, as Jung himself elaborates:

---

157 Jung, *Psychology & Alchemy*, op. cit. pp. 41, 95.
158 Ibid. p. 41.

"The horoscope is itself a mandala (clock) with a dark centre and leftward *circumambulatio* with "houses" and planetary phases. The mandalas of ecclesiastical art, particularly those on the floor before the high altar or beneath the transept, make frequent use of the zodiacal beasts or the yearly seasons."[159]

So, Jung establishes that the collective unconscious provides a universal pool of archetypal imagery from which our dreams are constructed, and that one of the most prominent of these symbols is the mandala, an image of the self and its quest for wholeness.

Delphi and Mandala Symbolism

In *The Archetypes and the Collective Unconscious,* Jung outlines some of the key characteristics of a mandala. Here are the first seven elements on the list, *The Formal Elements of Mandala Symbolism*:

- Circular, spherical or egg-shaped formation
- Circle elaborated into a flower or wheel
- Centre expressed by a sun, star or cross usually with four, eight or twelve rays
- The circles, spheres and cruciform figures are often represented in rotation
- The circle is represented by a snake coiled about a centre, either ring-shaped or spiral
- Squaring of the circle, taking the form of a circle in a square or vice versa
- Castle, city and courtyard motifs[160]

It is intriguing to compare this list with the chief characteristics of the architecture, layout and mythology of Delphi according to the research of Jean Richer, which include:

- the omphalos, a stone as a central organising point shaped like an egg
- axes of the cardinal directions
- a zodiac wheel
- a serpent
- symbolism of the sun and its path through the heavens

---

159   Ibid. p. 206.
160   C.G. Jung *The Archetypes and the Collective Unconscious*, (London, Routledge, 1991) Para. 646.

All these (and more) are encapsulated, metaphorically and physically, in the omphalos, or navel, at the heart of the temple complex.

It is apparent that the two lists are not just similar but essentially interchangeable or even identical. Jung's description of the formal elements of mandala symbolism is equivalent to the description of Delphi as understood in light of Richer's sacred geography. This is a highly unexpected and quite extraordinary correlation. Jung is describing a formalism of the collective subconscious, a highly specific cluster of related ideas that manifest universally in dreams, across age, culture, gender and location. His description of this dream object coincides with the key elements according to Jean Richer of the most important temple complex of the ancient Greeks, an oracular centre which has its origins in even earlier cultures. Why is it that the architecture of an ancient temple and a universal dream motif should share the same underlying form or pattern?

The Quaternity

> "The mystical side of alchemy, as distinct from its historical aspect, is essentially a psychological problem. To all appearances, it is a concretization, in projected and symbolic form, of the process of individuation."[161]

Now we turn from dream symbolism to alchemical practice. As we have touched on, the goal of the alchemical process, according to Jung, was the reconciliation of opposites, or the *coniunctio*.

Jung gives examples of some typical pairs of qualities that form these oppositions, or polarities, which must be overcome in the performance of the Work by the alchemist.

> "The factors which come together in the *coniunctio* are conceived as opposites, either confronting one another in enmity or attracting one another in love. To begin with they form a dualism: moist/dry, cold/warm."[162]

He then makes a deeply perceptive observation about the way the oppositions, or contrary forces, often manifest. In practice, Jung notes, we often find two pairs of such polar qualities occurring together – that is, four elements forming a crossed pair. He calls this arrangement the quaternity, or *quaternio*:

---
161   Jung, *Alchemical Studies*, (London Routledge and Kegan Paul, 1973) p. 105.
162   C.G. Jung, *Mysterium Coniunctionis*, op. cit. p. 3.

"...Often the polarity is arranged as a *quaternio* (quaternity) with the two opposites crossing one another, as for instance the four elements or the four qualities (moist, dry, cold, warm) or the four directions and seasons, thus producing the cross as an emblem of the four elements and symbol of the sublunary physical world."

The quaternity, therefore, is the four-fold figure made up of two crossed pairs of internal contrary forces in the self, which must be brought into harmony and balance.

The format leads to the idea of the cross or cruciform as a kind of diagram or schematic of the unconscious forces. Jung often illustrates his examples by using such a diagram of a cross, in which the two pairs of opposing or polar inner forces are combined in a single glyph, forming an archetypal map of this underlying central task of the unconscious.

Jung's brilliant insight that brings this material alive is this: he observed that the quaternity is simply another variation on the mandala form, and that it functions as a diagram or map of the quest for wholeness in alchemical practice, just as the mandala also does in dream symbolism. Alchemy in this sense, as in dream analysis, can be understood as a work of the unconscious in pursuit of the integration of contrary interior forces.

It is this longing for inner unity that is projected onto the material in the crucible, which then becomes the stage or site on which the work plays out. This striving towards wholeness is characteristic, therefore, both of the modern person in therapy and the alchemist in his laboratory in the Middle Ages, from which we can draw the conclusion that it is a universal goal of the human condition.

If the alchemist's genuine task is to resolve these inner tensions and seek a state of wholeness, then it can also be understood as a type of inner quest, or journey, to find that place of unity within the soul which is the ultimate goal of the work. Projecting this narrative outward, the work in the alchemical laboratory can be depicted in symbolic form as a voyage or journey in the wider world.

It follows that as the archetypal forces are characterised as two pairs of oppositions forming the four-fold or cruciform arrangement, so too will the world map on which such a symbolic journey takes place be conceived in terms of a four-fold geometric format, to reflect the same underlying template.

This may take the form, for example, of the four quarters of the earth, or the four cardinal directions, or other related expressions. If the mandala may be considered as a map of the centre and how to get there, then it is also a natural template for the cartography of sacred journeys. The mandala and the omphalos are two different expressions of the same underlying archetypal reality: the internal and external manifestations respectively of the self or centre. Richer even explicitly notes this identity in *Delphes, Délos et Cumes*. Referring to the omphalos he writes:

> "Figurations of this type, which are true mandalas, or templates for meditation, have several superimposed meanings."[163]

Landscape zodiacs may in turn be considered as elaborations or variations on the omphalos. From this we can conclude that the quaternity, the mandala, the omphalos and landscape zodiacs are simply different expressions of the same archetypal form of the sacred centre manifesting in different contexts.

The implication of Jung's insight for Richer is therefore that these forms, the mandalas, omphaloi, and landscape zodiacs, are one and the same. They emerge from the quaternity of polarities within the psyche, and are an expression of the yearning for wholeness, or unity, emerging from the dream state, from the collective unconscious and projected onto the world.

In summary, Jung shows that the mandala form and its equivalents are symbolic representations of the quest of the self for wholeness, by analysing its role both in modern dream study and in alchemy. He also establishes that the task of the alchemist is to strive to balance the complex of inner psychological forces which may be symbolised by the cross, or mandala.

According to the alchemists themselves, one means by which this can be accomplished is a circular motion around the centre. This is why the opus is frequently represented, or dreamed, as a journey around a representation of the world: it is the circular progress around the centre that brings about the integration of the two, or four, opposing forces that represent the inner tensions of the soul.

Jung quotes the alchemist Heinrich Kunrath, who expressed this notion succinctly, though in somewhat antiquated terms for modern ears:

---

163 Jean Richer, *Delphes, Délos et Cumes*, (Paris, Julliard, 1970). p. 66.

> "Through Circumrotation or a Circular Philosophical revolving of the Quarternius, it is brought back to the highest and purest Simplicity of the plusquamperfect Catholic Monad." [164]

This is the reason why the voyage of the hero around the world path functions as an allegory of the alchemical process. The same motif may also often be figured by other related formats signifying wholeness. For example, a circular journey around a landscape can be considered as an emblem of the course of the sun as it traverses the planetary houses in the heavens above.

Hence Jung can state:

> "The synthesis of the elements is affected by means of the circular movements in time (*circulatio, rota*) of the sun through the houses of the Zodiac."

This concept of the *circulatio* opens a rich vein of potential symbols by which the work can be represented. Jung observes:

> "It is to be noted that the wheel is a favourite symbol in alchemy for the circulating process, the *circulatio*. By this is meant firstly the *ascensus* and *descensus*, for instance the ascending and descending birds symbolizing the precipitation of vapours, and secondly the rotation of the universe as a model for the work and hence the cycling of the year in which the work takes place. The alchemist was not unaware of the connection between the *rotatio* and the drawing of circles."

That final sentence offers the possibility of an intriguing link between alchemy and geometry. The construction of a circle by the geometer with compass and pen and paper is analogous to the circular work of synthesis performed by the alchemist. The wheel, or circle, is even explicitly identified with the zodiac in some texts:

> "In the Manichean system the saviour constructs a cosmic wheel with twelve buckets – the zodiac – for the raising of souls. This wheel has a significant connection with the *rota* or *opus circulatorium* of alchemy, which serves the same purpose of sublimation."[165]

---

164 Jung, *Psychology & Alchemy*, op. cit. pp. 124, 384, 164.
165 Ibid. pp. 380, 186.

Jung quotes a passage from the alchemist Gerard Dorn who manages to combine nearly all these ideas into a single potent passage, which might even be considered a distillation of the contents of this book!

> "Enlarging on the idea of the *rota philosophica* Ripley says that the wheel must be turned by the four seasons and the four quarters, thus connecting this symbol with the *peregrinatio* and the quaternity. The wheel turns into the wheel of the sun rolling around the heavens, and so becomes identical with the sun-god or hero who submits to arduous labours and to the passion of self-cremation…"

Dreams and the Unconscious

> "Since alchemy is concerned with a mystery both physical and spiritual, it need come as no surprise that the composition of the waters was revealed to Zosimos in a dream. His sleep was the sleep of incubation, his dream 'a dream sent by God'."

We can now see why Richer was inspired in his work on sacred geography by reading these ideas in Jung. Once the identity between omphalos and mandala is admitted, it is only a small step to apply the motif of the world-journey around the mandala to the great legends of the Greek heroes moving around a landscape arranged around an omphalos. Thus, Richer found in Jung a rich theoretical basis for his discoveries in sacred geography.

This was no less true for his work on Nerval and indeed other poets and writers. After his encounter with Jung's ideas in the early 1960s, the *leitmotif* of Richer's analytical approach across all his research activities and fields is to identify the zodiac and planetary themes that are often running beneath the surface and informing any given author's work. It is no exaggeration to suggest that he applies this template universally to every topic to which he subsequently turns his attention.

For example, he produced a remarkable collection of essays on zodiac symbolism in several Shakespeare plays. It is a fascinating work, full of deeply insightful and very surprising observations. Yet nowhere does Richer display his sensitivity to the zodiac foundation underpinning the text more so than in the writings of Nerval, and in particular his short story *Sylvie*. As we will see, for Richer, Nerval is the example *par excellence* of the Artist as Alchemist. This is why his literary output may be read as an authentic work of alchemy in the precise sense of the Jungian understanding of the term.

It also helps illuminate the recurring theme of the dream that runs like a silver thread through these texts. Richer discovered the Greek alignments through a dream. *Le Serpent Rouge* is a poem about a dream. Nerval is the poet of the dream. Jung's work on mandalas emerges out of dream analysis.

I will now proceed to look in closer detail at the trail of references left by Richer in his books to Jung's alchemical works. All of this, to recall, is in order to explore Richer's portrayal of Nerval as the Grand Voyager of the Unknown, the archetypal heroic figure at the heart of *Le Serpent Rouge*.

Prometheus and the Mythology of Fire

Why would Nerval qualify to represent this archetype of the hero? To understand, we need to delve into the mythology surrounding the figure of Prometheus, the bringer of fire to mankind. In Chapter VII, entitled "Prométhée. Pandora", of his *Nerval: Expérience et Création*, Jean Richer examines in detail the role played by the cycle of ancient Greek myths about Prometheus and Pandora in Nerval's thought and writings. He relies heavily on an extended discussion of the myth of Prometheus in Jung's *Psychology and Alchemy*, found in Chapter 5: The Lapis-Christ Parallel, specifically in Part II: Evidence for the Religious Interpretation of the Lapis.

This section of Jung's book was obviously a favourite of Richer's because he cites these same passages of *Psychology and Alchemy* in several other places, from a selection of his books at different stages of his career. It becomes apparent that these key pages of Jung were crucial to Richer in the development of his ideas.

It is worth quoting in full the passage that opens Richer's chapter, as it draws together openly all the relevant themes: alchemy, Prometheus, Nerval, and even explicitly cites Jung's *Psychology and Alchemy*.

> "Myths as rich in possibilities as those of Prometheus and Pandora have given rise to diverse interpretations: this is why Prometheus is sometimes considered as a volcanic myth. And one does not need to have read the Vedas to perceive the analogy between man's internal fire, the solar spark suspended in the blood, and the supposed internal fire of the earth.

Precisely, the techniques of alchemy seem to have had as its object a taking charge of matter, man assuming responsibility and collaborating with God to accelerate evolution. For this purpose, according to scholars, the alchemists used several 'fires', the fire of the crucible and also their own 'fire'. That is why there can be a purely spiritual alchemy, and that alone is the one practiced by Nerval.

A work such as that of Jung, *Psychology and Alchemy*, shows that a considerable number of Nervalian symbols and myths are drawn from the field of alchemy because of their multiple senses of meaning.

On the other hand, a large number of stories or books having long held the attention of Nerval have an alchemical meaning: this is particularly so for the legend of Solomon and the Queen of Sheba (considered as an allegory of the soul in search of truth or the philosopher's stone), for *The Dream of Poliphile*, for Goethe's *Faust*."[166]

Richer, following Nerval, assimilates the fire of Prometheus with the internal fire in man. He identifies this with the "fire" of the alchemists, the inner spiritual furnace in which the true work takes place.

Thus, explains Richer, there is a purely spiritual alchemy, which may be distinguished from the alchemy of physical matter, and it is in this former sense that he considers that Nerval was indeed an alchemist. He mentions Nerval's deep engagement with certain alchemical works, including the legend of "King Salomon and the queen of Saba (Sheba)", and how these underpin many great myths and legends of humanity.

Richer has employed the narrative of Prometheus and his gift of stolen fire to mankind as an allegory of alchemy. He has established Nerval as an alchemist and named *Psychology and Alchemy* as the text by which to understand this. And he has tied all of this together by linking his chapter on Prometheus and Pandora to Jung's text on Prometheus and Pandora. Thus, Richer had already made it clear as early as 1963 that he considered Nerval to be an alchemist in precisely the sense in which Jung had understood the term.

After these opening remarks, Richer then quotes a long passage from Jung, taken from a work by Zosimos, a third-century AD Greek-Egyptian alchemist, who in turn was paraphrasing an account by Hesiod, the seventh-century BC Greek poet, from his work *Theogonies*. In the passage, Zosimos mentions the tale of Prometheus as part of a discussion

166 Jean Richer *Nerval : Expérience et Création*. op. cit. p. 257.

about Adam, the first man, who he also refers to as "Anthropos" or the "Son of God". Zosimos states that Adam has an essentially fourfold nature, because he has been constructed out of the four cardinal directions. This is reflected for example in an esoteric derivation of the name Adam itself in which each of its four Hebrew letters stands for one of the cardinal directions. He writes:

> "...with reference to his body, they named him symbolically after the four elements of the heavenly sphere."[167]

These complicated esoteric musings of Zosimos continue for several pages, and then draw to conclusion on page 368 of *Psychology and Alchemy*. Jung then elucidates Zosimos' conception of the "Son of God" and his relationship with Adam, the first man, in the following crucial passage:

> "In both cases, he (i.e., the Son of God) is identical with Adam, who is a quaternity compounded of four different earths. He is the anthropos, the first man, symbolized by the four elements, just like the *lapis*, which has the same structure. He is also symbolized by the cross, whose ends correspond to the four cardinal points. This motif is often replaced by corresponding journeys, such as those of Osiris, the labours of Herakles (footnote 75) the travels of Enoch and the symbolic *peregrinatio* to the four quarters in Michael Maier."

The *lapis* is the mysterious philosopher's stone which aids in the accomplishment of the Work. Its precise identity is never quite made clear anywhere in the writings of the alchemists, but it was said to be comprised of four elements. Thus, for Zosimos, Adam can be associated or even identified with the *lapis* because he embodies within himself the four cardinal directions.

This is the same passage in Jung to which Richer provided a footnote in the critical quotation from *Sacred Geography of the Greek World*, where he wrote:

> "A psychologist of the stature of Jung saw the hero as an image of the integrated man, who contains in himself the four elements and the four cardinal points."[168]

---

167   C.G. Jung, *Psychology and Alchemy*, op. cit. p. 363, 368.
168   Jean Richer. *Sacred Geography of the Greek World*. Op. cit., p. 103. The citation of Jung in Richer's quote has a footnote (18), directing the reader to Jung *Psychology*

By linking this quotation to the same passage in Jung to which he also footnoted in his 1963 book on Nerval, Richer is establishing a clear, if somewhat convoluted connection between all these related concepts: Nerval, alchemy, the hero, or integrated man, the *lapis*, and sacred geography.

These passages from Jung are dense and complex and hardly lend themselves to simplification or summary. Nevertheless, in order to follow Richer's understanding as expressed in his works, I will attempt to highlight the main points of the argument:

- Prometheus brought fire to mankind.
- The Alchemists employed two types of "fire" in their work: the fire that heated the crucible, and the internal spiritual fire.
- There is a spiritual alchemy, which works with this latter "internal" fire alone, and it is precisely in this sense that Nerval can be considered an alchemist.
- This relationship between the allegory of Prometheus and this inner, spiritual alchemy, is illuminated in Jung by the commentary of Zosimos.
- For Zosimos, the allegory relates to the formation of the first man, Adam, and his relationship with the Son of God, the Redeemer, both of whom have a four-fold nature.
- Adam, as the original man, can be compared to the *lapis*, the transforming stone of alchemy, which is also comprised of four elements.
- This four-fold structure is also echoed in the quaternity and provides the format for the mandala, or by equivalence, the omphalos.
- Finally, this motif of the quaternity, which provides the inner form of the "hero", the integrated man, Adam, the Son of God, the *lapis*, the mandala and the omphalos, can also be projected outwards as a journey around a four-fold space, such as the four quarters of the world.

Via this roundabout, yet rigorous route, Richer has woven an extraordinary tapestry of relationships, leading from Prometheus to Nerval as supreme alchemist, to the idea of the four-fold integrated man, and finally to the hero undertaking a ritual journey around a sacred landscape.

& Alchemy, page 355, note 75 in the first edition. This corresponds to page 368 in the second edition which I have used and referenced.

The Solar Hero

Richer has used the word "hero" to describe how Jung viewed the image of the integrated man, and yet that particular word does not occur in the passage in *Psychology and Alchemy* from which Richer has quoted.

Why then does Richer choose to employ the specific term "hero" in his restatement of Jung's discussion of Zosimos, when it is absent in the source material? The answer can be found if we turn a few pages forward in the same chapter to page 380, where Jung offers a sustained discussion of some themes from another alchemical source text, the "*Aurora consurgens*". A reference to a group of twelve stars in the original manuscript leads to the notion of a cosmic wheel with 12 buckets, which are the zodiac, for the raising of souls. This wheel is assimilated to the *rota* or *opus circulatorium* of the alchemists, another term for the transforming principle. This integrates the wheel with the four seasons, the quaternity and the *peregrinatio*.

The wheel can also be thought of as the wheel of the sun rolling through the heavens. Here Jung brings in the notion of the hero, the image of the sun on his journey, citing the alchemist, Michael Maier.

> "Michael Maier actually takes the *opus circulatorium* as an image of the sun's course:
> 'For while the hero, like a joyful giant, rises in the east and hastens to his sinking in the west, that he may forever return out of the east, he sets in motion these circulations…'." [169]

We have encountered the same metaphor earlier in this chapter:

> "The wheel turns into the wheel of the sun rolling around the heavens, and so becomes identical with the sun-god or hero who submits to arduous labours and to the passion of self-cremation…"

Richer has combined this notion of the hero as representative of the sun in its daily journey around the heavens, with the notion of the integrated man undertaking the symbolic journey to the four quarters of the world.

Here is the reason that Richer used the term "hero" when he referred to Jung's/Zosimos "four-fold" Adam, or first man. He wanted to clearly establish that this metaphor or motif of the four-fold man on

[169] C.G. Jung *Psychology and Alchemy*, op. cit. p. 380.

his symbolic journey can equally be extended to encompass the sun in its traverse around the heavens.

And why was this important? A potent clue is found on the next page of *Psychology and Alchemy*, where Jung notes:

> "The circle described by the sun is the "line that runs back on itself, like the snake that with its head bites its own tail, wherein God may be discerned".[170]

Here, Jung has added the *ouroboros*, the snake biting its own tail, to the chain of symbolic associations, as the image of the path taken by the sun, or hero, in its circuit. Everything culminates in this passage.

Here, at last, we have also found the textual source of Richer's notion, expressed on the first page of *Sacred Geography of the Ancient Greeks*, that the serpent represents the path of the sun through the zodiac.

> "Now the serpent not only is a symbol of the earth but also represents the path of the sun through the zodiac."[171]

This was the potent clue, cited again in his follow-up volume, *Delphes, Délos et Cumes*, that helped establish the link between Delphi, the Python and *Le Serpent Rouge*.[172]

This brings the trail full circle: it is the seed idea of *Le Serpent Rouge*. The red path is the red serpent, and both represent the path of the red sun as it makes it *peregrinatio* through the houses of the zodiac.

This is the distillation of a theme that runs through the entire alchemical writings of Jung, and from which Richer found deep inspiration for his understanding of both Nerval as alchemist and the sacred geography of the ancient world. Yet, perhaps nowhere is it more succinctly expressed than on this crucial page 368 and the facing page 369, in Jung's *Psychology and Alchemy*, in which the first man and the *lapis* are associated, via their common quaternity structure, with the motif of the allegorical journey to the four quarters of the earth of the hero, who is also an image of the sun.

An illustration appears on this double-page spread that shows Christ at the centre of a wheel of twelve apostles and twelve rivers, effortlessly invoking in concert with the accompanying text the notion of the zodiac in landscape with the omphalos/mandala/hero/sun at the focus.

---

170   C.G. Jung *Psychology and Alchemy*, op. cit. p. 381.
171   Jean Richer, *Sacred Geography of the Ancient Greeks*. op. cit. p.1.
172   The facing pages in *Psychology and Alchemy* in which this passage appears include two illustrations: on the left is a depiction of Helios, the sun-god in his chariot, and on the right is a depiction of the Prophet Elijah.

It is these two pages, more than any other in Jung, which encapsulate what Richer was trying to achieve in *Le Serpent Rouge*: a narrative based on a journey of testing around a central focal point, around a landscape conceived as a zodiac, after the form of the great works of ancient myth and literature, namely as alchemical and zodiacal allegory.

The Mystic Peregrinations of Michael Maier

In 1963, ten years after *Psychology and Alchemy*, Jung published *Mysterium Coniunctionis*, in which he expanded on many of the ideas from the earlier book. He investigates several texts written by alchemists that take the form of a mythical journey, or a quest, around a mandala-like form, such as the world-clock, or the horoscope, or a central mountain, or other equivalent.

In one extended section, entitled *Salt*, he discusses in some detail the journey, or "symbolic *peregrinatio* to the four quarters" of Michael Maier. At one point, he even provides a footnote of his own back to those crucial pages 368/369 of *Psychology and Alchemy*.

It is the very first paragraph of this section which contains the striking parallels with *Le Serpent Rouge*, discussed in Chapter Three, which began my interest in the relationship of Jung's work to the poem. These were based around a cluster of related terms: salty and bitter, seawater, baptism, the Red Sea/Red Serpent, which are the themes of his discussion that follows.

These passages of Jung culminate in an analysis of the *peregrinatio* of Michael Maier, an exemplary model of the alchemical opus presented as an allegorical journey to the four quarters of the world. Maier was a seventeenth-century German alchemist, physician and counsellor to Rudolf II. In 1617 he published a work entitled *Symbola aurea mensae*, also known as *The Peregrinations of Michael Maier*.

This word *peregrinatio* which is used to describe his journey is the Latin term which is the origin of both the English "pilgrimage", and the French *pèlerinage*. This is of course the same word employed in the third stanza of *Le Serpent Rouge* to describe the journey on which the narrator embarks, the *pèlerinage éprouvant*. This is discussed at great length in Jung's Salt section.

Symbola aurea mensae recounts a journey around the four quarters of the world, through the planetary houses, in which Maier encounters a sequence of archetypes: the monster, the maiden, the phoenix. If Nerval's *Sylvie* provides the literary prototype for *Le Serpent Rouge*, then it is these passages of Jung on Maier's voyage that provide a template for the alchemical foundations for the poem. Jung summarises Maier's travels in this manner:

> "During his mystic peregrination Maier reached the Red ('Erythrean') Sea, and in the following way: he journeyed to the four directions, to the north (Europe), to the west (America), to the east (Asia)."[173]

In this case, the four directions of the compass correspond to the four archetypal forces. Thus, for Jung, as Maier moves around the world to the south, north, east and west, he is encountering and reconciling his inner oppositions, figured as a four-fold arrangement of archetypal forces within his soul.

Here, as in other alchemical texts Jung discusses, the hero on the ritual journey will typically encounter various challenges that must be met and overcome, representing the struggle of the self. During the fourth and final segment of his travels, Michael Maier encounters a mysterious red-headed animal, the Ortus, as he crosses the Red Sea.

> "This journey is reminiscent of the voyage of the hero, one motif of which becomes evident in the archetypal meeting at the critical place (the "ford") with the Ortus, its head showing the four colours. There are other motifs too. Where there is a monster, a beautiful maiden is not far away, for they have, as we know, a secret understanding so that one is seldom found without the other."

This resonates with *Le Serpent Rouge* in which there is also an encounter with a mysterious red monster in the final segment of the journey, and a beautiful maiden is also not far away. The monster is of course the "*serpent rouge*" itself. The maiden is the "queen of the beautiful spring" whom the traveller meets in the Leo stanza, and who bears the names of Isis, Madeleine, and Our Lady of the Cross. She is the "goddess of a thousand names" according to Robert Graves, the archetype of the divine feminine – and the constant obsession of Nerval.

It is apparent that Richer has modelled the encounter with the "*serpent rouge*" in the poem on the monster/dragon/serpent met by Maier in the fourth quadrant of his *Grand Peregrination*. This is a recurring theme in the alchemical texts according to Jung. For example:

> "The encounter with the monster on the final fourth part of the journey towards integration is a frequent motif in

---

173 C.G. Jung, *Mysterium Coniunctionis* op. cit. p. 210.

> such stories. (...) The idea of transformation and renewal by means of a serpent is a well-substantiated archetype. It is the healing serpent, representing the god."

The reason for the encounter and why the serpent is a "well-substantiated archetype" is elaborated by Jung:

> "The dragon, or serpent, represents the initial state of unconsciousness, for this animal loves, as the alchemists say, to dwell in "caverns and dark places". Unconsciousness must be sacrificed; only then can one find the entrance into the head, and the way to conscious knowledge and understanding. Once again, the universal struggle of the hero with the dragon is enacted, and each time at its victorious conclusion the sun rises: consciousness dawns, and it is perceived that the transformation process is taking place inside the temple, that is, in the head."[174]

Again, Jung elucidates the connection between the circular motion that leads to the synthesis of the elements, the serpent encounter, and the transformation which leads to healing and change.

> "But if the life mass is to be transformed a *circumambulatio* is necessary, i.e., exclusive concentration on the centre, the place of creative change. During this process, one is "bitten" by animals; in other words, we have to expose ourselves to the animal impulses of the unconscious without identifying with them and without running away."[175]

The dragon/serpent motif is a fluid symbolic apparatus that lends itself readily to further interpretation. For example, the snake can be envisioned biting its own tail, leading to the famous *Ouroboros*, allowing the symbolism to join seamlessly to the idea of the wheel and the *opus circulatorium*, and more.

> "Time and again the alchemists reiterate that the opus proceeds from the one and leads back to the one, that is to a sort of circle like a dragon biting its own tail. For this reason, the opus was often called *circulare* (circular) or else *rota* (the wheel). Mercurius stands at the beginning and the end of the work; he is the *prima materia*, the *caput corvi*,

---

174 C.G. Jung, *Alchemical Studies*, op. cit. p. 89.
175 C.G. Jung, *Psychology and Alchemy*, op. cit, p. 145.

the *nigredo*, as dragon he devours himself and as dragon he dies, to rise again as the lapis. He is the play of colours in the *cauda pavonis* and the division into the four elements."

Psychology and Alchemy provides many illustrations of this principle from the alchemical texts, including an image from the Ripley Scroll showing two dragons biting each other's tail within a zodiac. Jung makes the following observation, which wraps up the entire thesis in a compact nutshell:

> "The circular figure, together with the *Ouroboros* – the dragon devouring itself tail first – is the basic mandala of alchemy."[176]

We find this confirmed in Richer's own writings. Here he is in the 1963 book on Nerval:

> "The image of the snake is a versatile image, which is what Jung calls an archetype of the unconscious. The ancients represented Cronos wrapped in the folds of the zodiacal serpent, an allusion to the annual revolution of the sun. Nerval also emits this image in relation to the tempting serpent of Genesis. But the writer certainly knew the cosmic sense of the image of the serpent *Ouroboros*, an allusion to the eternal return and cycles of creation, so often reproduced in books of alchemy."[177]

The entire thesis of this chapter is brought together and encapsulated in these compact passages: the snake, the zodiac, the revolution of the sun, Nerval, the Ouroboros, and books of alchemy. Here is Richer again on the same themes:

> "The large serpent ... is an image of devouring time, time that the annual race of the sun punctuates in the zodiac, and the snake represents this course."[178]

Finally, to round off this section, here is an intriguing quotation from *Psychology and Alchemy* in which Jung takes all this one step further, and even presents the path to wholeness as a serpentine path! One can well imagine Richer's delight in reading these words, given that he associates the "*Sentier Rouge*", the walking path around Rennes-les-Bains,

---
176 C.G. Jung, *Psychology and Alchemy*, op. cit. p. 126.
177 Jean Richer, *Nerval : Expérience et Création*, op. cit. p. 434.
178 Ibid., p. 194.

including the forestry track snaking up the flank of Pech Cardou, with the *"serpent rouge"*, as I have discussed in detail in Chapter Fourteen.

> "But the right way to wholeness is made up, unfortunately, of fateful detours and wrong turns. It is a *longissima via*, not straight but snakelike, a path that unites the opposites in the manner of the guiding caduceus, a path whose labyrinthine twists and turns are not lacking in terrors." [179]

The serpent is a multivalent symbol, perhaps the ultimate multivalent symbol, capable of effortlessly supporting all these different layers of interpretation. We have seen that it represents Pythia, or Python, the tutelary spirit of Delphi. It is also the physical path in the landscape around Rennes-les-Bains itself. It is the path of the sun, the Ouroboros and the monster that must be encountered in the fourth quadrant of the journey around the world map or mandala. It is also the symbol of the final stage of the alchemical process, the *rubedo*, and therefore there is only one colour which it can be in this potent allegorical scheme: the serpent must be red.

Jean Richer found the theoretical foundation for his ideas about the zodiac from the work of Jung on alchemy. He consciously employed the themes that Jung discusses at great length as the underlying archetypal content of the alchemical Great Work. Specifically, he constructed a voyage around a world arranged as a quaternity, or in this case, an elaboration of the cross format into a zodiac.

He had the hero move around the space, imitating the sun moving around the heavens through the houses. He has an encounter in the fourth and final sector with a beast, or dragon, or serpent, representing the final forces to be overcome in the quest to achieve the goal of wholeness. The meeting with the serpent would have been especially resonant for Richer, as the narrative of the poem was already based loosely around the founding myths of Delphi, including the struggle between Apollo and the serpent Python. The narrative employs alchemical motifs and references, as we have seen. It also adopts the formula that Jung identifies from the alchemical texts, of the journey around the four-fold world, culminating in the encounter with the monster, or serpent, in the final quadrant, marking the final stage, the *rubedo*, or redness, the Completion of the Great Work.

Richer has woven together so skilfully all these different layers. He has taken the walk of Boudet around his cromlech as the basic outline.

---

179 C.G. Jung, *Psychology and Alchemy,* op. cit. p. 6.

He has identified this with a zodiac, which is a genuine historical intervention in the local landscape. He has taken Nerval's *Sylvie* as an example of a story which unfolds around a zodiac template for inspiration. Then using elements taken from Jung on alchemy, he has constructed a narrative along the lines of the texts of the old alchemists, an account of the world journey, around the mandala or omphalos, in search of the balance of the contrary forces in the soul.

Chapter Seventeen
IMPRINT OF A SEAL

"Habientibus symbolum facilis est transitus."
"For those who have the symbol the passage is easy"
– Mylius, *Philosophia Reformata*[180]

IN THIS chapter, I will investigate possible sources of inspiration that might have given Boudet, and ultimately the parchment-makers, the idea of encoding a hexagram and inch grid as a concealed subliminal layer in a work of art. I will also try to establish whether there is any precedent for such a strange thing.

Hidden Layers in Mutus Liber

One of the most famous alchemical treatises is *Mutus Liber*, composed by an unknown author under the pen-name *Altus*, (or "High" in Latin) and published in La Rochelle in 1677.

Mutus Liber consists of fifteen plates that depict, almost in the manner of a modern comic book, a long and mysterious process undertaken by an alchemist and his *soror mystica*, or "mystical sister", his partner in the work. It involves unidentified substances, and certain operations that take place both outside in a country landscape and inside in an alchemical laboratory. The detailed visual instructions for this arduous work are interspersed with several plates showing visionary scenes. Strangely beautiful and utterly enigmatic, *Mutus Liber* is an irresistible book that has long fascinated students of esoteric lore.

Consider, first, the name. *Mutus Liber* can be translated as the "Mute Book", or the "Book without Words". Yet it is not quite wordless: there is a small amount of text on the first page, a phrase or two throughout the rest, and two words (repeated) on the final page. But the title

[180] Johann Daniel Mylius *Philosophia Reformata*, (Frankfort on the Main, 1622)

is apt: rather than relying on words, it conveys its message through images and symbols.

It is enough to spend a little time examining its pages to be convinced that the book is attempting to transmit some kind of information, or message. It is a book that speaks, but silently. What is the answer to this riddle? What kind of information, message or even speech can be communicated *without words*? The answer is geometry.

There is a secret concealed in the final page, Plate XV, of *Mutus Liber*.

In the image, shown at upper left in Figure 129, a man is floating in mid-air, supported by two cherubim, under a smiling sun. In his hands he holds a rope with the ends dangling down on either side of him. Below him, the alchemist and his *soror mystica* kneel in prayer. Behind them, hovering horizontally several feet off the ground, is a ladder.

There are more strange details: ribbons emerge from their mouths bearing the Latin words "*oculatus abis*". This might be loosely translated as: "provided with eyes, you leave", or perhaps "now that you can see, you depart". What is it that we are being challenged to see?

Look closely at the length of rope held between the hands of the Floating Man. It forms a perfect square with the two ends dangling down either side and the lower edge of the ladder. The position of the man's thumbs defines the corners accurately.

If we rule in this square defined by the rope and the ladder, the result is as shown at upper right of Figure 129.

This looks promising. What do we do next? Fortunately, the puzzle-maker has left us a clue. It is the ladder. We find that the length of the lower edge of the ladder corresponds impeccably to the side of the triangle which, together with a second triangle by symmetry, forms the hexagram on the square.

We complete the two triangles around the square, and the result is shown at lower left of Figure 129. The ladder signals the presence of a concealed hexagram. Careful inspection of the image reveals confirmation in various elements, including the arms of the kneeling couple.

If we now add the familiar outer hexagram to form the complete "Seal of Solomon" design, the result is shown in the lower right quadrant of Figure 129. Again, there are multiple confirmation signals. Notice that the left- and right-hand points of the outer hexagram fall on the borders of the image. The upper point of the outer hexagram falls on the forehead of the sun. The effect is subtle, rather than formulaic, but the intention of the artist is unmistakable. The composition has been arranged around a hidden Seal of Solomon.

IMPRINT OF A SEAL

Figure 129: Mutus Liber, Plate XV.
The hidden geometry reveals itself.

Figure 130: Mutus Liber, Plate XV. The hidden geometry in detail.
A reference grid has been added to the image, consistent with the correct size of the original. The grid squares across the full image have sides of one half of an inch. The grid squares within the central square have sides of one eighth of an inch. Note the spacing of the rungs of the ladder in particular.

Figure 131: Mutus Liber, Plate XV, later edition.
The same hidden geometry, though executed with slight variation on the earlier version. A reference grid has been added to the image. The grid squares have sides of one quarter of an inch.

What about the measure used to lay out the grid? High resolution scans of *Mutus Liber* are available online from the Library of Congress website and careful geometrical measurement confirms that the sides of the inner square are exactly three inches in length on the original.

As may be seen in Figure 130, the central square with sides of 3 inches can be subdivided into 24 squares per side of one-eighth inch measure. Notice that the rungs of the ladder in the central section are spaced by exactly two squares, or one-quarter inches.

It is evident that the engraver has taken meticulous care to ensure that the visible design conforms to the invisible underlying geometry. These are strong indications of intentionality. The hidden layer of *Mutus Liber* contains both a grid of inches, and a Seal of Solomon.

Remarkably, confirmation of this interpretation is provided in another version of *Mutus Liber* by a different artist. While the scenes depicted are broadly the same as in the version we have been looking at above, they are not quite identical but have been recast with some slight variation. Nevertheless, inspection of the final plate reveals that the very same concealed geometry is present. It has been implemented in a slightly different manner, but there can be no doubt that it also reflects meticulous attention to the hidden design, as shown in the accompanying Figure 131. We can observe that this time, the Floating Man is slightly higher, and his arms are held in a different posture. If anything, this results in an even more satisfying artistic display of the visible action superimposed on the invisible geometry.

The message is intended to be seen without being noticed, heard without being spoken. The language of The Book without Words, the Mute Book, *Mutus Liber*, is geometry. The anonymous author of *Mutus Liber* has incorporated the Seal of Solomon and the inch grid as concealed geometrical elements in Plate XV. These are the same elements that we find in the Boudet map and the parchments. It is apparent that Boudet either learned these techniques from having acquired an insight into *Mutus Liber*, or by some third party who must have. Boudet's reference to the Seal of Solomon symbol was an exercise within an existing, valid, historical esoteric tradition.

Now we can begin to trace the imprint of the seal from *Mutus Liber*, (1677), to Boudet (1886), to the parchments (1967). Were these connections known to the Team? Was de Sède, for example, aware of Boudet's debt to *Mutus Liber*? He has left some tantalising clues. In an earlier book, not directly concerned with the Rennes affair, he speaks more openly and refers to *Mutus Liber* by name.

"Some masters confide their secrets to the image: the *Liber Mutus* of Soulat by Maretz or the *Traité symbolique de la pierre philosophale* by Conrad Barchusen, are rebus albums that must be deciphered to rediscover the technique of laboratory operations. So again, here are seals, pentacles, and more or less abstract schemas that we could not understand without a key."[181]

De Sède knew in 1962 that *Mutus Liber* was a rebus album, that is to say, a book with messages hidden in pictures or images. In 1967 he used precisely the same language when he wrote that the analysis of the coded parchments concluded that the author had included rebuses. Then in 1978, in his introduction to the Bélisane edition of Boudet's *La Vraie Langue Celtique*, he pens the following:

"Let us therefore put into practice the wise advice of the old alchemists who placed at the top of their enigmatic books: '*Lege, lege, relege et invenies.*'" [182]

De Sède is being a little coy here in not naming the source of this latin phrase (translation: 'read, read, read again and discover'), but it is easily identifiable. It is a slightly abbreviated version of a quotation which is found in *Mutus Liber*. The text appears at the bottom of Plate XIV, the final scene before Plate XV with its hidden geometry. Having earlier declared that *Mutus Liber* is a rebus album, in his 1978 introduction de Sède quietly established a definite link between the alchemical text, the parchments and Boudet's book.

Curiously, Plate XIV from *Mutus Liber* also happens to be reproduced in Jung's *Psychology and Alchemy*. It appears on the penultimate page opposite a depiction of "The Phoenix as symbol of resurrection", intended as an allegory of the completion of the work.[183]

Geometrical Figures of the Old Rosicrucians

Rudolf Steiner, in his lectures on Rosicrucianism, provided some fascinating background material on the use of the Seal of Solomon as an esoteric symbol. It is worth quoting some passages at length. He speaks of certain books whose content consisted of "geometrical figures", intended to convey information in images rather than words

---

181   Gérard de Sède, *Les Templiers sont parmi nous* (Paris, Julliard, 1962). p. 126.
182   Gérard de Sède in Henri Boudet op. cit. p. xii.
183   C.G.Jung, *Psychology and Alchemy,* op. cit. p. 482 from Boschius *Symbolographia* (1702).

in a very curious manner. He does not give specific examples of such books, but we might not be too far away from the truth to consider that *Mutus Liber* emerged from just such a tradition.

> "A great number of them possessed — to speak in modern language — 'editions' of the geometrical figures of the old Rosicrucians. These they would show to those who approached them in the right way. When, however, they spoke about these figures — which were no more than quite simple, even poor, impressions — then the conversation would unfold in a strange manner. There were many people who, although they took interest in the unpretentious wise man before them, were at the same time overcome with curiosity as to what these strange Rosicrucian pictures really meant, and asked about them. But they received from these wise men, who were often regarded as eccentric, no clear and exact answer; they received only the advice: If one attains the right deepening of soul, then one can see through these figures, as through a window, into the spiritual world."[184]

These figures that one looks through, like a window, into the spiritual world: what did they consist of? Fortunately, Steiner has provided some more specific information:

> "We find in particular one symbol that played a great part for this little company of men. You get the symbol when you draw apart this 'Solomon's Key', so that the one triangle comes down and the other is raised up. The symbol thus obtained played, as I said, a significant part even as late as the nineteenth century, within this little community or school."

Steiner is speaking of the Seal of Solomon. He then goes into some detail as to the meaning and interpretation of the symbol:

> "The symbols by which he teaches them consist in certain geometrical forms, let us say for example a form such as this — (two intersecting triangles) — and at the points are generally to be found some words in Hebrew. It was impossible to find any direct connection with such symbols,

---

184 Rudolf Steiner, http://wn.rsarchive.org/Lectures/RosiModIn-it/19240106p01.html.

one could do nothing with them directly. And the pupils of this master knew through the instructions they received that what, for example, Eliphas Levi gives later on, is in reality nothing more than a talking around the subject, for the pupils were at that time still able to learn how the true meaning of such symbols is only arrived at when these symbols are rediscovered in the nature and being of the human organisation itself."

We shall have more to say on Eliphas Levi in a moment, but Steiner still has some remarkable information to convey about the meaning of the Seal of Solomon in these old Rosicrucian emblem books:

"The Master then made the members of his little circle of pupils take up a certain attitude with their bodies. They had to assume such a position that the body itself as it were inscribed this symbol. He made them stand with their legs far apart, and their arms stretched out above. Then by lengthening the lines of the arms downwards, and the lines of the legs upwards, these four lines came to view in the human organism itself.

A line was then drawn to unite the feet, and another line to unite the hands above. These two joining lines were felt as lines of force; the pupil became conscious that they do really exist. It became clear to him that currents pass, like electro-magnetic currents, from the left fingertips to the right fingertips, and again from the left foot to the right foot. So that in actual fact the human organism itself writes into space these two intersecting triangles.

The next step was for the pupil to learn to feel what lies in the words: 'Light streams upwards, Weight bears downwards.' The pupil had to experience this in deep meditation, standing in the attitude I have described. Thereby he gradually came to the point where the teacher was able to say to him: 'Now you are about to experience something that was practised over and over again in the ancient Mysteries.' And the pupil attained then in very truth to this further experience, namely that he experienced and felt the very marrow within his bones."[185]

---

[185] Rudolf Steiner, http://wn.rsarchive.org/Lectures/RosiModIn-it/19240112p01.html.

A trail of clues has been revealed. An underground stream has come to the surface. There is a secret hidden in some, at least, of these old alchemical manuscripts. It is geometrical, and it is subliminal. The goal is to arrange the hidden figure to conform to the underlying geometry but not to be overtly visible. It must be present, but subtle enough to pass beneath the viewer's level of conscious perception. This is subliminal communication.

Steiner's citation of Eliphas Levi in relation to this material is not an accident. An image from one of his works depicts a man with his arms reflected in a mirror. It has been engraved to encourage the eye to see the limbs and their reflections as two different coloured ribbons entwined to form the two interlinked triangles of the Seal of Solomon (Figure 132).

The symbol here is on open display yet hints at the means by which invisible layers can be concealed. It is a lesson in a technique by which certain geometrical forms or images can be embedded in works that depict something else entirely. This is of course exactly the concealed strategy that we have encountered in both the Boudet map and the parchments. This image from Levi is merely a glimpse into an elusive esoteric tradition, but it is enough to confirm its existence, as Steiner described. Boudet had evidently encountered these techniques and learned the secrets. He was working within an established esoteric practice.

Martinism

This symbol of the Seal of Solomon has a rich history that dates back to antiquity. It has represented many different meanings over time. However, within the context of French esoteric history, one group in particular has come to be associated with it, namely the Martinist Order. Martinism is a form of esoteric Christianity and high Christian mysticism, made up of many related groups and branches. It was first established around 1740 in France by Martinez de Pasqually, and later given further impetus in different forms in the late eighteenth century by his two students, Louis Claude de Saint-Martin and Jean-Baptiste Willermoz.

The Martinist Order re-emerged in Paris in or soon after 1886, when Augustin Chaboseau met Gérard Encausse (aka Papus), both of whom had been initiated in separate lineages from Saint-Martin. Louis-Claude de Saint-Martin (1743–1803), a French philosopher and Christian mystic popularly known as the Unknown Philosopher,

Figure 132: Image from Eliphas Levi.
Published in 1855, showing a "concealed" Seal of Solomon, made of interlaced black and white ribbons, encircled by the Ouroboros.

wrote many remarkable works of deep mysticism. His life story and his writings are a rich and fascinating subject, but it would be outside the scope of this present work to be able to give any adequate account of them.

In the Martinism of Louis-Claude de Saint-Martin, emphasis is placed on meditation and inner spiritual alchemy. In his words, he explained it as a silent "way of the heart", with the spiritual goal of attaining what he termed reintegration. He also instituted a body of rituals and knowledge designed to make contact with angelic spirits. This material was transmitted only by personal one-to-one initiation into the techniques and inner philosophy.

The symbol of Martinism is the Seal of Solomon, formed of two interweaved "ribbons" of contrasting colours, typically red and white or black and white. Various examples of the seal in different contexts and formats are shown in the attached figures.

There remain today many organisations around the world that claim affiliation and lineage with these original roots of Martinism, although there have been, as is inevitable, various splits and schisms and competing claims to be the genuine heirs of the tradition over the years. By the

1960s, when the parchments and related materials were created, the Order was highly active and their materials, including the seal, were widely known and available in such circles. In this context, the Seal of Solomon concealed in the parchments would have been instantly recognisable as the emblem of the Martinist Order.

There would seem to be little doubt that the Seal of Solomon design concealed in the parchments is intended as a reference to Martinism. But, as we have seen, the Seal of Solomon is also indicated in the Boudet map and on the title page. Is it possible he would have known that this was the Martinist emblem in 1886? Figure 134 shows a masthead from a newsletter of the Order in 1890, depicting the emblem.

Boudet, 1886 and the Martinist Order

Boudet's book, *La Vraie Langue Celtique*, carried the publication date 1886 on its title page. It is printed within the arabesque which, as we have seen, defined the concealed inch measure. Researchers of The Affair have long queried the veracity of this date because the historical records show that the printer named on the page, Francois Pomiés, went out of business in 1880. If it cannot be a genuine publication date from that printing house, then what else might it represent?

As it happened, that year was a highly significant one in the modern history of the Martinist Order. It was in 1886 that Augustin Chaboseau received his initiation and met Gérard Encausse soon after. They realised they had both received the same initiation via two different historical lines of transmission. This inspired them to launch the modern reorganisation of the order. Some sources also date this event to 1886, though most cite it as 1887 or 1888. [186]

The inception and the founding of the modern Martinist Order took place therefore over the years 1886, 1887 and 1888. The date on Boudet's book correlates exactly with this time frame.

It is not difficult to suggest some possible connections that would place Boudet in proximity to known Martinists. Firstly, Jules Doinel, who was one of the initial members appointed to the Council overseeing the activities of the Order, was associated with Carcassonne, and was involved in occult and ecclesiastical matters that would undoubtedly have brought him and Boudet into contact. Perhaps even more tantalising is evidence uncovered by a French journalist and researcher in recent years that shows Abbé Saunière's name written in the guestbook and attendance records of a Martinist lodge in Lyon.

---

186   http://omeganexusonline.net/rcmo/martinit.htm.

Figure 133: The sign of the Martinist Order.

Martinist Initiation

As I have noted above, there were two parallel paths of initiation in the history of the Martinist Order that united in modern times when Chaboseau met Encausse. Both lineages trace back to Saint-Martin himself. It comes as a surprising twist in our story to find that one of these lines passed through Henri Delaage, a journalist and author who was a close friend, collaborator and confidant of Gérard de Nerval, throughout the 1840s. They worked together on a book called *Le Diable Rouge*, or The Red Devil, which was published in 1851. Delaage had been initiated into the Martinist Order by his grandfather. This raises the question: is it possible that he passed the initiation by private instruction to Nerval?

Gérard de Nerval's work draws on an impressively wide and deep knowledge of many topics. His knowledge of spiritual matters, of the realms of the unconscious, the history of religions, and theology and esoteric lore, is hard won from both his reading habits and his own personal explorations of inner space.

But was he also perhaps an initiate, in some sense? Did he receive any instruction in esoteric or spiritual teaching through any group, or individual? Was he formally trained and initiated into any order or lineage? These questions engaged Richer, and he makes some comments on them in a 1957 monograph:

> "Nerval considered that he had experienced the equivalent of an ancient initiation and spoke and acted in the last months of his life as if, being 'vestal' or initiated, he was assured of the Reintegration, to use Martinès' vocabulary from Pasqually."[187]

In the extended passage from which this quotation is taken, Richer outlines his conclusions. There is nothing in the careful examination of the life and work of Nerval that would rule out the possibility that he had a human "master" from whom he received an initiation. On the other hand, there is no evidence that he belonged to any initiatory brotherhood either.

Nerval's spiritual development was his own personal work and private journey, and he did not draw his instruction from methodical teaching but from conversation and reading. He was subject to spontaneous episodes of personality disorder that left him vulnerable for the disasters which eventually overtook him. For this reason, Richer did not believe Nerval would be amongst those few who were capable of achieving self-initiation.

If Nerval considered himself in some sense initiated, then who might have been responsible for initiating him? Richer doesn't speculate in his text but is bold enough to caption a photograph of Henri Delaage with the words: "*Le maître (?) de Nerval*" (or "The master (?) of Nerval").

Richer describes how Nerval would use red ink for certain manuscripts with deep significance or importance. He then notes in the following extraordinary passages in the closing pages of the text of *Expérience vécue et création ésotérique*:

> "But we also know that any work of art is a reorganization of the world according to certain models that the sculptor, the musician or the poet carries within him; it is a form of alchemy. Now Nerval, who was a reader of Dom Pernéty, knew well that the colour red corresponds to the final stage of the work, that of fixation, also called Purple, *rubedo* or Phoenix. (...) To summarise the general impression that emerges from the body of these texts, it can be said that the use of red ink by the poet at this precise

---

187 Jean Richer, *Poètes d'Aujourd'hui : Gérard de Nerval,* op. cit. p. 108.

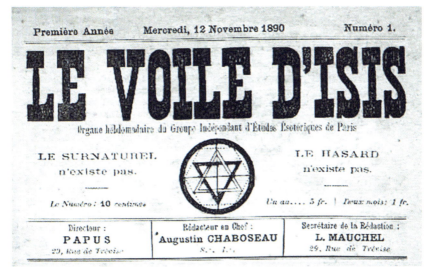

Figure 134: The seal on the first issue of Le Voile d'Isis, the Martinist newsletter, dated 12 November 1890, just four years after the publication date of La Vraie Langue Celtique.

moment in his career probably means that he considered that he had, in his own way, *accomplished the Great Work.*"[188]

Earlier, he had suggested a direct link between Nerval and Martinism:

"The second part of *Aurélia* is the story of a spiritual experience during which the poet thought he had obtained what the disciples of Martinès de Pasqually called a 'pass', that is to say a sign of friendship from God supposed to cause the certainty of salvation."[189]

It is not only Richer who has scrutinised Nerval's work for signs of an initiate. Sieburth remarks:

"Given its pervasive allusions to the subterranean Enlightenment counter-traditions of Swedenborgianism, Rosicrucianism, Mesmerism, Martinism and Freemasonry, *Les Illuminés* has often been taken as evidence of Nerval's deep initiation into the doctrines of esotericism and the occult."[190]

---
188 Jean Richer, *Expérience vécue et création ésotérique*, (Paris, Guy Trédaniel, 1987) p. 360.
189 Jean Richer, *Poètes d'aujourd'hui : Gérard de Nerval* (Paris, Editions Pierre Seghers, 1957) p. 93.
190 Richard Sieburth, *Gérard de Nerval Selected Writings*, op. cit. p. xxvi.

Richard Holmes ponders these same questions, as he notes how every element in Nerval's stories resonates with symbolic correspondences.

> "... mysterious spirit of these strange tales, in which each object contains a symbol. One could even say that he took from them certain occult meanings intended only for the neophyte, certain cabalistic formulae, and overtones of the Illuminati, which made one believe, at times, that he was writing directly of his own personal initiation. I would not be altogether surprised if, like Jacques Cazotte, the author of the *Diable Amoureux*, he had received a visit from some stranger making Masonic signs, who was quite confounded not to find in him a true member of the Secret Brotherhood."[191]

Holmes becomes consumed by the depths of Nerval's journey, and his search on the trail of his initiatory knowledge leads him to his novella *Aurélia*, written in the last years of his life, soon after *Sylvie*. He says of it that "it appears that Nerval's story partly describes an initiation essay based on alchemy, astrology and tarot". Holmes continues:

> "Indeed, the further I went into this labyrinth of signs and rituals the more I came to believe that *Aurélia* was the complete and literal statement of his life."

Holmes then quotes Nerval from *Aurélia*, in an extended passage which is worth citing in full, as it speaks eloquently to the poet's vision and experience of the world. Here he speaks openly of his inner trials as a Sacred Initiation.

> "From that moment, when I felt sure that I was being subjected to the trials of a Sacred Initiation, an invincible strength entered my soul. I imagined myself a Hero, living under the direct gaze of the gods. Everything in Nature took on a new aspect, and secret voices, warning and exhorting me, came from plants, trees, animals, and the humblest insects. The talk of my companions took on mysterious turns of meaning which I alone could understand, and formless, inanimate objects lent themselves to the calculations of my mind. From combinations of pebbles, from shapes in corners, from chinks or openings,

---

[191] Richard Holmes, *Footprints: Adventures of a Romantic Biographer*. op. cit., location 4544.

Figure 135: Henri Delaage, (1825-1882)
This photograph appears in a 1957 book by Jean Richer on Gérard de Nerval. with the caption: "l'ami et maître (?) de Nerval", or "the friend and master (?) of Nerval".

from the outlines of leaves, from colours, scents and sounds, I could see hitherto unknown harmonies springing forth. "How have I been able to live so long," I asked myself, "outside Nature, and without identifying with her? Everything lives, moves, everything corresponds ... Though I am captive now, here on earth, I commune with the chorus of the stars, and they take part in my joys and sorrows!"[192]

It is uncanny that, of all people, of all groups, Nerval should have such a close brush, if that is what it was, with an initiate of the Martinist Order, and then for his name to become involved posthumously in this whole affair so many years later.

---

192   Gérard de Nerval *Aurélia* in Sieburth op. cit. p. 306.

Chapter Eighteen
A WALK IN THE WOODS

"My dear sweet aunt, tell your son that he does not know you are the best of mothers and aunts. When I shall have triumphed over everything, you will have your place in my Olympus, just as I have my place in your home. Don't wait up for me tonight, for the night will be black and white."
- Final words of Gérard de Nerval, in a note written to his aunt.

WHEN I first attempted to read Umberto Eco's novel *Foucault's Pendulum* during the late 1990s, I was intrigued to come across many references that glanced at the Affair of Rennes-le-Château. I was particularly struck by one tantalising passage. It was about a *map hidden in a manuscript*.

"And, very circumspectly, he replied: 'In the manuscript, of course, there was also the map, or, rather, a precise description of the map, of the original. It's surprising; you can't imagine how simple the solution is. The map was within everyone's grasp, in full view; why, thousands of people have passed it every day, for centuries. And the method of orientation is so elementary that you just have to memorize the pattern and the map can be reproduced on the spot, anywhere. So simple and so unexpected.'"[193]

These lines struck a deep chord in me. When I launched a website on my work back in 1998, I even used this quotation as the opening text. I felt intuitively that there was something valuable hovering over

[193] Umberto Eco, *Foucault's Pendulum*, (London, Picador, 1990). p. 552.

this passage, though I did not yet know exactly how it might relate, if at all, to the questions surrounding the Rennes parchments.

Like many people who have set out to read it over the years, I suspect, I never did finish Eco's novel at the time, but in 2008 and 2009 when the parchment solution and the Boudet map had begun to reveal their secrets, I marvelled again at the prescience of his remark about the map in the manuscript. The idea of a manuscript as a "precise description of a map" seemed so uncanny in light of my discoveries. I wondered how much Eco really knew, so I resolved to attempt to read the book again, armed with my newfound insights on the role of Richer, and the solutions to the parchments and the maps.

On this second reading I found the novel only slightly less impenetrable, but I was aided this time in my comprehension by another of Eco's works, a much slimmer volume entitled *Six Walks in the Fictional Woods*. In this fascinating book, comprised of a series of lectures that he had delivered at Harvard in 1993, I found that Professor Eco had deposited certain keys that unlocked an unexpected hidden layer in *Foucault's Pendulum*.

In *Six Walks in the Fictional Woods,* Eco presents a characteristically brilliant examination of episodes from a range of fictional sources, loosely clustered around the theme of walks or journeys. He uses these to investigate the way that fiction interacts with reality, and explores questions around the relationship between the voice of the author, authorial identity and the experience of the reader.

Eco brings up *Foucault's Pendulum* in the opening pages of the book, and returns to it again later to discuss a crucial episode from the novel which we will look at in more detail shortly. He offers some fascinating remarks including the intriguing statement that "...the story was so thick with mysteries both true and false."[194]

On the very next page after introducing *Foucault's Pendulum*, Umberto Eco informs the reader that he has held a deep lifelong admiration of the work of Gérard de Nerval, and in particular, *Sylvie*.

> "In my subsequent lectures, I shall often refer to one of the greatest books ever written, Gérard de Nerval's *Sylvie*. I read it at the age of twenty, and still keep rereading it. When I was young, I wrote a very poor paper about it, and beginning in 1976 I held a series of seminars about it at the University of Bologna, the result being three

---

[194] Umberto Eco, *Six Walks in the Fictional Woods,* (New York Harvard University Press, 1994). p. 77.

doctoral dissertations and a special issue of the journal VS in 1982."[195]

In fact, the first of the "Six Walks" that Eco presents to explore his thesis is *Sylvie* itself. In an extended discussion in Chapter One, he analyses in detail the concept of the identity of the author in relation to Nerval's short story. He carefully distinguishes between the man himself, who was born Gérard Labrunie, his literary nom de plume, Gérard de Nerval, and the voice of the author within the story as the reader encounters him. He teases apart the subtle nuances that distinguish these three different aspects of the writer's persona, and how they relate to the experience of reading his work.

Even when he moves on to examine the other five walks in the lecture series, he returns time and again to Nerval and *Sylvie* as reference points for his discussion. I was intrigued to find Eco bringing *Foucault's Pendulum* and *Sylvie* into the same frame, and discussing them at length, side-by-side, in light of the references to Rennes-le-Château in the novel and the progress I had made linking Nerval, through Richer, to the Affair of Rennes. I wondered if there was perhaps more to it. This prompted me to redouble my efforts to firstly finish the book and secondly to think about it carefully.

Foucault's Pendulum

The events that make up the narrative of *Foucault's Pendulum* unfold over the course of a single night: 23/24 June 1984, in Paris. First published in 1988, the novel is a mystery thriller set in the world of arcane philosophy, esoteric books and the publishing industry. It's both a delicious send-up of historical conspiracy theories, and a serious education in hermetic lore.

It is a book that displays Eco's vast erudition to virtuoso effect – *Foucault's Pendulum* is an encyclopaedia of knowledge on the history of western esoteric thought. The novel is also huge in every sense of the word, the fictional equivalent of reading Jung: there is so much learning on every page, so many ideas in play that it becomes a major challenge for even the keenest and most knowledgeable reader to fully comprehend, or even finish.

The novel opens with a scene set in the Musée des Arts et Métiers, a museum of science and technology that houses the collection of the Conservatoire national des arts et métiers, established in 1794 in the deserted building of the Saint-Martin-du-Champs Priory on

---
195   Ibid. p. 11.

the rue Saint-Martin. Staying behind after closing time, the narrator, Causabon, conceals himself inside one of the exhibits where he waits patiently to witness a ritual he knows will take place at midnight in the central hall of the museum. During the evening, in a series of extended flashbacks, he recalls the long chain of events that brought him to this place.

It is futile to even attempt to reduce the plot to a manageable outline, so a brief and utterly inadequate overview will have to suffice. Causabon and two friends, Jacopo Belbo and Diotallevi, had set up a publishing house together, with the aim of specialising in books on obscure topics and unconventional ideas on the fringes of scholarship, the kinds of works usually rejected by mainstream publishers.

Business flourished. They received manuscripts of varying quality from aspiring authors all over the world. Amused and inspired in equal parts by this avalanche of material, they decide to invent their own imaginary version of a grand conspiracy. They come up with what they call The Plan, an intricate over-arching secret history of the world that interconnects everything. It involves a strange manuscript, a mysterious map and an explosive Secret. At the heart of The Plan is the knowledge of the existence of powerful underground currents of energy and how to direct their flow.

> "At the centre of the earth is a nucleus of fusion, something similar to the sun – indeed an actual sun around which things revolve, describing different paths. Orbits of telluric currents. The Celts knew where they were and how to control them."[196]

These "telluric currents" are streams of an invisible force that exert a potent influence over all life on the planet. They can be controlled by those who know the Secret, with a certain map that identifies key locations on the earth where the currents can be manipulated. These techniques were passed down from the Celts and jealously guarded by certain groups throughout history, including – naturally enough – the Knights Templar.

After waiting all evening, midnight arrives and Causabon watches on silently from his hiding place as the members of a secret society interrogate his friend Belbo, demanding he reveal the secret of The Plan. When he refuses to comply, a small riot breaks out, and Belbo ends up hanged by a wire from the eponymous Foucault's pendulum.

---

196 Umberto Eco, *Foucault's Pendulum*, op. cit. p. 444.

The pendulum, named after Léon Foucault, the nineteenth century French physicist, demonstrates the rotation of the earth with simple, direct evidence. The basic historic facts surrounding its origin and installation in the Musée des Arts et Métiers are explained in the first pages of the novel.

> "'It's Foucault's Pendulum,' he was saying. 'First tried out in a cellar in 1851, then shown at the Observatoire, and later under the dome of the Panthéon with a wire sixty-seven meters long and a sphere weighing twenty-eight kilos. Since 1855 it's been here, in a smaller version, hanging from a hole in the middle of the rib.'"[197]

So Causabon has witnessed the death of his friend by hanging from the original Foucault's pendulum, which had been re-installed in the museum of the Conservatoire national des arts et métiers in 1855. After this scene, Causabon escapes by an underground route through the Parisian sewers and emerges on the rue Saint-Martin. Then, in Chapter 115, he walks the length of the street in a fascinating extended episode that I will discuss in more detail later.

After this walk, Causabon takes a taxi to the Eiffel Tower, where the final scene of the novel plays out. This brief, utterly inadequate description of *Foucault's Pendulum* omits all the flashbacks, (which in fact comprise the bulk of the story), but at least sketches in the basic framework around which the narrative is woven and provides enough background detail for the discussion to follow in this chapter.

On one level, the novel can be considered as an exploration of the pitfalls and potential folly of succumbing to conspiracy theories. *Foucault's Pendulum* reads as a literary parody of topics like the Affair of Rennes, in which Eco is satirising the people and the publishing industries that spring up around such fringe ideas. This nexus between *Foucault's Pendulum* and the Affair of Rennes is confirmed by Eco himself as well as explicitly within the text of the novel. Rennes-le-Château is mentioned more than once. In chapter 18, for example, the narrator remarks: "I had recently read a book about the secret of Rennes-le-Château". Over the course of a paragraph, he then gives a detailed outline of Saunière's alleged discovery of manuscripts in the church, (without naming him) and subsequent events which followed.

The novel consists of 120 chapters divided into ten parts, named after the ten Sephiroth of the Tree of Life. Each chapter begins with a

---

[197] Ibid., p. 5.

quotation from a dazzlingly wide array of sources, some of whom have appeared in this book. Chapter 33, for example, opens with an excerpt by Papus from his monograph on Martines de Pasqually. Eliphas Levi and Saint-Martin are cited, as are Jacques Cazotte and Dom Pernety on the concept of whiteness in the Great Work. Chapter 13 begins with the famous phrase *"Et in arcadia ego"*, which could almost be the motto of the Affair of Rennes.

Chapter 85 has a paragraph from an extraordinary book by the French researcher Michael Lamy *The Secret Message of Jules Verne*, which explores the great French science fiction writer's surprisingly complex relationship with the Rennes-le-Château mystery.[198] Lamy's work also discusses Nerval and his relationship to Verne, so Eco is tacitly glancing at our poet with this epigram. If any doubt remains that the author is thoroughly familiar with the Rennes Affair, Chapter 66 opens with a quotation from *The Holy Blood and the Holy Grail* itself:

> "If our hypothesis is correct, the Holy Grail....was the breed and descendant of Jesus, the 'Sang real' of which the Templars were the guardians...At the same time, the Holy Grail must have been, literally, the vessel that had received and contained the blood of Jesus. In other words, it must have been the womb of the Magdalene."
> —M. Baigent, R. Leigh, H. Lincoln, *The Holy Blood and the Holy Grail*, 1982, London, Cape, xiv[199]

Then the action of the novel resumes with the following conversation between Diotavelli and Causabon:

> "'Nobody would take that seriously,' Diotallevi said. 'On the contrary, it would sell a few hundred thousand copies,' I said grimly. 'The story has already been written, with slight variations, in a book on the mystery of the Grail and the secrets of Rennes-le-Château. Instead of reading only manuscripts, you should look at what other publishers are printing.'"

Hence, without in any way diminishing or limiting the scope of Eco's literary intentions in the novel, it is fair to say that on one level at least, *Foucault's Pendulum* was partially inspired by Lincoln, Baigent and Leigh's 1981 book, *The Holy Blood and the Holy Grail* and, ultimately,

---
198 Michael Lamy, *The Secret Message of Jules Verne: Decoding his Masonic, Rosicrucian and Occult Writings*, (Rochester, Destiny Books, 2007).
199 Umberto Eco *Foucault's Pendulum* op. cit. p. 377.

by the Affair of Rennes itself. Several years after its publication, Eco was asked in an interview whether he had read *The Da Vinci Code*, to which he replied:

> "I was obliged to read it because everybody was asking me about it. My answer is that Dan Brown is one of the characters in my novel *Foucault's Pendulum*, which is about people who start believing in occult stuff."[200]

Of course, Dan Brown is not an actual character in *Foucault's Pendulum*, but a real-life representative of a certain literary type whom Eco is gently satirising.

Death of a Poet

The final chapter of the life of Gérard de Nerval had a sad and untimely ending. In the grey dawn of 26 January 1855, he was found dead, hanging from a window grating, in the rue de la Vieille Lanterne, a narrow alleyway by the Seine. On the wall above was the sign of a key, signifying the premises of a *serrurier*, or locksmith. He was just 47 years old.

A huge throng of those who loved him dearly in his life attended Nerval's funeral. His friends Théophile Gautier and Arsène Houssaye carried the expenses. There was enough doubt about his state of mind and intention that the police were persuaded to enter an open finding on his cause of death, which allowed him to be buried in consecrated ground in Père Lachaise Cemetery in Paris. In this discussion of *Sylvie* in *Six Walks in the Fictional Woods,* one of the first topics Eco introduces is the tragic circumstances of Nerval's death.

> "In *Sylvie* we must deal with three entities. The first is a gentleman who was born in 1808 and died (by committing suicide) in 1855."[201]

In the same paragraph, he gives the location where he died:

> "With Michelin guide in hand, many visitors to Paris still look for the rue de la Vieille Lanterne, where he hanged himself."

The reason why visitors look in vain for the rue de la Vieille Lanterne, Eco informs the reader, is that the small alleyway no longer exists. It

---
200   Deborah Solomon, "Questions for Umberto Eco" The New York Times, November 25, 2007.
201   Umberto Eco, *Six Walks in the Fictional Woods*, op. cit. p. 13.

was demolished in the late nineteenth century to make way for the Théâtre de la Ville, one of the two theatres built by Baron Haussmann at Place du Châtelet. Hence the "rue de la Vieille Lanterne" can no longer be found today on a map of Paris, or in the city itself.

Note that in this single passage, Eco has placed before the reader the year of Nerval's death by hanging, 1855, and the image of tourists wandering around looking for the site where he died, in the introduction to his text in which he has brought together Nerval's *Sylvie* and his own *Foucault's Pendulum* for a discussion of related themes and literary strategies shared by the two works.

Meanwhile, the climactic scene of *Foucault's Pendulum*, as we have noted, involves a man being hung from a pendulum of whose date of installation – 1855 – Eco had made sure to inform the reader in the opening pages.

Taken together, these snippets offer a surprising and dramatic new insight into the plot twist at the heart of the novel, the climax of the narrative action, namely, the hanging of Belbo from the Pendulum. It is apparent that in *Six Walks in the Fictional Woods*, Eco is assimilating the circumstances of Nerval's tragic demise in 1855 with the installation of Foucault's pendulum in 1855.

By this oblique means, he is quietly signalling that the hanging of Belbo on the pendulum, the central metaphor of his book, stands for the death of Gérard de Nerval himself! This is such an unexpected revelation that we must surely require confirmation before it can be accepted as the author's deliberate intention. Fortunately, Eco has provided it. It is the events which unfold after Causabon has witnessed the terrible scene in the *Conservatoire* that seal this interpretation, as we shall soon see.

A Walk in Paris

Driven from his hiding place by the horror following the midnight ritual, Causabon makes his escape from the museum through the sewers and emerges in the suburb of Port Saint-Martin, a kilometre or so to the north. He then walks back south towards the Seine, along the length of the rue Saint-Martin. Still in mild shock, he has no clear destination in mind, but nevertheless some hidden purpose seems to drive him forward.

In Chapter Four of *Six Walks in the Fictional Woods*, Eco brings in a discussion on this crucial scene in *Foucault's Pendulum*. It begins:

> "In chapter 115 of my book *Foucault's Pendulum*, the character Causabon...walks, as if possessed, along the entire length of the rue Saint-Martin."

In the novel, Causabon describes various locations that he passes, noting certain buildings, including the home of the famous alchemist Nicholas Flamel and other curiosities that drew his attention.

> "...display windows of Editions Rosicruciennes...church of Saint-Merri....Librairie la Vouivre..."[202]

He passes the Conservatoire, where he has just spent the night and continues:

> "...after a few blocks, on my left, the Conservatoire...I continue south towards the Seine. I have a destination, but I'm not sure what it is."

He also happens to pass by some interesting places that he does not explicitly mention, including the house, at number 168-170, in which Nerval was born in 1808 on this very Paris street. He keeps walking.

> "I continue along rue Saint-Martin, I cross rue aux Ours, broad, a boulevard, almost; I'm afraid of losing my way, but what way? Where am I going? I don't know."

Urged on without knowing why, he eventually arrives at the end of the rue Saint-Martin, then makes several random turns down sidestreets until he finally stops. Whatever his destination was, he has apparently now arrived at it. He finds himself in a small alley, the rue Nicholas Flamel, standing in front of a bookshop, the Librairie Arcane 22. Everywhere he looks, he is reminded of alchemy, the tarot and the esoteric:

> "At the corner, the Librairie Arcane 22, tarots and pendulums. Nicolas Flamel the alchemist, an alchemistic bookshop,"

He stands and directs his gaze along the alley. He is now looking directly at a prominent local landmark; the tall white tower named the Tour Saint-Jacques. Sometimes referred to as the "navel of Paris", this striking monument dates to 1508 and rises to a height of 52 metres, a narrow but richly embellished grand spire sitting in its own square.

---

202   Umberto Eco *Foucault's Pendulum*, op. cit. pp. 602-603

> "Rue des Lombards...at right angles, rue Nicholas Flamel, and at the end of that you can see, white, the Tour Saint-Jacques ... with those great white lions at the base, a useless late-Gothic tower near the Seine"

Causabon pauses for a moment while he muses on the significance of the Tour. He compares it to the Eiffel Tower and speculates on the role both of these monuments might have played in the grand secret at the heart of the Plan, the control of the telluric currents.

> "Maybe They began with the Tour Saint-Jacques, before erecting the Eiffel Tower. These are special locations. And no one notices. I go back towards Saint-Merri..."

He then turns and retraces his steps. After this, the action moves swiftly to the resolution of the main plot at the Eiffel Tower, and a few pages later, the end of the book.

Why did this turn out to be the unknown destination to which Causabon found himself compelled to walk? Why should this moment of gazing at the Tour Saint-Jacques have held such significance? There is no explanation supplied in the text for these crucial questions.

Umberto Eco devotes several pages in *Six Walks in the Fictional Woods* to some fascinating and insightful remarks about Causabon's walk in Chapter 115, but neither in the novel nor in his lecture series does he enlighten the curious reader on Causabon's purpose in walking to the rue Nicholas Flamel to stare at the Tour Saint-Jacques.

Using Google Earth to visualise the path of Causabon's walk and his precise location and orientation when he arrived at the rue Nicholas Flamel led me to the answer.

When Causabon directed his line of sight along the street towards the Tour Saint-Jacques, he was staring at the precise location, now the site of the Théâtre de la Ville, on the other side of the tower, where Nerval had taken his life, the former rue de la Vieille Lanterne. There is a plaque dedicated to Nerval installed in the square in which the Tour stands, the only public memorial to the poet in Paris, as it is the closest convenient position to the long-disappeared alleyway.

This was the destination and the reason for his walk. It was a kind of unconscious *pèlerinage*, or pilgrimage, ritual journey, in which, after witnessing the ritual in the *Conservatoire* which represented symbolically the death of Nerval, he traversed the length of the rue de St Martin, passing his place of birth on the way to the place of his literal death. Although Causabon was not aware of what he was doing, the

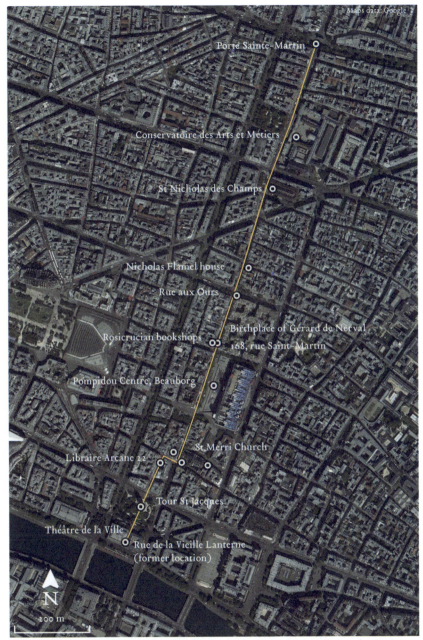

Figure 136: Causabon's walk in chapter 115 of *Foucault's Pendulum* by Umberto Eco. It begins at Porte Saint-Martin in the north, and proceeds south along rue Saint-Martin, and ends with him looking south towards the Tour Saint-Jacques from outside the bookshop Libraire Arcane 22. The rue de la Vieille Lanterne, now long gone, was on the site of what is now the Théâtre de Ville.

author of the novel, Umberto Eco, has arranged for his character to trace this route in homage to Gérard de Nerval.

In the introduction to *Six Walks in the Fictional Woods*, Eco described tourists wandering around the streets of Paris looking for the rue de la Vieille Lanterne. In this snippet, he deposited the clue that answers the riddle of the purpose of Causabon's walk. Just like the tourists, he was looking for the place of Nerval's death.

All of this confirms that Umberto Eco's novel *Foucault's Pendulum* contains a layer of hidden references to Gérard de Nerval.

The clues that arrive at this extraordinary conclusion were laid out in *Six Walks in the Fictional Woods*, a book in which Eco declares his love of Nerval's *Sylvie* which, as we now know, also played a pivotal role in the conception and creation of *Le Serpent Rouge* by Jean Richer.

Eco, Richer and Nerval

Professor Eco must surely have been aware of the work of Professor Richer, and conceivably might have even known him personally, as they both shared a lifelong fascination with Gérard de Nerval and had both published extensively on *Sylvie*.

Did Eco also know something about Richer's role in the Affair of Rennes? Is there anywhere in *Foucault's Pendulum* where we might detect a collegiate nod from Professor Eco to Professor Richer? If there is, the following passage would be a good candidate. It brings in many of the keywords from this book itself: the sun, the circular walk, the coiled serpent, rotary motion, myths and rites.

> "The sun is good because it does the body good, and because it has the sense to reappear every day; therefore, whatever returns is good, not what passes and is done with. The easiest way to return from where you've been without retracing your steps is to walk in a circle. The animal that coils in a circle is the serpent; that's why so many cults and myths of the serpent exist, because it's hard to represent the return of the sun by the coiling of a hippopotamus. ... And that's why the circle and rotary motion and cyclic return are fundamental to every cult and every rite."

Elsewhere in the novel, Eco discusses ancient landscape engineering:

> "The ground had been punctured and the deep strata tested, but the Celts and the Templars had not confined themselves to digging wells; they had planted their stations

Figure 137: The destination of Causabon's walk in chapter 115 of Foucault's *Pendulum* by Umberto Eco. Causabon directed his view south, towards the Tour Saint-Jacques, and the location of the rue de la Vieille Lanterne.

> and aimed them straight at the heavens, to communicate from megalith to megalith, and to catch the influence of the stars."[203]

He also talks about alignments and even grids connecting sacred places:

> "You know what the English leys are? If you fly over England in a plane, you'll see that all the sacred places are joined by straight lines, a grid of lines interwoven across the whole country, still visible because they suggested the lines of later roads."

---

203   Umberto Eco *Foucault's Pendulum*, op. cit. pp. 363, 444, 465

And how the knowledge was later forgotten:

> "It's a secret that was lost after the Roman invasion, but there are those who still know it."

It is not difficult to see all of this as an oblique reference to Richer's alignments, his maps, his recovery of sacred geography and perhaps even his contact and peripheral involvement with other networks.

Eco would not have to go far to find such ideas in Nerval's writings, either. He would have been very familiar with passages like the following example from Nerval's *Aurélia*, in which the narrator imagines himself sinking beneath the surface of the Earth in a characteristically dream-like vision. Then, he is carried along underground currents that form a vast, interconnected network. It is entirely possible that in this passage Eco even found the direct inspiration for his telluric currents.

> "I felt myself being buoyed along by a current of molten metal; a thousand similar streams whose hues varied with their chemical composition were criss-crossing the earth like the vessels or veins that wind through the lobes of the brain. From the pulse and flux of their circulation, I gathered these streams were made up of living beings in a molecular state, which only the speed at which I was travelling made it impossible to distinguish."[204]

In Eco's novel, the successful orientation of the map is provided by a pendulum. Finally, we learn how it all works, in language that would not be out of place in Nerval's writings themselves.

> "With the Pendulum, the exact point under the earth's concave vault where the telluric currents converged. (...) And now you see the beauty of the idea. The telluric currents become equated with the celestial currents. ... The Mystic Pole coincides with the Heart of the Earth. The secret pattern of the stars is nothing other than the secret pattern of the subterranean passages of Agartha."[205]

In light of such passages, it is intriguing to compare *Foucault's Pendulum* and *Le Serpent Rouge* side by side. Both are encyclopaedic works of esoteric fiction based around the Affair of Rennes. They are both huge in scope, despite their obvious difference in size.

---

204 Gérard de Nerval *Aurélia* in Sieburth op. cit. p. 272.
205 Umberto Eco, *Foucault's Pendulum*, op. cit. p. 514.

Both texts are also patterned on the author's reading of the temporal structure of Nerval's *Sylvie*. For Richer, the flow of narrative time in the story can be read as a cyclical motion around a clock-face or zodiac. When he composed *Le Serpent Rouge*, he used this motif as the template for the narrative of his poem. Eco, on the other hand, draws attention to the back-and-forth motion of the plot, and the use of flashbacks as a signature narrative device. When he wrote *Foucault's Pendulum*, he consciously employed this same motif as the template for the flow of action in his novel. He lays this out in *Six Walks in the Fictional Woods* where he discusses Nerval's treatment of time at length:

> "The fundamental mechanism in *Sylvie* is based on a continual alternation between flashbacks and flashforwards, and on certain groups of embedded flashbacks."[206]

Eco becomes so inspired by the possibilities for understanding the deep structure of Nerval's masterpiece of this analytical approach, he even provides the reader with a diagram of the unfolding of time through the work, echoing Richer's zodiac figure. He expanded on this in a later interview, explaining how he adopted the pendulum as the metaphor for the temporal structure of the novel. As a pendulum swings back and forward, so too does the narrative, from the present to the past, back to the present, then forward to anticipate the future, and repeat.

> "With *Foucault's Pendulum*, the oscillatory movement of the eponymous artifice obliged me to accept another temporal structure. (...) I built a kind of lace-up structure that recorded the returns to the past and the anticipations of the future."[207]

Furthermore, both texts are arranged on a system derived from the esoteric realms of astrology and tarot. *Le Serpent Rouge* is divided into thirteen stanzas named for the zodiac signs, while *Foucault's Pendulum* consists of ten parts named after the Sephiroth on the Tree of Life. Given all these parallels, we might not be too far off the mark to consider that *Foucault's Pendulum*, like *Le Serpent Rouge*, can also be understood as a *conte initiatique*: Eco's homage and secret tribute to the life and work of Gérard de Nerval.

---

206   Ibid., p29.
207   https://www.academia.edu/31426591/Le_voyage_formatif_et_cognitif_de_Causabon_dans_Le_Pendule_de_Foucault_dUmberto_Eco, cf. Umberto Eco, De la littérature, Paris, Grasset, 2003, p410.

The Hanged Man

If the central metaphor of Eco's novel is the Pendulum, in the opening pages, he establishes exactly what it represents when Causabon, peering through a periscope in his hiding place in the museum, remarks:

> "I realized that the periscope gave me a view of the outside as if I were looking though a window in the upper part of the apse of Saint-Martin—as if I were swaying there with the Pendulum, like a hanged man, taking his last look."[208]

This key unlocks the hidden compartment in *Foucault's Pendulum*. The hanging of Belbo on the Pendulum installed in the Musée des Arts et Métiers in Paris in 1855 represents the death of Nerval in 1855. There is an archetypal element to this. Eco is suggesting that in the closing act of his personal *conte intiatique*, Gérard de Nerval himself became the Hanged Man. Richer also sensed that there may have been a ritual element to Nerval's tragic suicide. In a 1957 essay he wrote:

> "The possibility of a mystical suicide, similar to that practiced by certain Cathars, is therefore not excluded. The choice of the place of the suicide, near the tower-talisman and the old Place de Grève, a sacrificial place, lends itself to reflection."[209]

The Riddle of the Three Named Authors

These observations offer the opportunity to resolve the final remaining riddle that I posed at the beginning of this book, namely, the strange business surrounding the three named authors of *Le Serpent Rouge*. Recall that the men named on the cover did not know each other and had nothing to do with the production of the poem or pamphlet. Their only connection was that they had all died by hanging on the same weekend in early March 1967.

Le Serpent Rouge was ready to be deposited in the Bibliothèque nationale at this time but had not been submitted to the library for inclusion. As Franck Marie determined, after the weekend of the three deaths, the pamphlet was somehow introduced into the collection with the paperwork backdated to create the impression that the order of events was the other way around.

---

208 Umberto Eco, *Foucault's Pendulum*, op. cit. p. 16.
209 Jean Richer *Poètes d'aujourd'hui : Gérard de Nerval* op.cit. p. 93.0

What was the motivation for creating this false impression? Here is a scenario that might account for it. Pierre Plantard was undoubtedly well-connected in Parisian affairs, including the Gendarmerie. By some means, during the month of March 1967, news of the three tragic deaths by hanging during the same weekend reached his ears as he was preparing the pamphlet for publication. He was struck by the uncanny co-incidence of the manner of the deaths which had taken place as he worked on a poem inspired by Gérard de Nerval, the Hanged Man.

Perhaps it was intended the poem appear anonymously, but now an idea occurred to him. He decided to introduce an additional element of macabre mystification to the project by attributing it to the three men: the "hanged men". This twist would baffle researchers if anyone should ever investigate the three men. Yet it left a glaring clue to the secret, right out in the open: the true "author" of the work was the Hanged Man, Gérard de Nerval.

This was Plantard's signature style: to conceal and reveal at the same time. It was also an opportunity that must have seemed too good to resist. He listed the three men as the authors, both on the cover of *Le Serpent Rouge* and on the blank deposit slip that one of his contacts within the Bibliothèque nationale had obtained for him. He copied the details from the death certificates, taking meticulous care to introduce exactly one error of a different kind into each of their details, as a signal that this was all deliberate and intentional, rather than casual and accidental. He returned the completed deposit slip to his inside contact at the library, who then backdated it with the official stamp and signature and inserted it into the catalogue files.

Sometime later, Jean Richer became aware of Plantard's crude attempt at adding intrigue to the project. No doubt he would have been incensed by such an insensitive and needlessly dark addition to his work, considering it to be in very poor taste.

With his academic career and reputation at stake, Richer insisted that his involvement in the Affair be kept a closely guarded secret amongst the participants. It is impossible of course to know the exact nature of the relationships between the various parties, but perhaps somehow Richer had sufficient clout and influence over the other participants that he was able to extract from them firm commitments never to reveal his role in the Affair. Yet all three - Plantard, de Sède and de Chérisey - could not resist leaving a trail of subtle clues that would one day lead back to Richer. If this is so, it might explain the reticence which all the Team members displayed in later years in their various writings to openly associate Richer with the poem. Nevertheless, they all sailed as close to the wind as they thought they might get away with, and each found a way to leave a trail of clues that might allow someone, one day, to put the pieces back together.

We will never know the exact sequence of events, but the identification of Nerval as the Hanged Man would certainly seem to be the underlying motivation for the misattribution of the poem to the three men. Notice also that within Eco's novel, three men, one of whom dies by hanging, compose the (false) Plan, although of course in reality it was the author himself who wrote it. In the case of *Le Serpent Rouge*, Plantard (falsely) assigned the authorship of the poem to three men, all of whom died by hanging, whilst in truth it was written by Jean Richer. And all of this was brought to light by a lecture series examining the relationship between author, voice and identity in such texts.

The Other Foucault

The title of Eco's novel obviously refers in the first instance to the pendulum invented by Leon Foucault, and we can now see that it also carries a symbolic association to the death of Nerval. But there is of course another Foucault: the influential twentieth century French philosopher, writer and professor, Michel Foucault.

The phrase "Foucault's pendulum", therefore, could also conceivably refer to a pendulum relating to the philosopher Michel Foucault. This might seem like a stretch, particularly as he is nowhere mentioned in the book. However, there is one very curious fact that compels us to take the possibility seriously: Michel Foucault died on 25 June 1984, the very day after the principal events of the novel take place! Eco and Foucault were both leading European philosophers of the latter decades of the twentieth century, so there is no question that the novelist could have been unaware of this connection.

Thus, there are three deaths implied in the symbolism of the novel's title, *Foucault's Pendulum*: the death of Belbo, the death of Nerval by hanging in 1855 (the year of installation of the pendulum) and now the death of Michel Foucault on the day after the principal events of the novel.

Linda Hutcheon explores this connection between *Foucault's Pendulum* and Foucault the philosopher in an insightful 1992 essay entitled *Eco's Echoes: Ironizing the (Post)Modern*.[210] She is cautious about pursu-

---

210 Linda Hutcheon, *Eco's Echoes : Ironizing the (Post)Modern*, Diacritics, Vol. 22, No. 1 (Spring, 1992), pp. 2-16 (15 pages).

ing this interpretation, or any specific reading of the text, too rigidly, noting that Eco makes it difficult for readers and critics alike to engage with his novels as "he self-reflexively ironizes the position not only of author but also of reader".

And she is also reticent about the idea of any neat solution to the multi-layered mysteries at the heart of the book, observing that "this is a novel in which there is no final Secret, or is the Secret simply kept silent?".

Nevertheless, she shows how it is possible to read it, at least on one level, as Eco's response to Foucault's philosophical outlook, and in particular his position on the relationship of older discourses of knowledge to modern rationalism and empiricism. To support her reading, she focuses on a course given by Eco in 1986 on hermetic semiosis, for which his prescribed reading list included Foucault's 1966 book *The Order of Things*.

> "In 1986 Eco gave a course on hermetic semiosis at the University of Bologna's Istituto di Discipline della Comunicazione, in which he studied the interpretative practice of seeing both the world and texts in terms of relations of sympathy and resemblance. His time frame ranged from prehistoric times to the present. Now, perhaps we can begin to see what all of this has to do with Michel Foucault.
>
> In *The Order of Things* (*Lets mots et les choses*), Foucault argued that this kind of thought was historically limited, a Renaissance paradigm which gave way to a modern scientific one. The epistemological space up to the end of the sixteenth century was one Foucault saw as governed by a rich 'semantic web of resemblance'. In this course Eco clearly wanted to challenge this temporal periodization, to argue that this kind of thought never really disappeared, that there was no final epistemic break. (...)
>
> In other words, the pendulum has continued to swing between the extremes of some form of reason and some form of mysticism, and this is one of the many meanings of the titular pendulum."[211]

In Hutcheon's reading then, the term "Foucault's pendulum" can be read as a reference to a historical movement back and forth or tension between earlier modes of mystical or sympathetic thought on

---

211   Ibid.

the one hand, and scientific materialism on the other. In her view, the novel represents an ironic rejoinder to Foucault's thesis, that hermetic thinking has been largely surpassed and overtaken within the scientific project and suggests instead the persistence into the modern world of such earlier modes of thought that are grounded in notions of sympathy and resemblance.

Given that we have now associated the symbolism of Foucault's pendulum in this book with the death of both Nerval and Foucault, we might consider these two writers as emblematic of the two modes of thought in the scheme. If so, Nerval would clearly represent the earlier mystical mode, and Foucault the latter scientific mode. We know where Eco's sympathies lie. Just as he argued in his 1986 lecture at the University of Bologna, he seems to be suggesting in *Foucault's Pendulum* that the magical mode has not completely lost its power to illuminate, if not explain.

Eco provokes us to consider the two writers in relation to each other and to compare their philosophical outlooks and programmes. If we do so, the domain in which the primary focus of the work of Nerval and Foucault overlap is not difficult to recognise. Both were deeply concerned with the problem of madness and its relationship to language, discourse and power.

This nexus between the poet and the philosopher has been considered before. In a 1971 essay, Shoshana Felman traced a particular thread of discussion in French literature running from Nerval to Foucault via Rimbaud and Artaud, on madness and its representation. Felman comes to an unexpected conclusion:

> "In fact, Nerval's poetic project surprisingly resembles Foucault's philosophical project. Nerval undertakes, as a poet, what Foucault will undertake as a historian and as a philosopher: to tell of madness itself, to write a history of madness, by trying to avoid the trap of 'what is commonly called reason'. Is this the triumph of Reason? Or, on the contrary, the refusal to believe that an 'unreason' exists, that there can exist, even in the madness, something radically foreign to the reason of things? Nerval, like Foucault, would like to go back to the origins of his being, to the zero point where Madness and Reason are not yet mutually exclusive, but on the contrary communicate in an enigmatic conjunction."[212]

---

212 Shoshana Felman « *Aurélia* » ou « *le livre infaisable* » : de Foucault à Nerval,

Felman is by no means alone in observing that Nerval was exploring the same themes as Foucault a century later. Richard Sieburth himself in his Introduction to the Penguin Classics edition of the Selected Works of Gérard de Nerval wrote:

> "In any event, he absolutely refused to concede that his condition deserved the clinical label of madness. Writing to the actress Ida Ferrier shortly after his release from Blanche's sanatorium, he provided an analysis of his predicament worthy of a Foucault: 'I was only allowed to be released and to mingle among reasonable folk once I had formally admitted that I had been sick—which took quite a toll on my pride and even on my honesty.' ... At stake, in short, was nothing less than the perennial battle of the Imagination against all these discourses—medical, penal or theological—that sought to strip it of its rightful sovereignty."[213]

And again later in the same volume, commenting on Nerval's short story *Aurélia*, almost a companion piece to *Sylvie*:

> "Throughout *Aurélia*, one can sense the intolerable pressures of the double-blind in which its author finds himself. On the one hand, he must publicly acknowledge his illness and denounce his madness as arrant self-delusion if he is to prove to his doctor that he has now regained his reason and thus deserves to be set free. On the other hand, to acquiesce to the moral and institutional legitimacy of the medical discourse that has committed him to bedlam is to renounce his sacred vocation as Promethean saviour and Orphic seer."

The problem that both Nerval and Foucault addressed might be described as follows: if madness is defined by the language of reason, how then can we speak of madness without resorting to a terminology that seeks to banish or even extinguish its claims to a voice? Or, to put it another way: if reason excludes madness, how can madness authentically represent itself to the world on its own terms? These questions lead naturally to others that are closely related. How do we deal with magical narratives and hermetic thought in a scientific, rational world?

Romantisme, 1971, 3. pp. 43-55
213   Sieburth. op. cit. pp. xix, 263.

We are brought back to Nerval's claim to have found a suitable voice in his life and art by which he was able to describe his journey in the realms of madness in language that does not succumb to reason's demands to dictate terms and definitions. How, then, did he claim to accomplish this?

Impossible Theories in Infeasible Books

Nerval categorically refused to accept the clinical definition of madness for his own experience, even when his continued detention in the hospital was made contingent on his acceptance of the diagnosis. The irony was not lost on him.

He bristled to be accused of madness and especially by his fellow writers. He was particularly stung by the words of Alexander Dumas, who mourned the passing of his friend's reason, and accused him in his folly of putting forth "impossible theories in infeasible books".

Nerval laid out his response to Dumas in an introductory essay that accompanied *Filles de Feu,* the collection of his stories in which *Sylvie* appeared. Here, he forcefully rejected any notion that he suffered from madness, instead defending his writings as part of a deliberate poetic strategy, an immersion in a mode of consciousness which, he claimed, made it entirely possible under certain conditions to transcend the limits of time. He fully accepted the reality of multiple incarnations and previous lives, and the ability to recover knowledge and memories from earlier ages. His poetic works were not the outcome of folly, he stated plainly, but composed in the true original spirit of poetry.

Richard Sieburth writes about Nerval's defence of his poetic methods, and identifies it with a specific historical tradition dating back to Plato:

> "Deeply wounded by Dumas' recent derision of his work, Nerval added a lengthy preface in which he attempted to clarify his creative method to his colleague–a method, he argued, that had nothing to do with madness but rather with a tradition of poetic anamnesis that reached back to Plato and was grounded in the Pythagorean doctrine of the transmigration of souls. To invent, he explained, (in a passage not lost on Proust), was always to re-remember, to experience the present as the suddenly rediscovered trace of all one's previous incarnations. "[214]

---

214 Richard Sieburth *Gérard de Nerval : Selected Writings*, op. cit. p. xxviii.

Anamnesis, a concept in Plato's epistemology, is the idea that humans possess innate knowledge, and that learning consists of rediscovering that knowledge from within. Each time the soul is incarnated its knowledge is forgotten in the trauma of birth. What one perceives to be learning, then, is the recovery of what one has previously known. Anamnesis is the closest that human minds can come to experiencing the freedom of the soul before it is encumbered by matter.

This, according to Sieburth, is the poetic tradition to which Nerval appealed in his letter to Dumas. Far removed from folly, he wrote of a mode of consciousness that permitted access to previous experiences of past lives and opened a door to knowledge from the ancient world.

Recall that in *Le Serpent Rouge*, the author credits the parchments of his friend with showing him the way when he was lost. He uses the phrase *fil d'Ariane*, invoking the thread which lead Theseus out of the labyrinth. Now the full implication becomes apparent, because it is found in the final paragraph of Nerval's letter to Dumas.

> "Once I was convinced I was writing my own story, I began to translate all my dreams, all my emotions, I moved on to this love for a fugitive star, who left me alone in the night of my destiny, I cried, I shuddered at the vain appearances of my sleep. Then a divine ray has him in my hell; surrounded by monsters against which I fought obscurely, I seized the thread of Ariadne, and from then on, all my visions became heavenly. Some day I will write the story of this 'descent into hell' and you will see that it was not entirely devoid of reasoning even if it always lacked reason."[215]

This paragraph must have resonated deeply for Richer. Here is Nerval speaking frankly of his own life as a *conte intiatique*. The story of the "descent to hell" is the legend of Orpheus, another of the classic world journey narratives. The second part of *Aurélia* begins "*Eurydice, Eurydice*", which hints that in this work, the companion piece to *Sylvie*, Nerval had kept his word. The monsters against which Nerval "fought obscurely", are the same ones that Jung describes, after Michael Maier. This is the encounter in the fourth quadrant with the monster/dragon/serpent, the unconscious content that must be faced and overcome, and the precursor for the motif of the *serpent rouge* itself.

215   Gérard de Nerval, *Filles du Feu : Lettre à Alexander Dumas*, in Nerval, Gérard de, *Œuvres*, (Paris. Bibliothèque de la Pléiade, Editors A. Béguin and J. Richer, 1961). Tome I p. 158.

Now we can understand what Jean Richer meant when he said that the parchments of his friend had been a *"fil d'Ariane"*. He was referring to Nerval's use of this term in his letter to Dumas, as part of his defence of his sanity and writings. This is the reason that the writings of Nerval showed Richer how it might be possible to access such states under certain conditions. Richard Sieburth summarised it this way:

> "The world of connections and intersections in which Nerval lived therefore supposed a constant recourse to the dreamlike, even in the waking state. What is called Nerval's madness seems to have come from a difficulty in distinguishing between, on the other hand, reminiscence, introverted feeling and, on the other hand, revelations of intuition or inner guide."[216]

Nerval pointed the way to techniques of consciousness which were capable of crossing boundaries of time and space, and even retrieving knowledge of an earlier era, based on his profound experiences of navigation in the dream-state. Here is the thread that Richer learned to follow. Would we be stretching this interpretation to suggest that Richer himself obtained some of his knowledge of the ancient world from following such techniques as the story of Nerval taught him? Not at all. He confirms it himself. In the introduction to *Delphes, Délos et Cumes*, he wrote:

> "I would not be so presumptuous as to try and explain 'rationally' what I have just reported which, in my eyes, only confirms what I am firmly convinced of: that all the religious and poetic experience of the peoples and individuals of the past are, in privileged moments, accessible to those who are able and know how to recapture it, in a particular state of consciousness."[217]

Here he speaks openly of a genuine process of transmission of knowledge across time, from ancient minds to the present. It was, after all, in a dream, on Mount Lycabettus in Athens that Richer's very first intuition of the landscape geometry arrived, as I described all the way back in the opening pages of this book.

If Jean Richer was able to recapture the poetic experience of peoples of the past, then we are in the presence of modes of transmission that

---

216   Jean Richer, *Gérard de Nerval : Poètes d'Aujourd'hui* : op. cit. p. xxx.
217   Jean Richer, *Delphes, Délos et Cumes*, op. cit. p. 16.

we currently do not understand, nor do we readily concede even exist. To modern ears, this sounds strange, yet the Ancient Greeks would have had no problem with the concept.

They conceived of a very different relationship between the dream state and waking life in comparison to our present understanding. This is conveyed succinctly in the healing technique known as incubation. The term referred to the practice of sleeping in the sacred enclosure of the temple in order to dream of being healed of an ailment or condition. By performing the cure or solution experienced in the dream, the issue would be resolved. Thus, the dream was a portal through which knowledge of self and healing arrived from higher realms. In Ancient Greece and modern France, the dream-state has been a bridge for the passage of true knowledge between worlds.

In Eliphas Levi's *Dogme et Rituel de la Haute Magie*, which by some uncanny coincidence was published in that same year of 1855, he described the tarot card of the Hanged Man and the meaning of its design. The gibbet from which the man is suspended, made from two trees, is in the shape of the Hebrew letter Tau. It also depicts a cross and a triangle formed by the position of his limbs and head. Levi remarks:

> "Now, the triangle surmounted by a cross signifies in alchemy the end and perfection of the Great Work, a meaning which is identical with that of the letter Tau, the last of the sacred alphabet. This Hanged Man is, consequently, the adept, bound by his engagements and spiritualized, that is, having his feet turned towards heaven. He is also the antique Prometheus, expiating by everlasting torture the penalty of his glorious theft."[218]

This extraordinary comment of Levi's offers the possibility to see in the circumstances of Nerval's passing the emblem of his accomplishment of the Great Work. If he did so, then this supreme alchemist, as Jean Richer did not hesitate to call him, had achieved the goal and attained reconciliation of the oppositions that lay at the heart of his own unique cosmology.

In the final tragic note he left for his aunt, the archetypal themes of Nerval's life and work are present: the mother and the gods of ancient Greece. Even the alchemical cannot help but break through to the surface. "The night will be black and white," he wrote in the poignant

---

[218] Eliphas Levi, *Dogme et Rituel de la Haute Magie Part I: The Doctrine of Transcendental Magic* Translated by A. E. Waite. (London, Rider & Company, 1896). Vol. I, Ch XII.

and heart-breaking final coda to his literary career. The journey was complete.

> "I've had enough of chasing after poetry; I believe that poetry lies at one's very door or perhaps in one's very bed. I'm still a man on the run, but I shall try to stop and wait."[219]

In the last analysis, Nerval was a man who inspired deep love and devotion from his friends during his lifetime and from a small but dedicated following of readers ever since. Umberto Eco was one of those readers, and in *Foucault's Pendulum* he has woven Nerval into the heart of the novel.

Whilst *Foucault's Pendulum* presents itself openly as a satire on the Affair of Rennes, it also fulfils a more subtle literary purpose: it is Umberto Eco's covert recognition of the importance of Nerval's *Sylvie*. The key to unlocking this relationship has been deposited by Eco in his *Six Walks in the Fictional Woods*.

In the final stanza of *Le Serpent Rouge*, it is revealed that the action has all taken place within a dream, the classic ending of a thousand fairy tales. In this case, though, it was no cliché. The dream was central to Nerval's thought, the kernel from which his entire imaginative inner world was generated. We have come a long way now from the rue Saint-Martin, and Causabon's walk, and perhaps we have gone further than even Eco anticipated. But this is where the trail has led. It was all a dream.

---

219 Richard Sieburth, *Gérard de Nerval Selected Writings*. op. cit. p. xxiii.

Chapter Nineteen
REASSEMBLING THE SCATTERED STONES

"When 700 years are fulfilled, the laurel will turn green again."[220]
Prophecy of Bélibaste, (1280 – 1321) the 'last Cathar'.

W E HAVE reached the other side of the woods and arrived at the destination that seemed impossibly far off when we first set out. It's time to catch our breath, reflect on the journey and take in the view.

The mystery of Rennes is laid bare. Refusing to be side-tracked by rumours of gold and treasure, we have followed a twisting, turning trail that has led instead to something far more valuable. A lost inheritance from our past has begun to reveal itself. The traces it has left are ephemeral, little more than a faint geometrical imprint gently laid over the landscape. Yet these slight clues point to work that is much more involved, elaborate and widespread. They are the remnants of a vast network of extraordinary alignments permanently marked in the mountains, valleys, churches and châteaux of the Pyrenees.

The discoveries provoke a host of obvious questions. Who built it? When was it built? Why was it built? In these pages, I have tried to focus on documenting what I have found, sticking to what I can prove and avoiding as much as possible, until now, any speculation, particularly on the origin of the Complex. It's time to loosen those restrictions.

Who built it?

Richer suggests that these practices had their origins in Babylon and were taken up by other civilisations around the Mediterranean, including the Phoenicians, the Hittites, the ancient Greeks, the Etruscans,

---
220   In the original Occitan: «*Al cap del set cents ans, verdejerà lo laurel*»

the Carthaginians and the Romans. In *Sacred Geography of the Ancient Greeks*, he wrote:

> "Faced with such astonishing systems of correspondences, one is almost forced to postulate the existence of 'great teachers', who at a rather remote date, situated at the very latest at the time of the 'return of the Heraclides' (that is, about 1,000BC) showed the inhabitants of the Greek world where to build their temples, and the general principles for selecting the site of a sanctuary."[221]

As these groups expanded their geographical spheres of influence they took their techniques with them, including to northern Africa, Gaul and the Iberian Peninsula. Presumably the knowledge was preserved in schools of astronomer-priests and passed on via oral transmission to local custodians and guardians of the geometry where it was established in the landscape.

When was it built?

The reader may have noticed that I have been rather vague in assigning historical dates to the different elements of the Complex. Conventional records are not much help. Exact dates of construction for most of the châteaux and churches which make up the physical elements of the geometry are often unknown. In any case, they were usually built on locations which were already significant, including peaks with valuable sightlines, and locations of strategic value, so we can be confident that the alignments and the geometry must predate the buildings which came later.

Richer's research led him to the conclusion that the early alignments and grids which framed the lands of ancient Greece were constructed in the period 2,050 to 1,800BC. The development of the system of zodiacal co-ordinates came later, around 1,000 to 800BC. Without another means of narrowing down the time frame for construction of the different stages of the Complex, it is difficult to be more specific. Fortunately, unlikely as it might seem, there is such a method.

Over my years of studying the landscape geometry, certain crucial observations have opened up an entirely different way of approaching this question of dating. The lines in the landscape were very often oriented to the rising and setting of major stars. These correlations between land and sky offer clues to techniques which might have been

---

221   Jean Richer *Sacred Geography of the Ancient Greeks* op. cit. p 68.

employed in the construction of the Complex, and perhaps even some potential insight into its purpose.

They also provide a methodology for dating the various stages of the geometry. Following these indications has enabled a detailed timeline of the development of the Complex to be assembled. It has not been possible to even begin to present this material in the current book, so the entire topic and the results which lead from it will be held over for what will hopefully be a follow-up volume.

Suffice to say for now that the results of applying these methods are broadly consistent with the estimates of date ranges that Richer provided, though they tend to push back the first signs of such activity in the Pyrenees as much as a millennium earlier, to around 3,000BC, and extend the later phases of construction to the first century AD.

This is an exceptionally long period, of up to three millennia, though it is not necessarily the case that activity was continuous throughout the entire era. The chronology which emerges from such considerations suggests that there were several bursts of intense construction work during this time, typically for around a century or two, separated by longer periods of stabilisation and maintenance of up to 700 to 1,000 years. I will refer to this extended period, from megalithic times to the Roman era, as the 'First Phase' of the Complex.

The First Phase

During this First Phase the main elements of the geometry were brought into being. In this book I have presented several examples, as summarised below, but there remain many more to be described.

A reference frame, consisting of a sequence of meridians passing through peaks and other prominent positions, was marked into the landscape of the Haute Vallée in the Languedoc during the First Phase. Some of these meridians were spaced at regular distances that resolve to whole values of key units, including the inch and the arcsecond.

The Sunrise Line must have been laid down at a very early stage of development of the wider system of alignments. It may be traced from the summit of Pic de Saint-Barthélemy, the highest and most holy mountain of this region of the Pyrenees and coincides with the alignment to the rising sun on August 24 each year as observed from the summit, a date preserved in local ritual and lore.

The Sunrise Line, in concert with its companion at right-angles, the Bugarach Baseline, established a system of coordinate axes that made possible a programme of high accuracy surveying and landscape interventions across a very wide area of territory, and in particular, to the north and east. It seems likely that a deep connection or exchange of some kind was established between ancient Greeks and the peoples of Occitanie in earliest times. Even the name Toulouse evokes the sacred enclosure, the *Tolos*, of the Greek temple. The practice of modelling city and landscape architecture after the pattern of the heavens was introduced into this region as it was around the Mediterranean.

These axes and meridians, together with their accompanying system of related measures, comprised the foundation layers of the Complex, on which subsequent alignments were then erected. These were in place by the beginning of the first millennium BC.

Later, the original site of Rennes-les-Bains was established on the banks of the River Sals. By the time of the arrival of the Romans, around the first century BC, the village had been laid out along the traditional format of ancient cities, with *cardo* and *decumanus* axes, and an omphalos at the crossing point. These axes were aligned to conform to the local terrain, with the primary *cardo* set down on a bearing of 15° to north. It passed through the summit of an impressive mountain peak at the northern end of the village that was accordingly given the name Pech Cardou, after the axis it helped to define.

The *cardo-decumanus* axes were then elaborated with a further four alignments equally spaced around the same central point, such that the six lines formed a twelve-fold division of the surrounding territory. This was a variation on the landscape zodiac format which was a defining feature and blueprint of ancient cities and cult centres across Europe and the Near East.

One of the zodiac lines in particular, the 45° alignment, was marked and surveyed with great care to create a calibrated reference instrument, the Arques Square, intended to enable very highly accurate geographical and astronomical observations and from these measurements, to derive knowledge about the earth. These techniques involved the sighting and timing of the transit of stars and other heavenly bodies across the field of calibrated meridians.

Why was it built?

The primary function of the Complex was to weave a reflection of the layout of the heavens into the shape of the earth to establish a harmonious relationship between the different realms. In our times we have lost all sense of how such a state of affairs could have any real effect or benefit, but evidently our ancient ancestors must have considered it to be a valuable and perhaps even essential goal, otherwise they would not have gone to such trouble.

These interventions—together with many others that have not yet been described—were predicated on an understanding of the dimensions and measures of the earth and observational astronomy. They comprise a co-ordinated system whose physical, historical and cultural traces remain to this day.

This was by no means its only purpose. One extremely practical application was as a communications system, enabling messages to be transmitted by line-of-sight signals across a vast and intricate network of peaks and other high points in the landscape. Curiously, this system integrated seamlessly with the underlying geometry of the Complex, though, again, to modern ears this feature seems to add no conceivable benefit if the use was simply to pass messages by fires at night or smoke signals by day.

There is a poetry to this ancient science, where geometry blends with myth to create archetypal forms. The zodiac becomes the symbol or even the signature of the cycles of becoming that have shaped this land. It denotes the cosmic axletree, the *axis mundi*, the wheel of the fates.

The installation of the zodiac must have been a decisive moment in the unfolding history of the Complex. As a reflection in the landscape of the order of the heavens, it marks and orients the local space. Further, because it is a zodiac, it organises time. And because it is a wheel, its nature is to turn.

And so, this wheel was inscribed and activated. The tides of history moved back and forth over the Pyrenees, and waves of people came and went. The Complex slowly became dormant and entered another hibernation phase. Yet, somehow, the deep knowledge survived and was never completely lost.

The Second Phase

There was a renewal or reactivation of the Complex in the Middle Ages, in the period from around the ninth to the thirteenth centuries AD. I call this the Second Phase. The Knights Templar were certainly aware of at least some aspects of the ancient geometry during their ascendancy in the ninth to twelfth centuries, as is proved by the location of Campagne-sur-Aude alone. They re-employed the signalling network between peaks that had been in active use from Celtic times, and built their châteaux on strategic highpoints in the system. Undoubtedly they also contributed to the re-marking of the geometrical design during this time. The Cathars, with their close association with the Templars and the châteaux throughout the area, might also have been privy to certain elements of the knowledge. Perhaps they played a role in ensuring its persistence into the modern era. If this was so, then certain legends from those times may take on a new significance.

It is recounted in a local guidebook that seven years after the final siege of Quéribus in 1255, a group of Cathars who had escaped and survived, met again at the ruins of the fortress château. They made a solemn vow to each other, that they would all reincarnate in seven hundred years, to continue their work and carry forward the memory and knowledge of the Cathar faith. Whether or not any of them managed to fulfil their pledge, somehow fragments and pieces of their worldview have survived. Streams may flow underground and then re-emerge at the surface.

Curiously, there is another piece of local Occitania lore from the era that also invokes the same time period of seven centuries. It is the saying quoted at the beginning of the chapter—"in seven hundred years, the laurel will turn green again". The phrase is attributed to Bélibaste, a pious Cathar who was born in Cubières in 1280 and died in 1321, burned at the stake at Villerouge-Termenès. He became known as the "last of the Cathars" and his words have left a lasting impression in local history and culture.

When he made his prophecy, in 1309, there were few Cathars left, and he must have known their hour had nearly passed. Nevertheless, he was bold enough to speak of a renewal at what must have seemed like a very far off date. The time has now arrived. If Bélibaste's words were to be fulfilled, perhaps we might look to see whether green shoots are appearing in our time. If so, the trail of texts and ideas that we have followed in this book may be signs of a revivification. Are we witnessing another activation phase in the recurring cycle?

Vehicle of Transmission: The Map

The Complex is an artefact of time whose emergence into historical awareness in our era is equally as remarkable as the pattern in the peaks itself. Somehow, the memory of this intervention in the landscape has propagated across centuries and millennia, ensuring that fragments at least have survived in the deep inventory of human collective consciousness.

This transmission can only have been accomplished with the aid of some kind of map, in the broadest sense of the term. What then is a map? It is a record of memories of excursions into a territory. A map can be a description, or a set of instructions, or even a poem or a song. It need not take physical form, on paper, or even on screen. It is, ultimately, a mental construct that may be passed from mind to mind. In this case, the language in which it has been recorded is geometry.

This "Complex" must have been described and defined by a map, a conceptual apparatus in the minds of the builders. The mental blueprint included precise details of measure, angle and geometry that were held as an image in inner visualisation. The geometry could then be encoded as a sequence of constrained gestures to be performed within certain defined domains. Such protocols permitted the materials to be communicated, and shared, and to function as plans and schematics of the construction.

The map was built up as a series of layers, each exquisitely related to the others in a seamless construction which seems to defy ordinary notions of historical development. It's a challenge to comprehend how any layer could have been devised before, or independently of, every other layer, let alone how the map format persisted in human memory and imagination for such lengthy periods. Yet, somehow, in some form, the knowledge survived. In more recent times, the map has been hidden in ingenious fashion in a sequence of manuscripts, from Boudet's enigmatic cartography in his strange book, to the mysterious parchments, to the astonishing poem of *Le Serpent Rouge*. I call this the Third Phase of the Complex.

The Third Phase

Abbé Henri Boudet, priest of Rennes-les-Bains, must have come into possession of detailed elements of this knowledge sometime prior to the 1880s, by what means I cannot say. He chose to use the term "cromlech" to refer to the ensemble of the twelve-fold division, or landscape zodiac. Evidently, he must have decided to encode the material in a book, and in a map, in order to ensure its preservation, even whilst, for unknown reasons, he was under some obligation not to explicitly reveal the details openly. To execute his plan, he conceived of creating an encrypted work of cartography with concealed geometry which would only become visible with the application of a key.

The key was concealed in the title on the map, *Rennes Celtique*. It defined the unit of measure to be employed, namely the inch, in the dimensions of the word Rennes, and determined a grid that was aligned with the edges of the letters of the title. The inch grid on the map corresponds in a remarkable manner with the meridians in the landscape. Boudet composed the title page of his book around the same hidden geometry, with the inch key being hidden this time in the ornate frame surrounding the publishing date.

At some moment before the early 1960s, Pierre Plantard had learned the hidden secret of Boudet's map, possibly from his grandfather, who had known the priest and been presented with a signed copy of his book. Together with certain others, he devised the Priory of Sion affair. As part of this elaborate prank at the intersection of literature, history, fiction and the esoteric, they came up with the idea of the parchments. These ingenious documents were conceived as recreations or reworkings of Boudet's map and included encrypted content which secretly documented their knowledge of the geometry and the inch grid. They spliced the parchments into an obscure narrative of a minor local mystery in the south of France, and the rest is, well, history.

Gérard de Sède was drawn into these affairs sometime around 1966, or before. He became acquainted with Professor Jean Richer and shared the incredible information about Boudet's book, map and hidden secrets with him. Richer was about to publish his research for the first time on his discovery of landscape zodiacs at Delphi and other ancient sacred sites in the Near East. They made a visit to Rennes-les-Bains in October 1966, and during this trip, I suggest, an event took place with far-reaching implications.

With Boudet's *La Vraie Langue Celtique et le Cromleck de Rennes-les-Bains* providing him with the crucial clues, Richer recognised that there was an ancient zodiac marked out as a twelve-fold division of the landscape around Rennes-les-Bains, based on the *cardo-decumanus* axes on which the plan of the ancient Roman town had been laid out. He realised that Boudet's cromlech was identical to this landscape zodiac, and that the priest had found in France in the 1880s a prime example of the zodiac landscape format that Richer had identified in Greece and elsewhere in the 1960s.

It is quite remarkable to consider that possibly the only person in modern times who was capable of recognising the presence of an ancient zodiac in the landscape should have been introduced to Rennes-les-Bains and the Boudet book, with its description of the cromlech, by people who happened to be in possession of the concealed content of its enigmatic map. There must have been some lively conversation around the dinner table amongst the group as these results were realised and digested.

During the visit to Rennes-les-Bains, the group decided to go for a walk. They planned to follow, as far as possible, the path laid out in Chapter VII of Boudet's book, describing a tour of the cromlech. In this manner, the group performed a circumnavigation, or better still, a circumambulation, around the zodiac of Rennes-les-Bains.

On completion of the walk, Richer composed a short poetic work, in which he encapsulated all this material into a single, cohesive, cryptic account of the cromlech. He mapped the route of their walk onto the zodiac of Rennes-les-Bains. He created a narrative, a *conte initiatique*, based around the notion of the heroic voyage around sacred space and inspired by his own literary hero, Gérard de Nerval. He drew on a rich tradition of scholarship including Goethe, Wirth, and Jung.

He called it *Le Serpent Rouge*. He wove various elements from his own literary and research work into its narrative, including the connections with Delphi, together with quotations from Boudet and other sources. The hidden message of the poem was Richer's insight that Boudet's cromlech was a landscape zodiac in the manner of ancient Delphi.

Plantard then took the poem and created the pamphlet of *Le Serpent Rouge*, as a vehicle for its dissemination. He added additional materials, typed it all out, pasted in the images and bundled it all together. It was early March 1967, and he was preparing for it to be published when news of three tragic deaths reached his ears. A mystery was set in motion.

On a summer family holiday to France in 1969, Henry Lincoln read *L'Or de Rennes*, the book by de Sède that launched the Affair to a wider public. He (and others) found evidence of geometry in the landscape of the area, alignments between churches, châteaux and peaks. It inspired him to make his three documentaries and to write various books, including *The Holy Place* in 1991. He became the mouthpiece by which the encoded messages hidden in the parchments were revealed to the world, via his 1974 film for the BBC's Chronicle series. The puzzle had been placed before the public, all part of an elaborate work of literary performance art created by Plantard and his Team.[222] Yet, beneath artfully concealed layers of blinds and double blinds, a genuine truth lay buried.

I stumbled on Lincoln's 1974 film on television that fateful Sunday afternoon in Adelaide as an impressionable fourteen-year-old. Without knowing it at the time, a seed had been planted in my mind. The topic burst into life for me in the early 1990s when I found and read *The*

---

222 See Appendix I for detailed discussion of the parchment text decoding.

Holy Place. I had taken my first steps on a path that was to lead to the discoveries in this book, not to mention an incredible life adventure.

I set out to tackle this strange riddle of geometry in landscape. Along the way, I was side-tracked by several puzzles which seemed to be entirely unrelated: *Le Serpent Rouge*, the Boudet book and map, the parchments. It took twenty years to discover that all of these yielded to the same simple solution. It was a symbol, or glyph. It was a map, hidden in a manuscript.

The secrets of the landscape geometry of the Two Rennes, and much more, were finally revealed through the lens of *Le Serpent Rouge*, Professor Jean Richer's astonishing prose-poem written in honour of Gérard de Nerval, a compact, dazzling, literary capsule fusing past, present and future into a single sublime work of alchemical art.

Epilogue
CODA TO A DREAM

As I put the finishing touches to this manuscript, I remembered a loose thread. It related to the very first alignment, the Apollo Line, from Delphi to Athens, Delos and Camiros that had come to Jean Richer in the dream on Mt Lycabettus.

As described in Chapter Five, after finding on Google Earth that the line as he had drawn it on his map in the early hours of the morning just missed the island of Delos by a very small margin, I proposed a slight variation. If the alignment begins at the Temple of Apollo in Delphi and continues to the peak of Mont Profitis Ilias on Rhodes, the second highest mountain on the island, rather than the monastery of Camiros on the coast as Richer had surmised, then with this minor alteration it now passes directly over the tiny island of Delos, birthplace of Apollo.

But I had never checked the effect of this slight change on the line at Athens. Previously the line did not seem to pass through any specific landmark in the city, but perhaps with the adjustment that too had changed. Now, I found that the line passed exactly through the highest peak in the modern city of Athens, Mt Attiko Alsos. To my astonishment, there was a monastery on the peak also named Moni Profitis Ilias, built as close as possible to the alignment from Delphi.

I searched in Google Earth for occurrences of the name Profitis Ilias close to Delphi. I quickly found one. It was another monastery, with almost the same name, Moni Profitis Ilia, located just a short distance to the west of the Temple Complex. It stands on a commanding location with views down the valley towards Delphi which lies just over the next ridge.

When I extended the Apollo Line beyond the Temple of Delphi, I found that it passed directly through the Monastery.

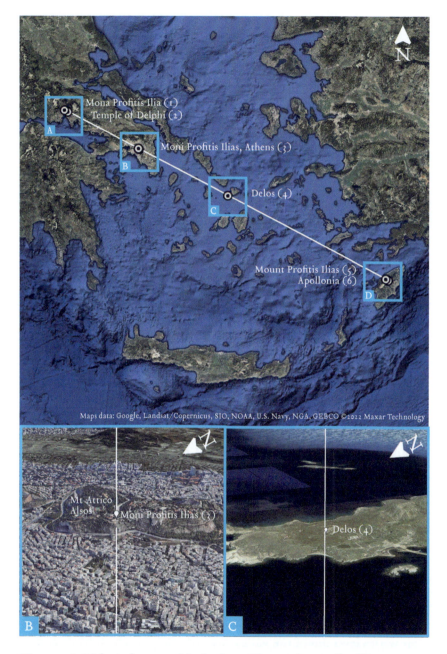

Figure 138: Richer's dream revisited. *Above:* The complete Apollo alignment. *Below left:* The alignment passing through the monastery of Moni Profitis Ilias, on Mt Attico Alsos, highest point in Athens. *Below right:* The alignment crossing the island of Delos, birthplace of Apollo.

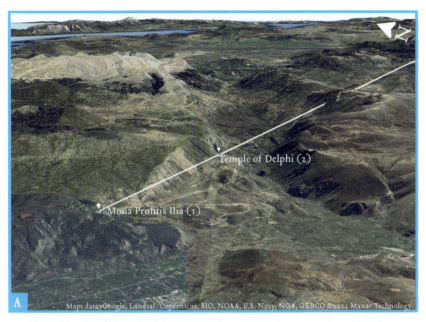

Above: The north-west end of the alignment at Delphi, passing through the Temple of Apollo in Delphi, and the monastery of Mona Profitis Ilia. *Below:* The south-east end on Rhodes, passing through Mount Profitis Ilias and the church in the village of Apollonia, site of an ancient temple to Apollo.

There were now six sites on the alignment that came to Richer in his dream, and the line passed precisely through each of these locations. In order from Delphi to Rhodes:

- Moni Profitis Ilia, Delphi: monastery on commanding peak
- Temple of Apollo, Delphi
- Moni Profitis Ilias: monastery on Mt Attiko Alsos, Athens
- Delos: island of the birthplace of Apollo
- Mount Profitis Ilias: second highest peak on the island of Rhodes
- Apollonia: church in village closest to peak on Rhodes.

The Apollo Line passes through the tiny island of the sun-god's birthplace, his most important temple, a village named after him, and three peaks all with the name of the Prophet Elijah attached, a figure who is also associated with Helios, solar symbolism, and Apollo.[223]

The precision and symbolic cohesion of the alignment validates in the strongest possible manner the genuine underlying accuracy of Richer's original insight in his divinatory dream. It has taken 63 years, and Google Earth, to fine-tune the line. With the minor corrections incorporated, full perfection is revealed.

The dream was real.

---

223 See also footnote 171 on page 375.

DELFICA

La connais-tu, Dafné, cette ancienne romance,
Au pied du sycomore, ou sous les lauriers blancs,
Sous l'olivier, le myrte, ou les saules tremblants,
Cette chanson d'amour qui toujours recommence ?...

Reconnais-tu le Temple au péristyle immense,
Et les citrons amers où s'imprimaient tes dents,
Et la grotte, fatale aux hôtes imprudents,
Où du dragon vaincu dort l'antique semence ?...

Ils reviendront, ces Dieux que tu pleures toujours !
Le temps va ramener l'ordre des anciens jours ;
La terre a tressailli d'un souffle prophétique...

Cependant la sibylle au visage latin
Est endormie encor sous l'arc de Constantin
— Et rien n'a dérangé le sévère portique.

- Gérard de Nerval

APPENDICES

Appendix 1
THE PARCHMENT TEXT DECIPHERMENTS

WHEN THE solution to decoding the parchment text was broadcast for the first time, as part of Henry Lincoln's 1974 BBC documentary *The Priest, the Painter and the Devil*, things were not quite as they seemed. A trick, or prank, or even perhaps a hoax was played on the viewing audience with profound implications for understanding the entire Affair.

In a nutshell, a *fake decoding* was presented that has fooled the world for fifty years.

In this Appendix, I will discuss and dissect the business of this trick in detail. The story is a little convoluted, but I hope the reward for persisting through the maze will be worth it.

I'll begin with a short summary of the decoding process with enough information to be able to grasp the trick. The full details are quite complicated, and some have been omitted that are not strictly necessary in order to follow the argument of this Appendix. There is, however, one critical element of the procedure that is essential to understanding the problem: the curious business of the letter "W".

The Decoding Process

The text of the large parchment consists of a passage from the Gospel of John, into which an additional 128 letters have been introduced. These 128 letters, interspersed within the Gospel text as every seventh letter, form the string that must be decoded to obtain the encrypted message. The first task, therefore, is to extract the 128-letter string by counting and recording every seventh letter from the text on the large parchment. For convenience, we arrange the 128-letter string thus obtained in a matrix of sixteen rows, each of eight letters.

We now need to perform four steps, according to certain defined alphabetic manipulations. At each step, every letter will be transformed into another. The string of 128 letters obtained after completion of the four steps must then be re-arranged, using a transposition involving a Knight's Tour on a double chessboard, to arrive at the target decoded message.

The final resulting message consists of recognisable French words but reads as garbled nonsense. Oceans of ink have been spilt, without success, attempting to make sense of this output. I will refer to this message as the POMMES BLEUES string, after the two words with which it concludes. [224]

The Letter W

There is a crucial detail that needs to be added to the above instructions. To perform the decoding correctly, the various manipulations of the letters must be performed with a 25-letter alphabet, without the letter W, as was used for the French language until the nineteenth century.

The decoding solution first presented to the world by Henry Lincoln in 1974, however, used the modern 26-letter alphabet, *including* the W.

This created an obvious problem. According to the story recounted in de Sède's 1967 book, *L'Or de Rennes* and repeated by Lincoln in his BBC documentary, the parchments were said to have been created by Abbé Bigou and deposited in the church of Rennes-le-Château around the year 1780, before being found by Abbé Saunière in the late 1890s.

If this account was historically accurate then any encoded message contained could only have been based on the 25-letter French alphabet in use in Bigou's time. Equally clearly, on the other hand, if the decoding relied on a 26-letter alphabet, as Lincoln put forward in 1974, then de Sède's narrative of the parchment origins must necessarily be false, and the parchments must be a modern concoction.

---

224   By the "POMMES BLEUES" message, I am referring to the final output of the complete process, which results in the 128-letter string: "BERGERE PAS DE TENTATION QUE POUSSIN GARDENT LA CLEF PAX DCLXXXI PAR LA CROIX ET CE CHEVAL DE DIEU J'ACHEVE CE DAEMON DE GARDIEN A MIDI POMMES BLEUES".

In 1991, in *The Holy Place*, Lincoln presented a revised version of the decoding, this time using the correct 25-letter alphabet. In commenting on this, he made the following statement about the effect of the letter W on the solution:

> "When de Sède sent the decipherment, he employed a normal 26 letter alphabet, and it was this method which was demonstrated in the 'Chronicle' film. I must thank a number of television viewers who wrote to me to point out that the letter W was not commonly incorporated in the French alphabet during the eighteenth century when, it is assumed, the cipher was devised. The removal of the letter W from the table produces a slight simplification in the decipherment process in its latter stages."[225]

The clear implication of Lincoln's statement was that the two versions of the decipherment based on the two different alphabets both produced the correct final result, with the only material difference between them being that the 25-letter process involved a slightly simpler final step. In other words, according to Lincoln: it didn't actually matter which version was used, as they both worked.

Thus, the reader was reassured that a *circa*-1780 date of origin of the parchments was not disqualified after all by this apparent problem. Lincoln did not speculate on the reasons why there would have been two different versions of the decoding process in circulation, nor why he had presented the "wrong" version in 1974.

I was intrigued by this comment when I first read *The Holy Place*, but as there was no way at the time to compare it with the solution from 1974, I could not take it any further. I tucked it away, however, and never quite forgot this odd quirk.

The 1974 BBC Documentary

When *The Priest, the Painter and the Devil* was first broadcast, domestic video-recorders were not yet available, so those who watched the original documentary were not able to review it carefully and check the results for themselves. If they had, the situation might have unfolded very differently, but as no one has been able to scrutinise this documentary properly in more than four decades, the problem described here has gone unnoticed in all that time. In the last year or two, however, Lincoln's original three documentaries have been uploaded in their

---

225   Lincoln, *The Holy Place*, op. cit. p. 163.

entirety to YouTube. For the first time since 1974, therefore, it is possible to review in detail the solution as originally presented and reconstruct what happened.

I did not know quite what to expect when I finally had the opportunity to sit down and watch the 1974 film again for the first time in 40 years, but I had no particular reason to doubt Lincoln's 1991 advice that the different alphabets both gave the correct result. I just wanted to check the claim and confirm it.

The decipherment process occupies around ten minutes of the documentary. In voice-over, Lincoln lays out the details of the full sequence of steps required to decode the hidden message, whilst a series of slides are shown that display the results of the manipulations with the letters as they are described.

On first viewing, I watched the decoding unfold on screen, noted the steps described and listened to Lincoln's voiceover. The procedure all seemed to go smoothly enough, and it arrived at the correct final string at the completion of the manipulations. There did not seem to be any obvious issues. On the face of it, the 26-letter version certainly appeared to work, just as Lincoln had later confirmed in 1991.

Before proceeding to a second viewing, this time with the intention of inspecting each step and carefully checking the details, I decided to prepare by working through the full sequence of manipulations for myself on paper. First, I would follow the instructions using the 26-letter alphabet as described in the film – the first time I had been able to check this work in nearly four decades. Then, I would compare the results with the slightly simplified 25-letter version as presented in Lincoln's 1991 book *The Holy Place*, to see for myself at last whether they both arrived at the correct final string as Lincoln had claimed.

After reviewing the output from the first step, which is based on a cryptographic technique known as a "Vigenère substitution", I realised that there was a problem. A big problem.

When I cross-checked the output obtained by using the 26-letter alphabet with that from the 25-letter alphabet version, I noticed that the two versions diverged sharply. Out of the full total of 128 letters, there were 68 that were different!

Having now stepped manually through the manipulations, the reason for this was obvious. The Vigenère substitution utilises a matrix made up of the letters of the alphabet, a 26 x 26 block for the 26-letter alphabet, or a 25 x 25 block for the 25-letter version. Because of the missing W, the output from the procedure using the 25 x 25 matrix was substantially different from that of the 26 x 26 equivalent. Immediately, I could sense that there must be something wrong with Lincoln's 1991 advice that the different alphabets had no impact on the final result. It was clear after just the first step that there was a very significant impact.

After the second and third steps, the errors continued to accumulate and the discrepancy between the two solutions grew wider.

In the fourth step, as Lincoln had noted, there is a slight difference in the procedure depending on which alphabet is being employed. Specifically, the 25-letter version calls for a shift along the alphabet of one letter, whilst in the 26-letter version, the stipulated shift is two letters.

This variation has the practical effect of cancelling out some of the errors that had accumulated in the 26-letter process, so that after the completion of the four steps, there was a total of 43 incorrect letters out of the full string of 128.

If I now performed the final Knight's Tour transposition on the result, those 43 errors in the 26-letter version had the effect of rendering the intended message (the POMMES BLEUES string) completely unintelligible. *Contra* Lincoln, it was obvious that the 26-letter procedure absolutely did not work at all! What was going on? How could this be? I had just watched his 1974 presentation using the 26-letter alphabet and as presented on the screen it had appeared to work flawlessly!

With my curiosity now piqued, I returned to YouTube for the second viewing. This time, armed with my results, I was ready to inspect very carefully the results at each step. I could hardly wait to see how the blatantly flawed outcome I had arrived at could be reconciled with the apparently perfect results presented in the documentary.

To begin, I freeze-framed the video at the moment where Lincoln displays the initial string for the first step of the sequence and compared this with the known correct initial string. These 128 letters had been taken, as per the decoding instructions, from the text written on the large parchment by reading off every seventh letter.

To my surprise, when I compared this 1974 initial string, as shown on screen with the initial string Lincoln published in 1991, there were seven letters that were different.

Admittedly, I already knew about three of these errors, because Lincoln had pointed them out in *The Holy Place*. It would seem that the scribe who wrote out the large parchment had made three transcription errors when he was creating it, presumably by copying some master original version of the coded string. The implication of this scribal error is that anyone setting out to decode the message by reading the letters from the parchment will have three incorrect letters in their initial string. These wrong letters will inevitably propagate through the four steps, resulting in three unavoidable errors in the final output message.

Having called attention to this, and noted it, Lincoln then corrects the errors for the purpose of displaying the fully correct solution in his 1991 account. [226] Regarding these transcriptions or "parchment-writing" errors, as I will refer to them, he writes that they:

> "...may or may not stem from slips by the original code-maker."[227]

This is an intriguing statement. Why would Lincoln doubt that these were anything but slips of the pen? If they did not stem from slips, what were they?

In any case, these three errors were indeed present in the 1974 version as was to be expected, but there were also an additional four errors, which were not. By careful comparison with copies of the parchment I was able to determine that each of these four arose from mistakes in reading, either from incorrectly reading the correct letter in the sequence, or by correctly reading the wrong letter.

This accounts for the seven errors that had crept into the initial string: there were three "parchment-writing" and four "parchment-reading" errors. The critical point here is that once a letter is wrong there is no way to retrieve the correct letter and get back on track as one proceeds through the decoding, without access to the original source code. The implication of the seven incorrect letters in the initial string was that there must unavoidably be seven incorrect letters in the output. Already this raised the obvious question: how then could the process shown in the documentary achieve an error-free result in the final output?

With all this in mind, I hit play and proceeded to review the first step, the Vigenère substitution with eight-letter keyword. In the voiceover, Lincoln gives instructions on how to perform this. The results are then displayed on screen. I paused the playback to check the result and counted 69 incorrect letters out of the total. However, according to my revised calculations, after taking into consideration the parchment-reading and parchment-writing errors there should have been

---

226 He does this by inferring the correct letters in the final string from their context within recognisable French words. He then converts this to the corresponding letter in the initial string by working through the steps in reverse.

227 Lincoln, *The Holy Place*, op. cit. p. 162.

70 errors. There was one less error than there should have been. Why the difference?

Detailed scrutiny revealed the very surprising reason. Two of the "parchment-reading errors" had reverted to the values they would be if they had originally been read correctly! After the Vigenère substitution with the 26-letter alphabet, one of these gave the same value as it would have with the 25-letter alphabet, and the other did not. Hence the overall error count decreased by one.

In addition, for completeness, I found there was another letter that had changed value, this time to the adjacent letter. In this case it would have been an error, and still was an error, so it did not alter the error count.

The key point to consider amidst this blizzard of confusing detail is that two of the "parchment reading errors" had reverted to the correct values that they would have been if they had been read correctly from the manuscript.

But how could this be? In order to be able to supply these details, the person who had provided these decoding instructions must have had the two correct values for these letters in their possession the whole time! This could only mean that these two apparent parchment-reading errors were not really genuine errors at all, but what I will call *fake errors*.

The apparent misreading of the two letters from the parchment was a *feigned move*. The act of replacing them with the correct values was a strong and crucial signal from the puzzle-makers, intended to clearly communicate that they were in control of the process, and the business of the errors was not a question of mistakes, or carelessness, but was *all part of the show*.

It slowly began to dawn on me that something very sophisticated was going on. These errors and their spontaneous corrections were not due to sloppiness or simple mistakes. They were deliberate and even meticulous in their level of attention to detail.

Then I recalled a comment made by Gérard de Sède in *L'Or de Rennes* reporting what he had been told by the cryptographic experts who he claimed had inspected the parchments and successfully decoded them. Amongst their three observations was this one:

> "Errors have been intentionally introduced so as to baffle any attempts at deciphering, the searcher being led along false trails."[228]

---

228 De Sède *L'Or de Rennes* op. cit. p. 103

I was beginning to understand what he meant. I had certainly been baffled by all of this and had to go to almost ridiculous lengths to work out what had happened and to have finally arrived at this conclusion. Now I realised that this was exactly the point. The errors were an inherent part of a very clever, if seemingly pointless, game.

I continued. After step two, consisting of a one-letter shift down the alphabet, I now counted 76 errors. An additional seven of the correct letters from the previous step had transformed into incorrect letters after the shift, because of the presence of the letter W. Things seemed to be going from bad to worse, but I was about to get an even bigger surprise.

Until this point, the results displayed on screen at least (nearly) matched the results of my calculations. When I performed the third step, however, (a second Vigenère substitution, with a 128-letter keyword), and compared it to the result displayed on screen, it made no sense.[229] Many letters were different from what I had expected, with no apparent rhyme or reason that I could discern at all, at least on first inspection.

I proceeded to the final step, described by Lincoln as a two-letter shift along the 26-letter alphabet. By my prior calculations, there should now be 43 incorrect letters out of the total of 128. The number had reduced because the two-letter shift (rather than a one-letter shift as is defined for the 25-letter alphabet version) had the effect of cancelling out many of the accumulated errors caused by the extra letter (W) in the Vigenère table.

And now, with my thanks to the reader for persisting this far, at last I arrive at the point of all this.

The result displayed on screen after the final step did not show 43 errors. On the contrary, as I had seen on first viewing, it displayed the *perfectly correct final 128 letter string*. None of the 43 errors appeared. Not one. Every single wrong letter had been corrected: not only the accumulated errors due to the effect of the W, but even the three "parchment-writing", the remaining two "parchment-reading" errors and the stray error. Every single one had been rectified!

How was this even possible? How had the documentary makers managed to conclude with a completely error-free final result, when the intermediate steps were riddled with wrong letters? After a few moments of head-scratching, followed by further careful comparison

229   The keyword consists of the 119 letters on the (genuine) Marie de Nègres gravestone, plus 9 letters (PS PRAECUM) from the fake one. This 128-letter keyword is also an anagram of the final output, the POMMES BLEUES string.

of the 1974 and 1991 versions, I suddenly realised the answer was simple. The final slide – which should have displayed the results of following faithfully the instructions in Lincoln's voiceover (and thus should have shown 43 errors) – had simply been swapped out for the correct final string.

It is crucial to note here that this correct target string can *only* be obtained using the 25-letter alphabet. It *cannot* be derived using the 26-letter alphabet, as I had shown by working through the process.

Therefore, the presence of the perfectly correct target string at the completion of the process proves that whoever had supplied this flawed 26-letter solution to Lincoln and the BBC had the correct 25-letter solution and the original master from which the parchments had been created the entire time. There was therefore no possibility that this 26-letter solution had been provided in good faith and there had simply been a miscommunication.

What about the previous string, the outcome after step three? I quickly worked out that it had been obtained by reverse engineering from the last slide, the correct target result. They had simply shifted all the letters in that replaced final string back along the alphabet by two letters.

This was a very clever move, I had to admit. It meant that if anyone had got as far as checking the results of the final two steps on their own, they would not have noticed any issue. The ending of the process looks fine. The problem is that these final strings do not follow on from the first two steps at all. The solution that was presented on screen is a hybrid. The results of the first two steps have been produced using the 26-letter alphabet, and are riddled with incorrect values and errors, but the results for the final two steps have been spliced-in and obtained using the 25-letter alphabet process.

To underscore this point: the 26-letter version simply does not work. If the instructions in the voice-over are followed correctly, and the correct results of this displayed on-screen, it produces a final string that has so many wrong letters (43 out of 128) that the target "POMMES BLEUES" message is incomprehensible as it is overwhelmed with errors.

But it has not been followed correctly and the results of following the instructions faithfully as given in the voice-over are not shown on screen. Instead, the results that should be displayed for the final two steps have been substituted with alternatives. It's a crude magic trick, an exercise in sleight-of-hand. The purpose is to create the illusion that the 26-letter procedure works correctly, when it does no such thing. When Lincoln claimed in 1991 that the 26-letter version worked, he was bluffing. It does not, and he must have known that it does not. It is impossible for these results to be obtained by accident, or by careless error or by any other innocent mistake.

I do not doubt that it has been as challenging to follow the argument of this appendix as it has been to write it, but this is precisely the point. These errors were introduced to baffle all attempts at following the decipherment, and they succeeded. It has taken more than 40 years for this trick to be uncovered.

Comparing the Parchment Decodings

Figure 139 has been created to assist the reader in following for herself these complicated manoeuvres and illustrate what happened as simply as possible. The first column, on the left, shows the steps of the correct decoding, free of any errors, as it would have been originally devised by the puzzle maker, (but before it was transcribed onto the parchment). As discussed, it employs a 25-letter version of the alphabet, lacking the letter "W".

The second column lays out the correct decoding as it was given in *The Holy Place* in 1991. It shows the initial string as it was written on the parchment. It also employs the 25-letter alphabet and reveals the three "parchment-writing" errors. The third column shows the decoding as it would proceed if the instructions given in the voice-over narration from the 1974 documentary are followed faithfully. The fourth and final column shows the decoding as it was displayed on screen in that documentary.

The four decodings proceed from top to bottom through the four steps, in each of which every letter of the 128-letter matrix is transformed into a different letter. The final block of 128 letters is the string that produces the "POMMES BLEUES" message (after one further step of rearrangement not shown at this stage, specified by a Knight's Tour on a double chessboard). Each of the versions (except the first) incorporates various "errors", as I've discussed. Far from being mere mistakes due to carelessness, these are in fact the vital key to understanding the entire point and purpose of the parchment codes.

I have colour-coded the squares to show the different types of errors that are present and to make it easy to track at a glance what is going on.

Squares in white indicate correct letters after each decoding step along the way.

Squares coloured gold in the final step mark letters which are correct at the conclusion of the process.

Squares coloured red, (or within the red border), blue or green represent incorrect letters as a result of the different categories of "errors", as we will see. Red denotes the errors made by the scribe in creating the parchment (the "parchment-writing" errors). Green represents the errors made in reading the letters from the parchment (the "parchment-reading" errors). Blue identifies letters that are incorrect as a result of the effect of the extra letter W in the Vigenère substitution and alphabet shift processes. The three squares surrounded by black borders in the second step of the fourth column are the three letters whose values were changed.

The string surrounded by the red border is the first of the two that have been swapped out for the versions derived from the correct 25-letter alphabet process.

Figure 140 shows the slide from the documentary with the substituted values. The output letters are in rows of 16. The first row, and the others in the smaller font, are the keyword. The second row, and the others in larger font, are the output that we are interested in. If we take the first of these larger font rows as an example, it represents to the viewer that the first ten letters of the output string after step three are:

V L J Q N Y L L Y Q.

However, if the instructions had been followed correctly these should read:

U L J Q O X L L X R.

The five false letters out of the ten shown have been altered so that after the next and final step, the two-letter shift down the alphabet, every letter will be correct to create the fully error-free 128-letter final string.

Conclusion

Whoever supplied this 26-letter "solution", which did not work, to Lincoln in 1974 knew exactly what they were doing. They could not have provided him with it in some kind of mix-up, or muddle, or confusion. Neither could it be a question of misunderstanding on Lincoln's side, as if perhaps the required alphabet had not been stipulated and he had wrongly assumed it was the 26-letter version, because the two procedures were different in the final step.

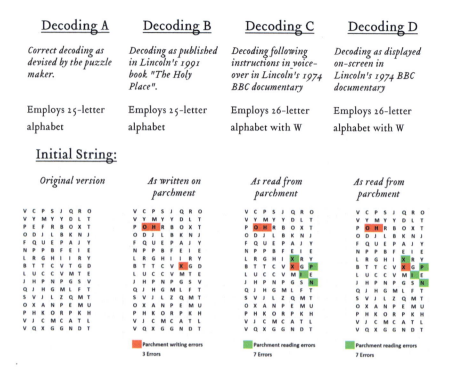

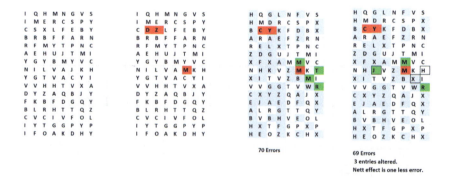

Figure 139: Anatomy of the false 1974 parchment decoding.
Side-by-side comparisons of the different versions of the parchment decodings, with "errors" colour-coded for reference. See text for full details. Red squares denote "parchment writing" errors. Green squares are "parchment reading" errors. Light blue squares show errors due to the extra letter W. Gold squares indicate correct letters in the final string.

THE PARCHMENT TEXT DECIPHERMENTS

457

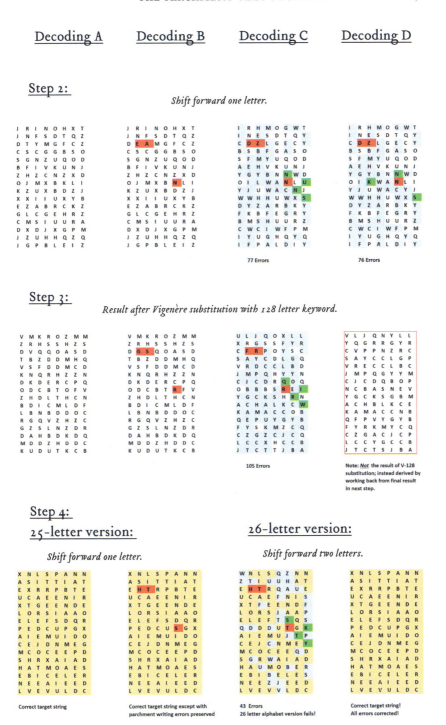

They (that is, the puzzle-makers) had instead knowingly created and supplied to Lincoln and the BBC a fake "hybrid" solution. It directs the use of the 26-letter alphabet, but in place of the error-strewn strings that would result from following the instructions faithfully all the way through with the wrong alphabet it simply displays the correct 128 letter strings for the outcome of the final steps, which can only be obtained by performing the entire sequence with a 25-letter alphabet. They simply swapped out the final two slides to create the false impression that the 26-letter method worked.

This inevitably raises the difficult question of Lincoln's involvement in this business. The slides displaying the results after each step were presumably created from materials supplied to Lincoln by de Sède. It is difficult to see how the documentary could have been assembled without the production team becoming aware of the fake results spliced into the on-screen depiction of the decoding process – someone must surely have checked them for accuracy. It is disconcerting to consider that a flagship BBC documentary series could have knowingly broadcast fake material, but the alternative, that they simply didn't check, is surely inconceivable.

In a book that he wrote years later in 1997, *Key to the Sacred Pattern*, Lincoln offers detailed recollection of all kinds of incidents and events surrounding his early involvement with the mystery, but there is no mention at all of the decoding included in the documentary, how he obtained it, or his subsequent reaction to the issue with the letter W.

He does, however, relate that after he first obtained the details of the solution from de Sède as early as 1971, he arranged a meeting with someone he describes as a "retired member of the British Intelligence code-breaking fraternity". He then writes:

> "I show him the coded parchments, the encoding system and the final message. He is intrigued and questions me closely on the provenance of the documentation and the possible dating of the cipher."

He leaves the materials with the cipher expert, and within a week, receives his reaction, which he reports as follows:

> "'It's one of the most complex ciphers I've ever encountered,' he tells me. 'The system is perfectly valid but laborious.'"[230]

---

[230] Henry Lincoln, *Key to the Sacred Pattern: The Untold Story of Rennes-le-Château*, (Gloucestershire, The Windrush Press, 1997), p. 44.

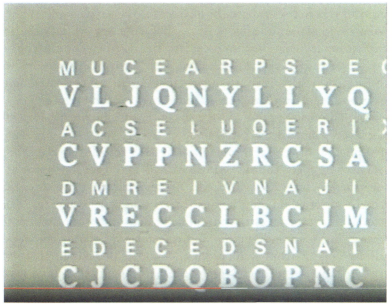

Figure 140: Proofs of a conspiracy: the doctored slide from Henry Lincoln's BBC Chronicle documentary The Priest, the Painter & the Devil (1974). Online at: https://www.youtube.com/watch?v=Uru8UiCxLnc. Timestamp 27:09/50:29

Something must obviously be wrong with this story, however, because if the cipher expert had been presented with the same version that was broadcast in 1974 and had carried out as thorough a job in checking the decoding as I had, he could not possibly have come to the conclusion that everything was "perfectly valid". He would have come to the conclusion that the method using the 26-letter alphabet did not work at all and would have had no means of correctly decoding it without the crucial detail that it required a 25-letter alphabet. So, the story does not make sense.

There seems no alternative except to suggest the somewhat uncomfortable conclusion that Lincoln and others at the BBC must have been knowingly involved in the prank, at least to some degree. Does this mean that the entire enterprise was nothing but a hoax? I don't think so at all. To ask whether the parchments are either "genuine", or a "fake" is, I suggest, to misframe the question. They are not necessarily what they purport to be, but this does not make them any less real, nor does it mean that they are merely a worthless deception.

We need to go beyond these simple dichotomies because there is something very subtle going on here. We are certainly dealing with a situation that is much more sophisticated than a simple hoax. A more insightful way to think about the business of the parchments might be to view them as akin to a work of performance art, a kind of serious joke that makes sense within a certain strand of twentieth-century French literature, even if it does not necessarily translate very well into English. If this is so, then the story of the parchments' apparent discovery by Saunière was intended to mask their true origin and purpose. The entire notion of the message encoded in the text is a false outer layer, a disguise, a distraction.

The purpose of the parchments was not as a vehicle for concealing a secret text message. The "POMMES BLEUES" output string was not a prize to be won by cracking the code. No one ever had to decode the text to obtain it, nor could they, because this was not even possible. Rather, it was an integral part of the ensemble of materials out of which this narrative has been put together and was never a separate component at all. In fact, far from being hidden, it was published in the *Dossiers Secrets* in 1964, several years *before* the parchments themselves were first shown publicly in 1967.

So, if the parchment text was not a code to be deciphered, then what was it? I suggest that it can be better understood as an exercise in *constrained writing*. The notion of generating text by adhering to a set of formally designed rules is a particularly French literary tradition of the twentieth century.[231]

Georges Perec is an example of an author who worked under such deliberately self-imposed constraints. He wrote an entire novel, for example, which does not include the letter "e" (*La Disparition* (1969), translated into English as *A Void*). In another, (*Life: A User's Manual* (1978)) he arranged the plot so that it played out according to the moves of a Knight's Tour on a 10 x 10 chessboard, a strategy that offers a conceptual link directly to the parchment decoding instructions.[232]

Under this reading, the parchments are the output of a sequence of elaborate manipulations of text that have been designed to fascinate, distract and deflect attention. The goal was to create conditions that would encourage the widespread dissemination of the parchments, to encourage scrutiny of their enigmatic and mysterious surfaces, and to put off for as long as possible the correct extraction of the genuine content. The "solution" itself, the "POMMES BLEUES" string, was

---

231   See for example the activities of the Oulipo group. (Hat-tip to Steve Mizrach who was the first to notice this.
232   This is not to suggest that Perec had any involvement in the creation of the parchment, but only that these kinds of ideas were prevalent in French literary circles of the time.

precisely *not* the point. It was merely the output of the sequence of manipulations. It was the manipulations themselves that were the purpose of the exercise.

A Serious Comedy of Errors

The puzzle-makers have played an elaborate game with the presence of errors in the process, intentionally designed and inserted into the text with great care. Amazingly, de Sède laid this out plainly enough all the way back in 1967 in *L'Or de Rennes*. Here is the full list of the conclusions from the cryptographic experts who de Sède claimed to have examined the parchments.

> "At the conclusion of a highly technical study, they arrived at the following deductions:
>
> 1. The texts have been well encoded by means of a double key substitution, and then by a transposition effected by means of a chequer board.
> 2. In addition to the coding proper, the author has added rebuses.
> 3. Errors have been intentionally introduced so as to baffle any attempts at deciphering, the searcher being led along false trails."[233]

The first is correct, but it is impossible for any cryptographer to have determined the details of the encoding. The experts must have been provided with both the original texts and the instructions for decoding.

The second is direct confirmation of the solution presented in this book: there are indeed "rebuses" in the parchments, in addition to the encoded texts themselves. A rebus is a picture puzzle. De Sède was hinting at the concealed Seal of Solomon design.

The third deduction confirms the existence of intentional errors and provides an explanation behind the strategy. They have been introduced to "baffle attempts at deciphering", and to lead researchers down false trails. Like banana-skins artfully strewn along the pavement to slow down pursuers, they are designed to impede progress, to lead astray, to confuse.

From this we can infer that the entire process of the encoding and decoding of the text message was designed and intended as a *distraction*, with as many layers, twists and trapdoors as possible. The puzzle-maker is in no hurry for the Seal of Solomon to be identified; indeed,

[233] De Sède *L'Or de Rennes* op. cit. p. 103

the longer it takes, the better. The genuine content concealed in the parchments is not encrypted in the text; it is the hidden geometry.

At every stage, errors were deliberately and systematically introduced, to the extent that we might even consider it as a signature textual manoeuvre. Recall the deposit slip for *Le Serpent Rouge* at the Bibliothèque nationale, in which each of the three alleged authors had precisely one incorrect detail in their names and addresses. We can now see that the presence of these carefully calibrated errors could even be considered as the characteristic signature of the Affair, almost like a watermark certifying authenticity, the identifying mark present at every stage of the performance of a meticulously choreographed magic trick.

This element carries over into the communication of the solution to Lincoln, which was an integral part of the execution of the literary performance art of the parchments. According to the self-imposed rules of this formal game, even the act of passing the solution to Lincoln and his team must therefore also involve at least one deliberate, traceable error introduced into the idealized "correct" response.

In the context of this high-concept madness, to satisfy this requirement they came up with the idea of the 26-letter alphabet. Even though the correct decoding proceeds in accordance with the 25-letter alphabet, they set out to convince the audience that it was based around the 26-letter alphabet. They were creating yet another layer of befuddlement, deliberate distraction and precisely choreographed confusion.

Of all the actions surrounding the appearance of the parchments, the spliced fake solution in the Lincoln documentary broadcast by the BBC in 1974 is perhaps the most audacious, and sheds significant light on the motivations behind it all. Consider that the Plantard team had everything to lose by daring to propose a blatantly false solution to the documentary team, and on the face of it, little to gain. They risked exposing the entire business as an obvious modern concoction.

This tells us that there was a deeper strategy at play. It was not about the final output, but the elaborate process itself. The goal was not a message to be decoded or a problem to be solved. It was the elaborate performance of a carefully staged literary ritual in the public space. Now we can begin to glimpse the broad outlines of the strange game that the Team were playing. The errors they introduced are the trail by which we can recreate the path of the puzzle-maker.

This strategy has nothing to do with any quest for buried treasure. The account of Saunière and his apparent discovery of riches was appropriated and exploited for the purpose by the Team. It was a cover-story, a hook they used by which to draw in a curious audience. The underlying purpose was much more subtle and interesting. The intent was to perpetuate the mystification for as long as possible. They wanted to create layers of sticky fascination to engage people in thinking about it, examining it, discussing and arguing about it.

The parchments were planned and created within the context of this complex multi-faceted game as a vehicle for the dissemination of a subliminal image, the emblem of the Martinists. The aim was to ensure that these materials were distributed as widely as possible. To use the modern term, they wanted the parchments to go viral so that the hidden symbol would be viewed subliminally by as many people as possible. By any measure, their plan must have succeeded beyond their wildest expectations.

This is the reason for the baffling complexity of the text encoding, and the deliberate mystification of the introduced errors. While generations of seekers and searchers and dreamers have obsessed about the "POMMES BLEUES" message and turned it this way and that to try to find a glint of a clue which will lead them to golden treasure, they have been unwittingly fulfilling the goals of the puzzle-maker: to spread the images far and wide.

Appendix II
LIST OF FIGURES

Figure 1: Richer's dream. 3
Figure 2: The 45° alignment. 23
Figure 3: The imposing tower of the Château d'Arques. 24
Figure 4: Ground plan of the Château d'Arques compound.
Image: Lucien Bayrou and Marienne Roques, *Le Château d'Arques*, (Carcassonne, Centre d'Archéologie Médiévale du Languedoc, 1988), p.18. 25
Figure 5: The La Pique Meridian. 28
Figure 6: The Pech Cardou meridian. 31
Figure 7: The two meridians of Pech Cardou and La Pique, with the 45° alignment. 33
Figure 8: The two meridians of Pech Cardou and La Pique, in perspective view, looking towards the south. 34
Figure 9: The two meridians of La Pique and Pech Cardou viewed in perspective looking north. 35
Figure 10: The Sunrise Line and the Bugarach Baseline (purple) with the two meridians and the 45° alignment. 38
Figure 11: Front cover of Le Serpent Rouge. 45
Figure 12: *Le Serpent Rouge*, page 2. 46
Figure 13: *Le Serpent Rouge*, page 3. 47
Figure 14: *Le Serpent Rouge*, page 4. 48
Figure 15: *Le Serpent Rouge*, page 5. 49
Figure 16: *Le Serpent Rouge* Deposit slip at Bibliothèque nationale, Paris.
Image: *Franck Marie, P.F...? le Serpent Rouge Preuves* (Malakoff, Editions S.R.E.S. Vérités Anciennes, 1978) p.19. 57

Figure 17: Comparison of Sagittarius stanza of *Le Serpent Rouge* with the opening of Jung's Part 5. Sal in *Mysterium Coniunctionis*. 65
Figure 18: The Fa Triangle. 71
Figure 19: The La Pique Meridian: the view looking due south from the summit of Le Sarrat Rouge. 75
Figure 20: The side profile of Le Sarrat Rouge. 76
Figure 21: Summit of La Pique. 76
Figure 22: View looking south towards Les Crêtes d'al Pouil. 77
Figure 23: Looking south towards Col du Vent. 77
Figure 24: The Pech Cardou meridian. 79
Figure 25: Panorama from Bézu, looking north-east. 82
Figure 26: Bézu meridian. 84
Figure 27: Meridians of Lavaldieu and Rennes-les-Bains. 85
Figure 28: The Sighting Ridge: Looking south over the sighting locations for the five meridians shown in Google Earth. 87
Figure 29: All five meridians on Google Earth, viewed from directly overhead. 88
Figure 30: All five meridians on Google Earth, perspective looking north. 89
Figure 31: View from Pic de Saint-Barthélemy, along the Sunrise Line, at dawn on August 24. 92
Figure 32: The Sunrise Line. Extends from Pic de Saint-Barthélemy in the west to Château de Villerouge-Termenès in the east. 93
Figure 33: Front cover of first edition of Jean Richer's *Géographie Sacrée du Monde Grec*.(Paris, Hachette, 1967). 99
Figure 34: The main axes of Greece through Delphi, as depicted in Figure 10 of Jean Richer's *Aspects ésotériques de l'œuvre littéraire* (Paris, Dervy-Livres, 1980), p. 145. 103
Figure 35: The Grand Alignments, depicted in Carte 1 of Jean Richer's *Géographie Sacrée du Monde Grec* (Paris, Hachette, 1967), p. 26. 104
Figure 36: The Zodiac of Delphi according to Prof. Jean Richer. 106
Figure 37: The cover of Jean Richer's *Delphes, Délos et Cumes* (Paris, Julliard, 1970). 113
Figure 38: Photograph of Gérard de Nerval by Félix Nadar. Image: Public domain. From Gallica Digital Library and under the digital ID btv1b10547584p/f1 121
Figure 39: The ancient Roman *cardo* and *decumanus* axes of Rennes-les-Bains as depicted in the map on page 59 of Jacques Rivière and Claude Boumendil's *Histoire de Rennes-les-Bains*, (Cazilhac, Bélisane, 2006). 135

LIST OF FIGURES

Figure 40: The ancient Roman *cardo* and *decumanus* axes of Rennes-les-Bains shown on Google Earth, perspective view from Bézu. 137

Figure 41: The ancient Roman *cardo* and *decumanus* axes of Rennes-les-Bains with the 45° alignment on Google Earth. 138

Figure 42: The axes of Rennes-les-Bains, the 45° and the alignment from Rennes-les-Bains, over Rok Nègre, to Château Blanchefort. 139

Figure 43: Completing the alignments at right-angles to the Blanchefort and 45° lines. 141

Figure 44: The title page of *La Vraie Langue Celtique* by Abbé Henri Boudet, priest of Rennes-les-Bains. 145

Figure 45: Boudet's cromlech as two circles, 16 and 18 kilometres in circumference, centred on Rennes-les-Bains church. 151

Figure 46: The route described in Chapter VII of Boudet's *La Vraie Langue Celtique*, shown on his map and Google Earth. 153

Figure 47: The path of the walk in *Le Serpent Rouge*. 155

Figure 48: The *Le Serpent Rouge* path on the zodiac of Rennes-les-Bains. Overhead view on Google Earth. 159

Figure 49: The *Le Serpent Rouge* path on the zodiac of Rennes-les-Bains .Perspective view looking north. 160

Figure 50: View south along the Pisces alignment of the *Le Serpent Rouge* zodiac in the landscape. 161

Figure 51: The Virgo sector, with the four hoof prints of the horse on the stone, depicted on the IGN map. 163

Figure 52: The St Sulpice meridian and the Scorpio segment. 163

Figure 53: The serpentine path of *Le Serpent Rouge* visible on the south flank of Pech Cardou. 171

Figure 54: The path of Le Serpent Rouge is shown in red. 172

Figure 55: The serpentine path of *Le Serpent Rouge* just visible on the south flank of Pech Cardou. 173

Figure 56: Photograph of *"le serpent rouge"* from a vantage point on the west side of the valley of the River Sals. 174

Figure 57: A plate from Jean Richer's *Delphes, Délos et Cumes*, (Paris, Julliard, 1970), p.64, showing a stone carving at Delos of a serpent coiled around an omphalos between two palm trees. 175

Figure 58: 'Leucate' on the coast in both Greece and France. 181
Above: Leucate shown due west of Delphi as depicted on a map of Greece from Jean Richer's *Aspects ésotériques de l'œuvre littéraire* (Paris, Dervy-Livres, 1980). 181
Below: Google Earth map of south-west France showing Cap Leucate due east of Rennes-les-Bains, with detail from IGN map. 181

Figure 59: Place-names and alignments depicted in Carte 13 of Jean Richer's *Géographie Sacrée dans le Monde Romain* (Paris, Guy Trédaniel, 1985), p.192.   183

Figure 60: Locations on the meridian of Gaul, as listed in Jean Richer's *Géographie Sacrée dans le Monde Romain* (Paris, Guy Trédaniel, 1985), p. 336.   184

Figure 61: Selection of place-names and alignments depicted in Carte 20 of Richer's *Géographie Sacrée dans le Monde Romain* (Paris, Guy Trédaniel, 1985), p.345.   185

Figure 62: Map of Rennes-les-Bains in Pierre Plantard's Introduction in the 1978 reprint of Henri Boudet's *La Vraie Langue Celtique*.
Image: Pierre Plantard, *La Vraie Langue Celtique et Le Cromleck de Rennes-les-Bains* (1886), (Paris, Belfond, 1978), p. 39.   191

Figure 63: *Le Zodiaque de Saint Emilion*
Image: Gérard de Sède, *Saint-Emilion insolite* (Villeneuve de la Raho, Pégase, 1980), p.42.   202

Figure 64: "*Une seule page pour un livre fantôme*", or "A single page from a "phantom book".
Image: Jean-Luc Chaumeil, *Le Testament du Prieuré de Sion: Le Crépuscule d'une Ténébreuse Affaire* (Villeneuve de la Raho, Pégase, 2006), p. 112.   203

Figure 65: Boudet's map in the 1978 Belfond edition of *La Vraie Langue Celtique*, with Pierre Plantard's introduction.
Image: Henri Boudet, *La Vraie Langue Celtique et Le Cromleck de Rennes-les-Bains* (1886), as reprinted by Belfond Press, Paris, 1978.   211

Figure 66: Boudet's map with inch grid, generated using the title size and position as reference, extended over the full map.   213

Figure 67: The secret key to unlocking the concealed inch grid of the map is deposited in the position and measure of the title.   215

Figure 68: The cover and title page of *La Vraie Langue Celtique* with superimposed grid of inches, further subdivided into squares of 1/8 inches.   217

Figure 69: The cover/title page of *La Vraie Langue Celtique* with grid of inches, subdivided further into squares of 1/4 inches.   219

Figure 70: Superimposing Boudet's map with its concealed grid on Google Earth reveals the three inner meridians align with grid.   221

Figure 71: Boudet's cromlech of 16 kms in circumference is represented by a circle of eight inches diameter on a 1:25,000 scale map.   227

Figure 72: Dimensions of Boudet's cromlech on the 1:25,000 scale map and in the landscape, and Earth dimensions.   231

LIST OF FIGURES 469

Figure 73: De Chérisey's version of the hidden grid in Boudet's map. Image: Philippe de Chérisey, *Un Veau à Cinq Pattes* (Paris, France Secret, 2008), pp. 135, 337. 237
Figure 74: This Figure shows the results of following de Chérisey's instructions to rule a grid on the Boudet map. 239
Figure 75: Reproductions of the Two Parchments 245
Figure 76: Parchment Geometry Step 1: 249
Figure 77: Step 2: 249
Figure 78: Step 3: 251
Figure 79: Step 4: False Step 5: 251
Figure 80: True Step 5: Step 6: Step 7: 253
Figure 81: Step 8: 253
Figure 82: Step 9: Complete the hexagram. 254
Figure 83: Geometry of the hexagram: A circle drawn within a 4x4 square generates two hexagrams, shown in red. 255
Figure 84: Superimposing the hexagram geometry of Figure 83 on the geometry of the small parchment. 257
Figure 85: Superimpose the hexagram geometry on the large parchment. 259
Figure 86: The geometry of Figure 83 can be extended naturally to create two further 'outer' hexagrams. 260
Figure 87: Superimposing the extended geometry accounts for the position of the lower target marker, the first line of the lower text, and the SION device. 261
Figure 88: Alleged gravestone of Dame Marie de Nègres (d.1781). Image: Pierre Jarnac, *Mélanges Sulfureux : Les mystères de Rennes-le-Château*, (Couiza, C.E.R.T. Centre d'Etudes et de Recherches Templières, 1995) 263
Figure 89: Superimposing the parchment geometry on the Marie de Nègres gravestone reveals the hidden design. 265
Figure 90: The secret scale on the Marie de Nègres gravestone. 266
Figure 91: The two parchments, correctly sized and positioned, back-to-back, with the hidden geometry shown. 267
Figure 92: The two parchments, at correct relative sizes and positions, back-to-back, without geometry. 269
Figure 93: Uncial and semi-uncial script 271
Figure 94: More semi-uncials: The letters on the covers of both books measure one half inch in height. Above: Jean Richer's *Géographie Sacrée du Monde Grec.*(Paris, Hachette, 1967). Below: *La Vraie Langue Celtique et Le Cromleck de Rennes-les-Bains*, (Paris, Belfond, 1978) 273

Figure 95: Boudet's map with concealed inch grid and his cromlech depicted as two circles of radii 4 and 4 ½ inches, centred on Rennes-les-Bains Church. 275

Figure 96: Superimposition of the small parchment onto Boudet's map, with grid, on Google Earth. 277

Figure 97: Trails of confirmation.
Above: Extract from the *Dossiers Secrets*.
Image: Pierre Jarnac, Mélanges Sulfureux : Les mystères de Rennes-le-Château, (Couiza, C.E.R.T. Centre d'Etudes et de Recherches Templières, 1995).
Below: Marginalia from the manuscript of de Chérisey's essay *Pierre et Papier*.
Image: As reproduced in Jean-Luc Chaumeil, *Le Testament du Prieuré de Sion : Le Crépuscule d'une Ténébreuse Affaire* (Villeneuve de la Raho, Pégase, 2006). 278

Figure 98: Page from the Dossiers Secrets.
Upper right: the fake and genuine Marie de Nègres gravestones.
Bottom left: coat-of-arms with Seal of Solomon made from triangles of two colours, echoing the hidden design of the parchments.
Image: Pierre Jarnac, *Mélanges Sulfureux : Les mystères de Rennes-le-Château*, (Couiza, C.E.R.T. Centre d'Etudes et de Recherches Templières, 1995) 279

Figure 99: Page from Gérard de Sède's *La Rose-Croix*, showing the Rose des Vents on the Croix des Saisons.
Image: Gérard de Sède, *La Rose-Croix*, (Paris, Editions J'ai lu, 1978), p.37. 281

Figure 100: "Here is the proof that I knew the secret of the seal of SALOMON." 283

Figure 101: The Arques Square. 289

Figure 102: The Arques Square, looking north, perspective view. 291

Figure 103: Google Earth images showing the co-ordinates of the two endpoints of the 45° alignment. 292

Figure 104: Dimensions of the Arques Square . 293

Figure 105: Zooming in on Campagne-sur-Aude. 295

Figure 106: The geometry of the Arques Square, showing Campagne-sur-Aude on the midline. 296

Figure 107: The Arques Square, looking east. 297

Figure 108: Geometry template of the Arques Square. 297

Figure 109: The alignment of six churches or villages on the 330° bearing from the centre of the Arques Square. 301

LIST OF FIGURES 471

Figure 110: The alignment from the centre of the Arques Square to Villeneuve-lès-Montréal at 330° runs at a tangent to the boundary of the circular village of Cailhavel. 303
Figure 111: Sightlines through Campagne-sur-Aude. 305
Figure 112: Woodcut engraving from Leonard Digges *A geometrical practise, named Pantometria* (London, Abel Ieffes, 1591) 307
Figure 113: Geometry of the Arques Square on 1:25,000 map. 309
Figure 114: The Measures of the Arques Square: Combining the cromlech, the 8 x 8 inch grid, the four grid meridians, the Arques Square and the Boudet map. 311
Figure 115: The field of the meridians, in red and the Sunrise Line and Bugarach Baseline, in purple. 313
Figure 116: Girbes de Bacou is the intersection of the Bugarach Baseline and the Lavaldieu meridian on longitude 2°18′00″ E. 315
Figure 117: Point Q is the intersection of the Bugarach Baseline with the Sunrise Line. 317
Figure 118: R is the intersection of the Sunrise Line and the Lavaldieu meridian. 319
Figure 119: Clay tablet from Susa, Iran, capital of the Elam and Achaemenid Empire.
Image: *Mémoires de la Mission Archéologique en Iran*. Tome XXXIV Mission de Susiane, E.M.Bruins et M.Rutten, 1960 321
Figure 120: The geometry of the Susa tablet. 322
Figure 121: Superimposing the Susa tablet geometry on the landscape geometry reveals they are identical, at a scale of 1:10,000 inches. The tablet is a map of the landscape geometry. 323
Figure 122: The distance between Girbes de Bacou and Bugarach is divided into four lengths, each of 150,000 inches. 325
Figure 123: Grid geometry of the Bugarach Baseline. 327
Figure 124: The bearing of 130° from Campagne-sur-Aude. 329
Figure 125: To acceptable accuracy, Caudiès-de-Fenouillèdes is due east of Pic de Saint-Barthélemy, and due south of Arques Church. The three locations form a large [7:24:25] triangle. 329
Figure 126: Crucuno, Brittany. 47°37′30″N; 3°07′18″ E. This megalithic stone rectangle is laid out as a [3-4-5] triangle in units of 10 megalithic yards and oriented to the compass angles. 331
Figure 127: Hypothetical geodetic squares, A, B, C and D, laid out on the surface of the Earth. 333
Figure 128: The zodiac structure of Nerval's *Sylvie* (1853), according to Jean Richer, as it appears in a 1966 paper *Saggi e ricerche di*

letteratura francese, (Pise, t. VII, 1966, pp. 201-37), and then later in 1971 in *Nerval au royaume des archétypes Octavie Sylvie Aurélia* (Archives des lettres modernes, no. 130, 1971, (11) V, archives nervaliennes no 10), then finally in 1987 in his *Gérard de Nerval: Experience vécue et création ésotérique Gérard de Nerval : Expérience vécue et création ésotérique*, (Paris, Guy Trédaniel, 1987), p.253. 345

Figure 129: *Mutus Liber*, Plate XV.
Image: Jean-Jacques Manget *Bibliotheca chemica curiosa, seu rerum ad alchemiam pertinentium thesaurus instructissimus*. (Genevae, Chouët, 1702). Public domain. 385

Figure 130: *Mutus Liber*, Plate XV hidden geometry in detail. Half inch grid added for reference. Inner Square is 3 inches x 3 inches. 386

Figure 131: *Mutus Liber*, Plate XV, later edition. 387

Figure 132: Image from Eliphas Levi. Public domain. 393

Figure 133: The sign of the Martinist Order.
Image: © User: Ohjay / Wikimedia Commons / CC-BY-SA-3.0 395

Figure 134: The seal on the first issue of *Le Voile d'Isis*, the Martinist newsletter, dated 12 November 1890, four years after the publication date of *La Vraie Langue Celtique*. 397

Figure 135: Henri Delaage, (1825-1882). Atelier Nadar.
Image: gallica.bnf.fr / Bibliothèque nationale de France. Digital ID btv1b530508392 399

Figure 136: Causabon's walk in chapter 115 of *Foucault's Pendulum* by Umberto Eco. 411

Figure 137: The destination of Causabon's walk in chapter 115 of *Foucault's Pendulum* by Umberto Eco. 413

Figure 138: Richer's dream revisited.
Above: The complete Apollo alignment.
Below left: The alignment passing through Athens.
Below right: The alignment crossing Delos. 438

Figure 138: (continued)
Above: The north-west end of the alignment at Delphi.
Below. The south-east end of the alignment on Rhodes 439

Figure 139: Parchment decodings with colour-coded "errors". 456

Figure 139: (continued) Parchment decodings with colour-coded "errors". 457

Figure 140: Proofs of a conspiracy: the doctored slide from Henry Lincoln's 1974 BBC documentary *The Priest, the Painter & the Devil* Online at: https://www.youtube.com/watch?v=Uru8UiCxLnc. Image taken at time stamp 27:09/50:29 459

LIST OF FIGURES

Cover Images:

Carte d'État-major de la France, Feuille Quillan S.O. 1/40 000.
Old military map of France, Quillan S.W. sheet, 1/40,000.
Dépôt de la Guerre (Q1270687) IGN, 1866

Excerpts from IGN map 2347OT Quillan from Institute Géographique Nationale credit and copyright ©IGN 2022. Fair use in accordance with guidelines.

All Google Earth maps credit to Google and the data providers as named on individual images, according to terms of use.

All photographs taken by the author.

For all images not in the public domain, every effort has been made to obtain permission to reproduce from the copyright holder. For any omissions, please contact the publisher who will be pleased to make appropriate arrangements in any future editions.

To view high resolution Google Earth Studio videos of many of the landscape images shown in Figures in this book, please visit the author's website at simonmmiles.com or his YouTube channel at https://www.youtube.com/@SimonMilesresearch/videos.

Appendix III

THE TEXT OF LE SERPENT ROUGE

Original French text in italics, followed by English translation.

1. *Verseau* /Aquarius

Comme ils sont étranges les manuscrits de cet Ami, grand voyageur de l'inconnu, ils me sont parvenus séparément, pourtant ils forment un tout pour lui qui sait que les couleurs de l'arc-en-ciel donnent l'unité blanche, ou pour l'Artiste qui sous son pinceau, fait des six teintes de sa palette magique, jaillir le noir.

How strange are the manuscripts of this Friend, great traveller of the unknown, they came to me separately, yet they form a whole for he who knows that the colours of the rainbow produce a white unity, or for the Artist who makes black appear by mixing the six colours of his magic palette.

2. *Poissons* /Pisces

Cet Ami, comment vous le présenter? Son nom demeura un mystère, mais son nombre est celui d'un sceau célèbre. Comment vous le décrire ? Peut-être comme le nautonnier de l'arche impérissable, impassible comme une colonne sur son roc blanc, scrutant vers le midi, au-delà du roc noir.

This Friend, how would one introduce him? His name remained a mystery, but his number is that of a famous seal. How can one describe him? Perhaps like the pilot of the imperishable ark, impassive like a column on his white rock scanning towards the south, beyond the black rock.

3. Bélier/Aries

Dans mon pélérinage éprouvant, je tentais de me frayer à l'épée une voie à travers la végétation inextricable des bois, je voulais parvenir à la demeure de la BELLE endormie en qui certains poètes voient la REINE d'un royaume disparu. Au désespoir de retrouver le chemin, les parchemins de cet Ami furent pour moi le fil d'Ariane.

On my initiatory pilgrimage, I tried to clear a path with my sword through the dense vegetation of the woods, I wanted to reach the abode of the sleeping BEAUTY in whom certain poets see the QUEEN of a lost kingdom. Desperate to find the way, the parchments of this Friend were for me, the thread of Ariadne.

4. Taureau/Taurus

Grâce à lui, désormais à pas mesurés et d'un œil sur, je puis découvrir les soixante-quatre pierres dispersées du cube parfait que les Frères de la BELLE du bois noir échappant à la poursuite des usurpateurs, avaient semées en route quant ils s'enfuirent du Fort blanc.

Thanks to him, henceforth with measured steps and a sure eye, I am able to discover the sixty-four scattered stones of the perfect cube that the Brothers of the BEAUTY of the black wood, escaping the pursuit of the usurpers, had strewn along the way as they fled from the white Fort.

5. Gémeaux/Gemini

Rassembler les pierres éparses, œuvrer de l'équerre et du compas pour les remettre en ordre régulier, chercher la ligne du méridien en allant de l'Orient à l'Occident, puis regardant du Midi au Nord, enfin en tous sens pour obtenir la solution cherchée, faisant station devant les quatorze pierres marquées d'une croix. Le cercle étant l'anneau et couronne, et lui le diadème de cette REINE du Castel.

Gathering together the scattered stones, working with the square and compass to put them back in order, looking for the line of the meridian going from East to West, then looking from the South to the North, finally in all directions to obtain the desired solution, stopping in front of the fourteen stones marked with a cross. The circle being the ring and crown, and it is the diadem of this QUEEN of the Castle.

6. Cancer/ Cancer

Les dalles du pavé mosaïque du lieu sacré pouvaient-être alternativement blanches ou noires, et JESUS, comme ASMODEE observer leurs alignments, ma vue semblait incapable de voir le sommet où demeurait cachée la merveilleuse endormie. N'étant pas HERCULE à la puissance magique, comment déchiffrer les mystérieux symboles gravés par les observateurs du passé. Dans le sanctuaire pourtant le bénitier, fontaine d'amour des croyants redonne mémoire de ces mots : PAR CE SIGNE TU le VAINCRAS.

The stones of the mosaic paving of the sacred place could be alternately white or black, and JESUS, like ASMODEUS, observing their alignments, my eyes seemed incapable of seeing the summit where the fabulous sleeping one remained hidden. Not being HERCULES with magic power, how to decipher the mysterious symbols carved by the observers of the past. In the sanctuary however the font, fountain of love of the believers, reminds us of these words: BY THIS SIGN YOU SHALL CONQUER him.

7. Lion/ Leo

De celle que je désirais libérer, montaient vers moi les effluves du parfum qui imprégnèrent le sépulcre. Jadis les uns l'avaient nommée : ISIS, reine des sources bienfaisantes, VENEZ A MOI VOUS TOUS QUI SOUFFREZ ET QUI ETES ACCABLES ET JE VOUS SOULAGERAI, d'autres: MADELAINE, au célèbre vase plein d'un baume guérisseur. Les initiés savent son nom véritable : NOTRE DAME DES CROSS.

From her that I wanted to free, rose towards me the fragrance that permeated the sepulchre. Once some called her: ISIS, queen of the beneficial springs, COME TO ME ALL YOU WHO SUFFER AND WHO ARE OVERWHELMED AND I WILL COMFORT YOU, others: MADELEINE, with the famous jar full of healing balm. The initiates know her true name: NOTRE DAME DES CROSS.

8. Vierge/ Virgo

J'étais comme les bergers du célèbre peintre POUSSIN, perplexe devant l'énigme : «ET IN ARCADIA EGO...» ! La voix du sang allait-elle me rendre l'image d'un passé ancestral. Oui, l'éclair du génie traversa ma pensée. Je revoyais, je comprenais ! Je savais maintenant ce secret fabuleux. Et merveille, lors des sauts des quatre cavaliers, les sabots d'un cheval avaient laissé quatre empreintes sur la pierre, voilà le signe que DELACROIX avait donné dans l'un des trois tableaux de la chapelle des Anges. Voilà la septième sentence qu'une main avait tracée : RETIRE-MOI DE LA BOUE, QUE JE N'Y RESTE PAS ENFONCE. Deux fois IS, embaumeuse et embaumée, vase miracle de l'éternelle Dame Blanche des Légendes.

I was like the shepherds of the famous painter POUSSIN, baffled by the riddle: "ET IN ARCADIA EGO..."! Would the voice of the blood show me the image of an ancestral past. Yes, a flash of genius crossed my mind. I saw, I understood! I knew now this fabulous secret. And, miraculously, when the four horsemen jumped, one horse had left four hoofprints on the rock, there was the sign that DELACROIX had given in one of the three paintings in the chapel of Angels. There was the seventh maxim which a hand had written: DELIVER ME FROM THE MIRE AND LET ME NOT SINK. Twice IS, embalmer and embalmed, miraculous vessel of the eternal White Lady of Legends.

9. Balance/ Libra

Commencé dans les ténèbres, mon voyage ne pouvait s'achever qu'en Lumière. A la fenêtre de la maison ruinée, je contemplais à travers les arbres dépouillés par l'automne le sommet de la montagne. La croix de crête se détachait sous le soleil du midi, elle était la quatorzième et la plus grande de toutes avec ses 35 centimètres ! Me voici donc à mon tour cavalier sur le coursier divin chevauchant l'abîme.

Begun in darkness, my journey could only end in Light. At the window of the ruined house, I gazed at the mountain peak through the trees stripped bare by autumn. The cross of the crest stood out under the midday sun, it was the fourteenth and the tallest of all at 35 centimetres! Here I am therefore, in my turn, rider of the divine charger astride the abyss.

10. Scorpion/ Scorpio

Vision céleste pour celui qui me souvient des quatre œuvres de Em. SIGNOL autour de la ligne du Méridien, au chœur même du sanctuaire d'où rayonne cette source d'amour des uns pour les autres, je pivote sur moi-même passant du regard la rose du P à celle de l'S, puis de l'S au P ... et la spirale dans mon esprit devenant comme un poulpe monstrueux expulsant son encre, les ténèbres absorbent la lumière, j'ai le vertige et je porte ma main à ma bouche mordant instinctivement ma paume, peut-être comme OLIER dans son cercueil. Malédiction, je comprends la vérité, IL EST PASSE, mais lui aussi en faisant LE BIEN, ainsi que xxxxxxxx CELUI de la tombe fleurie. Mais combien ont saccagé la MAISON, ne laissant que des cadavres embaumés et nombres de métaux qu'ils n'avaient pu emporter. Quel étrange mystère recèle le nouveau temple de SALOMON édifié par les enfants de Saint VINCENT.

A heavenly sight for he who remembers the four works of Em. Signol around the Meridian line, in the very choir of the sanctuary from which this source of love for one another radiates, I turn around looking from the rose of the P to that of the S, then from the S to the P . . . and the spiral in my mind becoming like a monstrous octopus expelling its ink, the darkness absorbs the light, I am dizzy and I put my hand to my mouth, biting my palm instinctively, perhaps like OLIER in his coffin. Curses, I understand the truth, HE PASSED, but he too doing GOOD, as did xxxxxxxx HE of the flowery tomb. But how many have sacked the HOUSE, leaving only embalmed corpses and numerous metals that they had not been able to take? What strange mystery is concealed in the new temple of SOLOMON built by the children of Saint Vincent?

11. Ophiuchus/ Ophiucus

Maudissant les profanateurs dans leurs cendres et ceux qui vivent sur leurs traces, sortant de l'abîme où j'étais plongé en accomplissant le geste d'horreur : « Voici la preuve que du sceau de SALOMON je connais le secret, que xxxxxxxxxx de cette REINE j'ai visité les demeures cachées. » A ceci, Ami Lecteur, garde toi d'ajouter ou de retrancher un iota ... médite, Médite encore, le vil plomb de mon écrit xxxx contient peut-être l'or le plus pur.

Cursing the profaners in their ashes and those who follow in their footsteps, emerging from the abyss into which I had plunged while making the gesture of horror: "Here is the proof that I know the secret of the Seal of SOLOMON, that xxxxxxxxxxx of this QUEEN I have visited the hidden residences." To this, Dear Reader, be careful not to add or remove one iota ... meditate, Meditate again, the base lead of my writing xxxx contains perhaps the purest gold.

12. Sagittaire/ Sagittarius

Revenant alors à la blanche coline, le ciel ayant ouvert ses vannes, il me sembla près de moi sentir une présence, / les pieds dans l'eau comme celui qui vient de recevoir la marque du baptême, me retournant vers l'est, face à moi je vis déroulant sans fin ses anneaux, l'énorme SERPENT ROUGE cité dans les parchemins, / salée et amère, l'énorme bête déchaînée devint au pied de ce mont blanc, rouge en colère.

Returning then to the white hill, the sky having opened its floodgates, I seemed to sense a presence near me, / with my feet in the water like he who has just been baptised, turning towards the east, facing me I saw, endlessly unwinding its coils, the enormous RED SERPENT cited in the parchments, salty and bitter, the huge raging beast became red with anger at the foot of this white mountain.

13. Capricorne/ Capricorn

Mon émotion fut grande, "RETIRE-MOI DE LA BOUE» disais-je, et mon réveil fut immédiat. J'ai omis de vous dire en effet que c'était un songe que j'avais fait ce 17 JANVIER, fête de Saint SULPICE. Par la suite mon trouble persistant, j'ai voulu après réflexions d'usage vous le relater un conte de PERRAULT. Voici donc Ami Lecteur, dans les pages qui suivent le résultat d'un rêve m'ayant bercé dans le monde de l'étrange à l'inconnu. A celui qui PASSE de FAIRE LE BIEN!

My emotion was great "DELIVER ME FROM THE MIRE" I said, and I awoke immediately. I have omitted to tell you in fact that this was a dream that I'd had this 17th JANUARY, feast day of Saint SULPICE. Afterwards, my confusion persisting, I wanted after giving it some thought to tell it to you like a fairy-tale by PERRAULT. Here then, Dear Reader, in the pages that follow, is the result of a dream that lulled me from the strange to the unknown. Let he who PASSES DO GOOD.

Appendix IV
CHRONOLOGY OF KEY TEXTS

Year	Author	Title
1797	Goethe	*Märchen*, or *The Green Snake and the Beautiful Lily*
1853	Nerval	*Sylvie*
1888	Boudet	*La Vraie Langue Celtique et le Cromleck de Rennes-les-Bains*
1935	Wirth	*Le Serpent Vert*
1947	Richer	*Gérard de Nerval et les Doctrines Esotériques*
1953	Jung	*Psychology & Alchemy*
1963	Jung	*Mysterium Coniunctionis*
1963	Richer	*Gérard de Nerval : Expérience et Création*
1967	de Sède	*Le Trésor Cathare*
1967	de Sède	*L'Or de Rennes*, or *Le Trésor Maudit de Rennes-le-Château*
1967		*Le Serpent Rouge*
1967	Richer	*Géographie Sacrée du Monde Grec*
1970	Richer	*Delphi, Délos et Cumès*
1974	Lincoln	*The Priest, the Painter & the Devil* Documentary
1977	de Sède	*Signé : Rose + Croix; l'énigme de Rennes-le-Château*
1978	Boudet	*La Vraie Langue Celtique*, with Plantard introduction
1978	Boudet	*La Vraie Langue Celtique*, with de Sède introduction

1978	de Sède	*La Rose-Croix*
1981	de Chérisey	*Circuit*
1982	Lincoln, Baigent, Leigh	*The Holy Blood and the Holy Grail*
1983	Richer	*Géographie Sacrée dans le Monde Romain*
1986	Lincoln, Baigent, Leigh	*The Messianic Legacy*
1987	Richer	*Gérard de Nerval : Expérience Vécue et Création Esotérique*
1988	Eco	*Foucault's Pendulum*
1988	de Sède	*Rennes-le-Château : le dossier, les impostures, les phantasmes, les hypothèses*
1991	Lincoln	*The Holy Place*
1994	Eco	*Six Walks in the Fictional Woods*
2006	Chaumeil	*Le Testament du Prieuré de Sion : Le Crépuscule d'une Ténébreuse Affaire*
2009	de Chérisey	*Un Veau à Cinq Pattes* (written in 1984)

Appendix V
BIBLIOGRAPHY

Ambelain, Robert, *Martinism: History and Doctrine* (1946), Editions Niclaus, Paris.

Andrews, Richard & Paul Schellenberger, *The Tomb of God* (1996), Little, Brown & Co., London.

Baigent, Michael, Richard Leigh and Henry Lincoln, *The Holy Blood and the Holy Grail* (1982), Jonathan Cape, London.

Baigent, Michael, Richard Leigh and Henry Lincoln, *The Messianic Legacy* (1986), Jonathan Cape, London.

Bayrou, Lucien *Le Château d'Arques* (1988), Centre d'Archéologie Médiévale du Languedoc, Carcassonne.

Boudet, Henri, foreword by Pierre Plantard, *La Vraie Langue Celtique et Le Cromleck de Rennes-les-Bains* (1886), 1978, Belfond Press, Paris.

Boudet, Henri, foreword by Gérard de Sède, *La Vraie Langue Celtique et Le Cromleck de Rennes-les-Bains* (1886), 1978 edition reprinted 1986, Belisane Press.

Boudet, Henri, *The True Celtic Language and the Stone Circle of Rennes-les-Bains* (1886), English translation 2008, Les Editions de l'œil du Sphinx, Paris.

Cahiers de la Société Gérard de Nerval No 15 : Jean Richer 1915–1992 (1992).

Chaumeil, Jean-Luc, *Le Testament du Prieuré de Sion : Le Crépuscule d'une Ténébreuse Affaire* (2006), Pégase, Paris.

De Chérisey, Philippe, *Un Veau à Cinq Pattes* (2008), France Secret, Paris.

De Sède, Gérard, *La Rose-Croix* (1978), Editions J'ai lu, Paris.

De Sède, Gérard, *Les Templiers sont parmi nous* (1962), Julliard, Paris.

De Sède, Gérard, *Le Trésor Cathare*, (1966), Julliard, Paris.

De Sède, Gérard, *Signé : Rose + Croix : l'énigme de Rennes-le-Château* (1977), Librairie Plon, Paris.

De Sède, Gérard, *The Accursed Treasure of Rennes-le-Château* (1967). Translated (2001) by Bill Kersey from *L'Or de Rennes*, DEK Publishing, Surrey.

De Sède, Gérard, *Rennes-le-Château: The dossier, the impostures, the fantasies, the hypotheses* (1998). Translated by Bill Kersey from French original (2006), DEK Publishing, Surrey.

De Sède, Gérard, *Saint-Emilion insolite* (1980), Pégase, Villeneuve de la Raho.

Digges, Leonard *A geometrical practise, named Pantometria diuided into three bookes, longimetra, planimetra, and stereometria, containing rules manifolde for mensuration of all lines, superficies and solides.* London, Henrie Bynneman, 1571.

Doumayrou, Guy-René, *Essai sur la Géographie sidérale des Pays d'Oc et d'ailleurs* (1975), Union Générale d'Editions, Paris.

Eco, Umberto, *Foucault's Pendulum* (1988), Picador, London.

Eco, Umberto, *Six Walks in the Fictional Woods* (1994), Harvard University Press, New York.

M. Eliade. M, *The Myth of the Eternal Return: Cosmos & History*, Volume 4 of Bollingen Series Mythos Series, (1971) Princeton University Press.

Graves, Robert, *The White Goddess* (1997, first published 1948), Faber & Faber, London.

Higgins, Godfrey, *Anacalypsis an Attempt to Draw Aside the Veil of the Saitic Isis or an Inquiry into the Origin of Languages, Nations and Religions*, London, Longman, 1836.

Holmes, Richard, *Footprints: Adventures of a Romantic Biographer* (1985), Harper Perennial, London.

Jarnac, Pierre, *Mélanges Sulfureux : Les mystères de Rennes-le-Château*, (1995), C.E.R.T. Centre d'Etudes et de Recherches Templières, Couiza (Aude).

Jung C.G., *Alchemical Studies: Collected Works of C.G. Jung, volume 13* (First published 1967, reprinted 1973), Routledge and Kegan Paul, London.

Jung C.G., translated by R.F.C. Hull, *Mysterium Coniunctionis: An Inquiry into the Separation and Synthesis of Psychic Opposites in Alchemy*, (1970), Routledge London.

Jung C.G., *Psychology & Alchemy, Collected Works of C.G. Jung, volume 12*, Second Edition, (1968), Routledge, London.

Jung C.G. *The Archetypes and the Collective Unconscious, Collected Works of C.G. Jung, volume 9 Part 1*. (1991), Routledge, London.

Kiess, Georges, *Les Tours à signaux au pays de Rhedae*, in Association Terre de Rhedae Bulletin No 9, Décembre 1995.

Michael Lamy, *The Secret Message of Jules Verne: Decoding his Masonic, Rosicrucian and Occult Writings*, Destiny Books, Rochester, 2007.

Lincoln, Henry, *Key to the Sacred Pattern* (1997), Windrush Press, Witney, Oxfordshire.

Lincoln, Henry, *The Holy Place: The Mystery of Rennes-le-Château – Discovering the Eighth Wonder of the Ancient World* (1991), Jonathan Cape, London.

Lincoln, Henry, *The Templars Secret Island: The Knights, the Priest and the Treasure* (2000), Windrush Press, Witney, Oxfordshire.

Lincoln, Henry, BBC Chronicle documentary: *The Lost Treasure of Jerusalem* (1971). Available 24.6.22 at https://www.youtube.com/watch?v=rkMaGAoHDxs

Lincoln, Henry, BBC Chronicle documentary: *The Priest, the Painter & the Devil* (1974). https://www.youtube.com/watch?v=Uru8UiCxLnc

Lincoln, Henry, BBC Chronicle documentary: *The Shadow of the Templars* (1977). https://www.youtube.com/watch?v=-bYKaXJVt6k

Marie, Franck, *P.F...? le Serpent Rouge Preuves* (1978), Editions S.R.E.S. Vérités Anciennes, Malakoff.

Mémoires de la Mission Archéologique en Iran. Tome XXXIV Mission de Susiane, E.M.Bruins et M.Rutten, 1960

Nerval, Gérard de, *Œuvres, Tome I* and *Tome II* (1961), Bibliothèque de la Pléiade, Editors A.Béguin and J.Richer, Paris.

Nerval, Gérard de, translated by Richard Sieburth, *Selected Writings* (1999), Penguin Classics, London.

Nerval, Gérard de, *Voyage en Orient*, Tome 1 and 2 (1964 edition), Julliard, Paris.

Olhagaray Pierre, *Histoire des comptes de Foix, Béarn et Navarre*, Paris, 1629

Pennick, Nigel, *Secret Games of the Gods: Ancient Ritual Systems in Board Games* (1989), Samuel Weiser Inc, Maine.

Picknett, Lynn and Clive Prince, *The Sion Revelation: The Truth about the Guardians of Christ's Sacred Bloodline* (2006), Touchstone Books, New York.

Bill Putnam & John Edwin Wood *The Treasure of Rennes-le-Château* (2003), Sutton Publishing, Stroud, Gloucestershire.

Quehen, Réné and Dominique Dieltiens, *Les Châteaux Cathares ...et les autres* (1983), Montesquieu Volvestre, France.

Robin, Jean *Le Royaume du Graal : Introduction au mystère de la France* (1992), Guy Trédaniel, Paris.

Richer, Jean *Aspects ésotériques de l'œuvre littéraire* (1980), Dervy-Livres, Paris.

Richer, Jean, *Delphes, Délos et Cumes : Les Grecs et le Zodiaque* (1970), Julliard, Paris.

Richer, Jean, *Géographie Sacrée dans le Monde Romain* (1985), Guy Trédaniel, Paris.

Richer, Jean, *Géographie Sacrée du Monde Grec* (1967), Hachette, Paris.

Richer, Jean *Gérard de Nerval : Expérience vécue et création ésotérique* (1987), Guy Trédaniel, Paris.

Richer, Jean, *Gérard de Nerval et les Doctrines Esotériques* (1947), Griffin D'Or, Paris.

Richer, Jean, *Nerval : Expérience et création,* (1963), Hachette, Paris.

Richer, Jean, *Nerval : Expérience et création, deuxième édition* (1970), Hachette, Paris.

Richer, Jean, *Poètes d'aujourd'hui : Gérard de Nerval* (1957), Editions Pierre Seghers, Paris.

Richer, Jean, *Iconologie et Tradition : Symboles cosmiques dans l'art chrétien* (1987), Guy Trédaniel, Paris.

Richer, Jean, *Nerval au royaume des archétypes : «Octavie» «Sylvie» «Aurélia»*, Archives Nervaliennes, no. 10 in Archives des lettres modernes no. 130 (1971).

Richer, Jean, translated by Christine Rhone, *Sacred Geography of the Ancient Greeks: Astrological Symbolism in Art, Architecture and Landscape* (1994), SUNY Press, New York.

Rivière, Jacques and Claude Boumendil, *Histoire de Rennes-les-Bains* (2006), Belisane, Cazilhac (Aude).

Skinner, J. Ralston, *The Source of Measures* (1875, reprinted 1982), Wizards Bookshelf, San Diego.

Steiner, Rudolf, Rosicrucianism and Modern Initiation. GA 233A. III. The Time of Transition. 6 January 1924, Dornach, https://rsarchive.org/Lectures/RosiModInit/19240106p01.html

Steiner, Rudolf, Rosicrucianism and Modern Initiation. GA 233A . V. Occult Schools in the 18th and First Half of the 19th Century, 12 January 1924, Dornach, https://rsarchive.org/Lectures/RosiModInit/19240112p01.html

Wirth, Oswald, *Goethe Le Serpent Vert : Conte Symbolique traduit et commenté par Oswald Wirth* (1935, reprinted 1977), Collection Mystiques et Religions, Série B, Dervy-Livres, Paris.

Wood, David, *Genisis* (1985), The Baton Press, Tunbridge Wells.

Wood, David, & Ian Campbell, *Geneset* (1994), The Baton Press, Tunbridge Wells.

Les Dossiers Secrets

Les Dossiers Secrets was a series of privately printed pamphlets deposited in the Bibliothèque nationale, Paris, between 1964 and 1977.[234] The series included the following publications (date, author and title are shown):

January 1964   Henri Lobineau
Généalogie des rois mérovingiens et origine de diverses familles françaises et étrangères de souche mérovingienne, d'après l'abbé Pichon, le docteur Hervé et les parchemins de l'abbé Saunière, curé de Rennes-le-Château

August 1965   Madeleine Blancasall
Les descendants mérovingiens ou l'énigme du Razès wisigoth

March 1966   Antoine l'Ermite
Un trésor mérovingien à ... Rennes-le-Château

June 1966   Eugène Stublein
Pierres gravées du Languedoc

November 1966   S. Roux
L'affaire de Rennes-Le-Château

20 March 1967   Pierre Feugère, L. Maxent et G. de Koker
Le Serpent Rouge

27 April 1967   Henri Lobineau
Dossiers secrets

October 1967   Nicolas Beaucéan
Au pays de la Reine Blanche

July 1977   Jean Delaude
Le Cercle d'Ulysse

---

234   The collection of *Les Dossiers Secrets* was reprinted as *Mélanges Sulfureux: Les mystères de Rennes-le-Château*, with texts re-assembled by Pierre Jarnac, and published by C.E.R.T in 1994.

EXPANDED TABLE OF CONTENTS

Prologue: A Dream in Athens 1

Introduction 5
   The Affair of Rennes   8

PART ONE
Identification

Chapter One: On the Path 15
   The Map   17
   The 45° alignment   21
   Château d'Arques   21
   Meridians in the Mountains   24
   La Pique Meridian   27
   The Pech Cardou Meridian   29
   Units of Measure   32

Chapter Two: *Le Serpent Rouge* 41
   The Poem   42
   Where Does the Action of the Poem Take Place?   50
   What is the Nature of the Zodiac in the Landscape?   50
   The Riddles   51
   Who Wrote the poem?   55
   Summary   58

Chapter Three: First Inklings 61
   The Salty and the Bitter   63

Chapter Four: Sightlines 67
   The Cathar Châteaux and Sighting Lines   *70*
   Visiting the Meridians   *73*
   Three More Meridians   *80*
   The "Sunrise Line" and the Pic de Saint-Barthélemy   *87*
Chapter Five: Sacred Geography 95
   Sacred Geography of the Ancient Greeks   *96*
   Landscape Grids and the Zodiac of Delphi   *99*
   Delphi as Navel or Sacred Centre   *103*
   The Riddle in the Dream Resolved   *104*
   The Path of the Hero   *107*
Chapter Six: Converging Circles 111
   The Python of Delphi   *114*
Chapter Seven: The Number of the Famous Seal 119

PART TWO
Orientation

Chapter Eight: The Zodiac of Rennes-les-Bains............................ 131
   The Roman Cardo and Decumanus Axes   *132*
   A Discovery in the Landscape Around Rennes-les-Bains   *136*
Chapter Nine: The Cromlech of Rennes-les-Bains..................... 143
   La Vraie Langue Celtique and Le Serpent Rouge   *146*
   What is the Cromlech?   *147*
   A Tour Around the Cromlech   *150*
   Allocating Zodiac Signs to the Twelve-fold Division in the Landscape *156*
   Summary   *164*
Chapter Ten: Delphi, Apollo and the Python 167
   Delphi and the Serpent Python   *172*
   Pech Cardou and Mount Parnassus   *177*
   Serpent and Symbol   *183*
Chapter Eleven: Confirmation from the Team 187
   Pierre Plantard's Zodiaque de Rennes-les-Bains   *187*
   Elements for a Bibliography   *190*
   Phillipe de Chérisey   *192*
   Gérard de Sède   *195*
   Summary   *202*

EXPANDED TABLE OF CONTENTS

PART THREE
Solution

Chapter Twelve: Geographic Cryptography 209
  Doctored From the First Page to the Last   214
  The Map-Grid and the Meridians   218
  The Scale of Boudet's Map   220
  Dimensions of the Cromlech   224
  The Sixty-Four Stones   226
  The Earth as Chequerboard   228
  The Cromlech as World Map   229
  De Chérisey's Map Grid   235
  Summary   239

Chapter Thirteen: The Riddle of the Parchments 241
  The Target for the Correct Solution   243
  The Solution to the Small Parchment   247
  Geometry of the Hexagram   255
  The Large Parchment   256
  Geometry of the Seal of Solomon   258
  The Lower Target Marker   259
  The Marie de Nègres Gravestone   262
  The Uncial Script   268
  Maps and Parchments   273
  Meridians on the Parchment   276
  Summary   282

Chapter Fourteen: The Arques Square 287
  The Dimensions of the Arques Square   290
  The Templar Headquarters of Campagne-sur-Aude   294
  The Alet-les-Bains Cathedral Alignment   299
  Compass on Campagne-sur-Aude   300
  Geometrical Construction of the Square   304
  The Meridians and the Width of Arques Square   308
  The Two Angles   316
  More Pythagorean Triangles   318
  The Susa Tablet   320
  A Fourth Pythagorean Triangle   320
  Pythagorean Triangles in Ancient Landscape   330
  Purpose of the Arques Square   331
  Summary   334

PART FOUR
Transmission

Chapter Fifteen: Grand Voyager of the Unknown 339
- *Sylvie* — 342
- *The Zodiac Structure of Sylvie* — 344
- *When Did Richer Discover the Zodiac in Sylvie?* — 347
- *Sylvie and Le Serpent Rouge* — 348
- *Conte Initiatique* — 349
- *Goethe's Märchen* — 349
- *Le Serpent Vert* — 350
- *Nerval's Life as Archetypal Story* — 352
- *The Itinerary of the Noble Voyager* — 353

Chapter Sixteen: Dreams, Alchemy and the Omphalos 357
- *Alchemy and the Quest for Wholeness* — 359
- *The Stages in the Work and their Colours* — 359
- *Dream Analysis and Mandala Symbolism* — 362
- *Delphi and Mandala Symbolism* — 364
- *The Quaternity* — 365
- *Dreams and the Unconscious* — 369
- *Prometheus and the Mythology of Fire* — 370
- *The Solar Hero* — 374
- *The Mystic Peregrinations of Michael Maier* — 376

Chapter Seventeen: Imprint of a Seal 383
- *Hidden Layers in Mutus Liber* — 383
- *Geometrical Figures of the Old Rosicrucians* — 389
- *Martinism* — 392
- *Boudet, 1886 and the Ordre Martiniste* — 394
- *Martinist Initiation* — 395

Chapter Eighteen: A Walk in the Woods 401
- *Foucault's Pendulum* — 403
- *Death of a Poet* — 407
- *A Walk in Paris* — 408
- *Eco, Richer and Nerval* — 412
- *The Hanged Man* — 415
- *The Riddle of the Three Named Authors* — 416
- *The Other Foucault* — 418
- *Impossible Theories in Infeasible Books* — 422

Chapter Nineteen: Reassembling the Scattered Stones 427
- *Who built it?* — 427

When was it built?	*428*
The First Phase	*429*
Why was it built?	*430*
The Second Phase	*431*
Vehicle of Transmission: The Map	*432*
The Third Phase	*433*
Epilogue: Coda to a Dream	437

APPENDICES

Appendix I: The Parchment Text Decipherments	445
The Decoding Process	*445*
The Letter W	*446*
The 1974 BBC Documentary	*447*
Comparing the Parchment Decodings	*454*
Conclusion	*455*
A Serious Comedy of Errors	*461*
Appendix II: List of Figures	465
Appendix III: The Text of *Le Serpent Rouge*	475
Appendix IV: Chronology of Key Texts	480
Appendix V: Bibliography	484

AUTHOR'S BIOGRAPHY

Simon Miles is an independent author, researcher and speaker.

After a successful business career in the field of scientific lasers, he became a full time writer in 2007.

Originally from Australia, he now lives in the UK, near Manchester, with his wife Judith, and their cat Leo.

This is his first book.